SPIDERS OF
NORTH AMERICA

SARAH ROSE

Princeton University Press
Princeton and Oxford

Published by Princeton University Press
41 William Street, Princeton, New Jersey 08540
99 Banbury Road, Oxford OX2 6JX

press.princeton.edu

All Rights Reserved
ISBN (pbk.) 978-0-691-17561-4
ISBN (e-book) 978-0-691-23706-0

Library of Congress Control Number: 2021951317

British Library Cataloging-in-Publication Data is available

Editorial: Robert Kirk, Abigail Johnson, and Megan Mendonca
Production Editorial: Karen Carter
Text Design: D & N Publishing, Wiltshire, UK
Cover Design: Ruthie Rosenstock
Production: Steven Sears
Publicity: Matthew Taylor and Caitlyn Robson

Cover image: Female *Argiope aurantia*. Image courtesy of the author.

This book has been composed in IBM Plex Sans and IBM Plex Sans Condensed

Printed on acid-free paper. ∞

Printed in Italy

10 9 8 7 6 5 4 3 2 1

CONTENTS

2 Sensing web weaver guild: *Sphodros rufipes*

2 Sensing web weaver guild: *Ariadna bicolor*

3 Sheet web weaver guild: *Frontinella pyramitela*

3 Sheet web weaver guild: *Agelenopsis* sp.

4 Orb weaver guild: Araneidae

4 Orb weaver guild: Araneidae

5 Space web weaver guild: *Dictyna coloradensis*

5 Space web weaver guild: *Hentziectypus globosus*

6 Ambush hunter guild: *Mecaphesa asperata*

6 Ambush hunter guild: *Loxosceles reclusa*

7 Ground active hunter guild: Gnaphosidae

7 Ground active hunter guild: *Schizocosa* sp.

8 Other active hunter guild: *Pisaurina mira*

8 Other active hunter guild: *Eris* sp.

9 Spider hunter guild: *Mimetus* sp.
with *Theridion* prey

9 Spider hunter guild: *Mimetus* sp. guarding
egg sac of different spider species

FOREWORD

Humans have a collective ambivalence toward spiders. We admire their industrious nature as they spin their intricate snares. We marvel at the symmetry and artistry of orb webs. On the other hand, we are frightened of the spinners themselves. Spiders can be large. They may appear suddenly and unexpectedly. They can run rapidly in any direction on too many legs. Cobwebs are an embarrassing reminder that we don't pay attention to details in our housekeeping.

Spiders figure in creation myths of indigenous peoples, were an inspiration to historical figures such as Robert the Bruce, and are today the subjects of intense research and experimentation. We have fed them illicit narcotics to record the effects on their web-spinning behavior, sent them into orbit to measure the effects of zero gravity, and harnessed the properties of their silk for our own uses. As I write this, new studies are emerging that suggest spiders may outsource some cognitive functions to their webs. Another study indicates that the pressure exerted by spider predators may increase plant diversity in certain habitats.

Somewhere between irrational fear and scholarly pursuits there is a place for nonscientists seeking a better understanding of spiders. That is the audience for this book. In the case of arachnids, familiarity breeds fascination, rather than contempt. Knowing which spiders live around you, and how they fit in the larger picture of outdoor (and indoor) ecology, can alleviate fears and foster an appreciation that lasts a lifetime.

Sarah Rose has compiled what may be the most thorough and easy-to-use field guide to common North American spiders yet created. By including images of webs, egg sacs, and other "signs" of spiders, as well as immature stages of some species, she has improved greatly on previous efforts. Distribution maps give a good indication of where a particular species can be found, though our collective knowledge of spider geography is a work in progress. This is one way you, the reader, can help. Uploading your images of spiders to such online platforms as iNaturalist, BugGuide.net, Project Noah, and similar sites can greatly enhance the science of arachnology.

Please enjoy this book, make good use of it, and consider becoming a citizen scientist. Recruit family, friends, and neighbors to the cause of conservation through enhancing the biodiversity of yards, gardens, farms, and parks everywhere. Spiders are wildlife, too, and in fact among the most important and impactful of all animals.

Eric R. Eaton

Author, *Wasps: The Astonishing Diversity of a Misunderstood Insect*
Lead author, *Kaufman Field Guide to Insects of North America*

ACKNOWLEDGMENTS

First and foremost, I would like to thank Princeton University Press. They have been great to work with, and I am grateful that they wanted to create a field guide to spiders, as that is something that has been lacking for North America.

When I was young, my parents encouraged me to ask questions and then to seek the answers. Often, an answer would not just be provided to me, but a quest would be undertaken to find the answer. This really shaped who I am as a person and a researcher. It would probably have been easier for my parents just to answer my questions or brush them off; rather, they gave me a set of skills and knowledge of the places to look for the answers or research projects that could be done to help resolve those questions. Each week, my family took a trip to the library, and I presented my list of burning questions to the librarian for assistance in finding the right book to aid me in my quest for answers. I always thought, and still think, that librarians must be some of the smartest people on the planet. They never led me astray and always seemed to point me in the right direction, even as a PhD student working on my dissertation. I would like to thank all those librarians, teachers, park staff, and anyone else who aids people in their quest for answers. To all those who never told me that I asked too many questions but kindled that flame of curiosity, thank you so very much.

If I were to list everyone who assisted in the creation of this field guide, there would not be many pages left for the spiders. I am also fearful that, after publication, I would realize that I had forgotten a name and then regret that for all eternity. So thank you to everyone who assisted me in any way. I sincerely appreciate all of you. That said, there are a few people I feel compelled to name. First would be Dr. Richard A. Bradley. Dr. Bradley has been a valuable mentor to me since my undergraduate days, and I now consider him a dear friend. He has never made me feel stupid for asking questions (even if I should have known the answers to those questions). The Ohio State University Museum of Biological Diversity has been very helpful in allowing me access to specimens to help confirm (or dispute) my identifications. Many of the reference specimens that I collected while working on this book have been donated to the museum. I would like to thank all the members of the American Arachnological Society, which is filled with people who love to share their knowledge and help other people. I have always appreciated the inclusiveness I have felt in this society. I would like to thank Eric Eaton, a friend who always promotes others. I would like to thank the anonymous reviewers who provided invaluable feedback that assisted me in developing the text for this book. I want to thank my parents, both of whom died while I was working on this book. They were supportive and always believed in me. They are missed tremendously, but they would be pushing me forward, saying that the work needs to go on. My son has also been a huge support during this process.

I do need to thank specifically the people below, who contributed photographs and spiders, assisted with spider identification, or allowed me to search their home or property for spiders. Thank you!

John Abbott
Alice Abela
R. J. Adams
Melody Albright
Wendy Alward
Erin Armstrong
Lyn Atherton
Kyron Basu
Danielle Bennett
Amy Bianco
Jason Bond
Clayton Bownds
Deb Bradley
Richard A. Bradley
Margarethe Brummermann

Scott Bundy
E. Christine Butler
Cade Campbell
Meghan Cassidy
Carmen Champagne
Lou Coticchio
Jillian Cowles
Jason M. Crockwell
Michael Davis
Laura Dirkx
Michael Draney
Sebastian Echeverri
Lynette Elliott
Michael Ferro
Diona Fredo

Alyssa Fuller
Mike Graziano
Christian Grismado
Jeff Gruber
Kevin Hall
Kat Halsey
Toby Hays
Marshal Hedin
Chad Heins
Solomon Hendrix
Andy Hoffman
Jeff Hollenbeck
David and Laura Hughes
J. C. Jones
Thaddeus Charles Jones

Rudo Kemper
Joseph T. Lapp
Joel Ledford
John Liebeskind
David Lightfoot
Annika Linqvist
Isaac Lord
Chandra Lowe
Tammy MacKenzie
William Mason
John R. Maxwell
Ann Mayo
Sean McCann
Rowan McGinley
Marc Milne
Graham Montgomery
Célio Moura

Tom Murray
Eric Neubauer
Tony Palmer
Jay Paredes
L. Brian Patrick
Laura Paxson
Arlo Pelegrin
Kevin Pfeiffer
Jen Phillips
Diane Placto Brooks
Erin Powell
Lisa Powers
Darius A Przygoda
Michael Rexman
Joanne Ritterspach
Johnstone
Jesse Rorabaugh

Jenn Rose
Nina Sandalin
Danielle Siddle
John Sloan
Michelle Sloan Boss
Natalie Stalick
Ian Stocks
Aaron Stoll
David Timmons
Leonard Vincent
Gerald Wegner
Kevin Wiener
Justin Williams
Susan Wise-Eagle
Ian Wright

This publication would not have happened without the resources available on the World Spider Catalog, the Symbiota Collections of Arthropods Network (SCAN), BugGuide.net, iNaturalist, Flickr, and several Facebook groups that focus on spiders (and sometimes other arthropods). Thanks to all the "Bugbook" friends out there.

Lastly, I need to acknowledge the spiders. They are vital members of our planet and beautiful, fascinating creatures. I dedicate this book to those spiders that lost their lives in the name of science.

PHOTOGRAPHY CREDITS

Any photographs not credited here were taken by the author.

John Abbott: *Ctenus hibernalis* (p.446). **Alice Abela**: *Badumna longinqua* (p.62 (*6a*); p.89; p.90); *Diaea livens* (p.309); *Ebo pepinensis* (p.460); *Lutica nicolasia* (p.421); *Nigma linsdalei* (p.225, 2 images); *Segestria pacifica* (p.60); *Tarsonops systematicus* (p.585; p.586, 2 images); *Tinus peregrinus* (p.481). **R. J. Adams**: *Holocnemus pluchei* (p.238). **Meoldy Albright**: *Euryopis* sp. (p.26 (*41a*)). **Wendy Alward**: *Mastophora phrynosoma* male (p.173). **Erin Armstrong**: *Cyrtophora citricola* (p.157, 3 images); *Selenops* sp. (p.297 (*3b*); p.300). **Lyn Atherton**: *Scoloderus nigriceps* (p.186, 4 images); *Aptostichus pacificus* (p.41; p.43, 2 images); *Blabomma californicum* (p.85, 4 images); *Cybaeus reticulatus* (p.87, 2 images); *Cybaeus signifier* (p.62 (*5b*); p.64 (*10b, 11b, 13a*); p.84, 2 images; p.88, 4 images); *Eratigena duellica* (p.63 (*8b*); p.69; p.76, 5 images); *Ero canionis* (p.589, 4 images; p.590, 4 images); *Ozyptila pacifica* (p.317, 2 images); *Pimoa altioculata* (p.65 (*14a*); p.121, 2 images; p.122, 4 images); *Rhomphaea fictilium* male (p.271); *Usofila pacifica* (p.63 (*8a, 9b*); p.122; p.123, 5 images). **Amy Bianco**: *Orchestina saltitans* (p.406; 409, 2 images). **Jason Bond**: *Antrodiaetus montanus* (p.42); *Aphonopelma eutylenum* (p.40 (*6a*); p.53, 3 images; p.601, urticating hair); *Aphonopelma hentzi* (p.54, 2 images); *Aptostichus simus* (p.40 (*5b*); p.47, 2 images); *Bothriocyrtum californicum* female (p.50); *Cyclocosmia truncata* (p.51); *Myrmekiaphila tigris* (p.49, 2 images); *Schizocosa crassipes* (p.387); *Sphodros* sp. (p.7 (*15b*)). **Clayton Bownds**: *Allocyclosa bifurca* (p.130); *Castianeira amoena* (p.332). **Richard A. Bradley**: *Araneus cavaticus* (p.132); *Araneus guttulatus* female (p.136); *Araneus niveus* female (p.140); *Araneus thaddeus* male (p.146); *Cybaeus giganteus* (p.87); *Dolomedes vittatus* (p.9 (*18a*)); *Episinus amoenus* (p.253, 2 images); *Heteropoda venatoria* male (p.578); *Hogna baltimoriana* (p.370, 2 images); *Hogna carolinensis* (p.371, 4 images; p.372, 4 images); *Hypselistes florens* (p.108, 2 images); *Latrodectus geometricus* lower left and right female, and female ventral (p.258, 3 images); *Menemerus bivittatus* male (p.516, 2 images); *Misumena vatia* (p.312, 2 images); *Neoscona domiciliorum* (p.183, 2 images); *Neriene litigiosa* (p.113, 2 images); *Ocepeira* sp. (p.32 (*55a*)); *Ocrepeira ectypa* web (p.185); *Octonoba sinensis* (p.213, 4 images); *Oecobius navus* (p.41 (*8a*); p.58, 2 images; p.59, 2 images); *Philodromus marxi* male (p.463); *Piratula minuta* (p.383, 2 female with egg sac images, 2 male images); *Scytodes thoracica* (p.575, 4 images); *Sergiolus capulatus* (p.353); sweep netting and sorting (p.13 (*26a*); p.14 (*26b*); *Trochosa sepulchralis* (p.402, 4 images; p.403, 2 images); *Varacosa avara* (p.404, 3 images); Xiphosura (p.3 (*3*)); *Xysticus texanus* (p.326, 2 images). **Diane Placto Brooks**: *Sphodros niger* (p.45, 2 images). **Margarethe Brummermann**: *Syspira tigrina* (p.448). **Illustration by Steve Buchanan**, from *Common Spiders of North America*, courtesy of American Arachnological Society: *Theotima minutissima* (p.120). **Scott Bundy**: *Theotima minutissima* (p.63 (*9a*); p.120, 2 images). **E. Christine Butler**: *Araneus miniatus* female (p.139); *Lupettiana mordax* (p.432); *Wulfila* sp. (p.31 (*53a*)). **Cade Campbell**: *Hypochilus pococki* web (p.217 (*4a* right); p.228; p.229). **Meghan Cassidy**: *Araneus detrimentosus* (p.134, 3 images); *Arctosa littoralis* (p.365, 3 images); *Cesonia bilineata* (p.343, 3 images); *Falconina gracilis* (p.337; p.338, 3 images); *Gasteracantha cancriformis* right female, and juvenile (p.162, 2 images); *Marpissa pikei* (p.515, 6 images); *Metazygia wittfeldae* (p.175, 3 images); *Neoscona oaxacensis* juvenile (p.184, 3 images); *Phidippus carolinensis* female (p.536); *Salticus austinensis* (p.559, 3 images). **Carmen Champagne**: *Sergiolus tennesseensis* (p.354; p.355, 2 images). **Lou Coticchio**: *Deinopis spinosa* (p.298, eye configuration; p.299, upper male); *Eriophora ravilla* right female (p.158); *Latrodectus bishopi* (p.257); *Loxosceles rufescens* (p.306, 3 images). **Jillian Cowles**: *Anyphaena catalina* (p.428); *Apollophanes punctipes* (p.459, 4 images); *Calilena arizonica* (p.73, 2 images); *Castianeira occidens* (p.335, 3 images); *Chalcoscirtus diminutus* (p.489, 3 images); *Cinetomorpha bandolera* (p.407, 4 images); *Colonus hesperus* (p.490, 3 images; p.491, 2 images); *Colonus sylvanus* lower female and upper male (p.493); *Curicaberis abnormis* (p.576, 2 images); *Darkoneta* (p.231, 2 images); *Dendryphantes nigromaculatus* (p.494, 3 images); *Diguetia canities* (p.218 (*8a*); p.227, 4 images); *Diguetia imperiosa* (p.218 (*8a*); p.226); *Emertonella taczanowskii* (p.249, 3 images); *Escaphiella hespera* (p.327 (*2b*); p.406; p.408, 4 images); *Euryopis scriptipes* (p.255); *Filistatoides insignis* (p.55); *Gertschosa* (p.345, 2 images); *Habronattus geronimoi* (p.500, 6 images); *Habronattus hallani* (p.501, 5 images); *Habronattus oregonensis* (p.502, 5 images); *Habronattus pugillis* (p.503, 4 images; p.504, 4 images); *Habronattus pyrrithrix* (p.504, 2 images; p.505, 7 images); *Habronattus ustulatus* (p.506, 2 images); *Hamataliwa grisea* (p.450, 2 images; p.451, 4 images); *Hesperocosa unica* (p.369, 2 images); *Hexurella* (p.62 (*4b*); p.66, 2 images); *Homalonychus selenopoides* (p.360, 2 images); *Isaloides yollotl* (p.309, 2 images); *Kibramoa paiuta* (p.329 (*7a*); p.414; p.415, 4 images); *Lauricius hooki* female on rock (p.582); *Loxosceles arizonica* (p.303, 3 images); *Loxosceles sabina* (p.306); *Megahexura fulva* (p.61 (*2b*); p.62 (*4a*); p.67, 2 images; p.68); *Messua limbata* female (p.517, 2 images); *Metacyrba floridana* (p.517, 2 images; p.518, 2 images); *Metacyrba taeniola similis* (p.519, 4 images); *Metaphidippus chera* (p.520, 3 images); *Micaria gosiuta* (p.348; p.349, 2 images); *Micaria longipes* (p.349, 2 images); *Micaria nye* (p.350, 3 images); *Micaria pulicaria* (p.351, 2 males); *Micaria pulicaria* ant model (p.351); *Neoanagraphis chamberlini* (p.361); *Neopisinus cognatus* (p.263, 2 images); *Nodocion eclecticus* (p.351, 2 images; p.599, scutum); *Olios giganteus* (p.425 (*9a*); p.575; p.579, 4 images); *Orthonops icenoglei* (p.584 (*2a*); p.586, 3 images); *Oxyopes apollo* (p.453, 4 images); *Oxyopes scalaris* female top left (p.455); *Oxyopes tridens* (p.456; p.457, 2 images); *Ozarkia* (p.231; p.232, 2 images); *Palpigradi* (p.4(*4*)); *Peckhamia americana* (p.525, 3 images); *Phidippus apacheanus* (p.529, 4 images); *Phidippus ardens* (p.530, 4 images); *Phidippus asotus* (p.530; p.531, 3 images); *Phidippus californicus* (p.534, 6 images); *Phidippus carneus* (p.535, 4 images; p.536, 2 images); *Phidippus comatus* (p.539, 6 images); *Phidippus octopunctatus* (p.542, 2 images; p.543, 3 images); *Phidippus phoenix* (p.545, 5 images); *Phidippus tigris* (p.550, 3 images); *Phidippus tux* (p.551, 4 images; p.552, 2 images); *Phidippus tyrrelli* (p.552, 4 images; p.553); *Platycryptus arizonensis* (p.554, 2 images; p.555); *Platycryptus californicus* (p.555, 4 images); *Plectreurys tristis* (p.329 (*6a*); p.415; p.416, 3 images); *Salticus palpalis* (p.559, 2 images; p.560, 2 images); *Sassacus papenhoei* (p.562, 2 images; p.563, 2 images); *Sassacus vitis* female (p.563, 2 images); Schizomida (p.3 (*2*)); *Selenops actophilus* (p.300, eye configuration; p.301, 4 images); *Septentrinna bicalcarata* (p.338, 2 images); Solifugae (p.6 (*12*)); *Sosippus californicus* (p.394, 2 images; p.395, 2 images); *Strotarchus beepbeep* (p.438); *Synema viridans* (p.319, 4 images); Thelyphonida (p.4 (*5*)); *Tidarren sisyphoides* (p.292, 4 images); *Tivyna moaba* (p.226, 2 images); *Trechalea gertschi* (p.579, eye configuration; p.580, female with egg sac); *Trogloneta paradoxa* (p.232; p.234, 4 images); *Zelotes monachus* (p.357). **Jason M. Crockwell**: *Chinattus parvulus* male (p.490). **Michael Davis**: *Zoropsis spinimana* (p.583, 3 images). **Laura Dirkx**: *Araneus diadematus* upper right female, lower left female, and female ventral (p.135, 3 images); Grackle with spider prey (p.29 (*46*)); Hummingbird nest (p.29 (*47*)); *Tetragnatha viridis* female (p.207). **Sebastian Echeverri**: Mesothelae (p.6 (*14*)).**Lynette Elliott**: *Castianeira descripta* (p.333). **Michael Ferro**: *Microdipoena guttata* (p.219 (*10a*); p.232). **Diona Fredo**: *Trichonephila clavata* (p.187). **Alyssa Fuller**: *Coras medicinalis* (p.75); *Hypochilus pococki* (p.217 (*4a* left); p.228 upper and lower right; p.229, 3 female images). **Mike Graziano**: *Geolycosa turricola* burrow entrance (p.367). **Cristian Grismado**: *Myrmecicultor chihuahuensis* (p.330 (*13a* right); p.405, upper 2 images). **Jeff Gruber**: *Euophrys monadnock* (p.496, 4 images). **Kevin Hall**: *Philodromus dispar* female (p.461).

Kat Halsey: *Latrodectus geometricus* (p.259, 2 images); *Neoscona oaxacensis* female (p.184, 2 images).
Toby Hays: *Gertschanpis shantzi* (p.124 (*2a, 3b*); p.125; p.126, 5 images). **Marshal Hedin**: *Antrodiaetus unicolor* female at burrow (p.44); *Aphonopelma* sp. burrow and eye configuration (p.39 (4b); p.53);
Aptostichus burrow (p.47, 2 images); Bothriocyrtum californicum burrow (p.39 (4a); p.50); *Calisoga longitarsis* (p.40 (*6b*)); p.52, 3 images); *Calponia harrisonfordi* (p.584 (*2a*); p.585, 2 images); *Diguetia canities* retreat (p.226); *Euagrus chisoseus* (p.61 (*2a, 3b*), 3 images; p.65, 2 images; p.66, 2 images);
Habronattus americanus (p.497, 2 images); *Hexurella rupicola* (p.67); *Latrodectus geometricus* upper left female (p.258); *Latrodectus Hesperus* (p.260, 3 images); *Loxosceles deserta* (p.303); *Metaltella simoni* female bottom right (p.91); *Microhexura Idahoana* (p.61 (*3a*); p.68, 2 images; p.69, 3 images);
Microhexura montivaga leg (p.61 (*3a*); p.68); *Sergiolus montanus* (p.353); *Titiotus shantzi* (p.427 (*13b*);
p.581; p.583, 2 images); *Trogloraptor marchingtoni* (p.217 (*5a*); p.295; p.296, 2 images; p.598,
pseudosegmented); *Yorima angelica* (p.89, 2 images). **Chad Heins**: *Emblyna annulipes* (p.223); *Theridion pictum* (p.289); *Titanoeca americana* male, and subadult male (p.295, 2 images). **Solomon Hendrix**:
Habronattus decorus female (p.499). **Andy Hoffman**: *Antrodiaetus unicolor* female (p.44); *Tigrosa aspersa* female with young (p.397). **Jeff Hollenbeck**: *Decaphora cubana* (p.577, 3 images); *Steatoda grandis* female (p.276). **David and Laura Hughes**: *Araneus guttulatus* female with egg sac (p.136);
Syspira longipes (p.448) *Trypoxylon* nest (p.30 (*50*)). **J. C. Jones**: *Tigrosa georgicola* female and female ventral (p.397). **Thaddeus Charles Jones**: *Neotama mexicana* (p.41 (*9b*); p.57, 3 images). **Rudo Kemper**: *Zosis geniculata* (p.215). **Joseph T. Lapp**: *Gasteracantha cancriformis* male (p.162);
Misumessus lappi (p.314, 3 images; p.315, 2 images). **Joel Ledford**: *Calileptoneta helferi* (p.218 (*9a*);
p.229; p.230, 3 images). **John Liebeskind**: *Neoscona* sp. (p.28 (*42*)). **David Lightfoot**: *Myrmecicultor chihuahuensis* (p.330 (*13a* left); p.405, lower; p.406, 2 images). **Annika Lindqvist**: *Myrmekiaphila comstocki* (p.48, 2 images). **Isaac Lord**: *Anapistula* (p.124 (*3a*); p.190; p.191, 2 images). **Chandra Lowe**:
Trechalea gertschi (p.424 (*6a*); p.579, female with young; p.580, female with young, male, 2 images;
p.581, 2 images). **Tammy MacKenzie**: *Araneus cavaticus* (p.133, 3 images); *Habronattus decorus* upper right male (p.499). **William Mason**: *Metellina mimetoides* (p.198, 2 images). **John R. Maxwell**: *Sphodros rufipes* male (p.39 (*3b*); p.45; p.46). **Sean McCann**: *Phidippus johnsoni* (p.540, 5 images); *Philodromus dispar* male (p.461, 2 images); *Plexippus paykulli* male (p.558, 3 images); *Rhomphaea fictilium* female
(p.271, 2 images); *Syspira* sp. (p.447). **Rowan McGinley**: *Schizocosa rovneri* (p.392). **Graham Montgomery**: *Leptoctenus byrrhus* (p.447). **Célio Moura**: *Cithaeron praedonius* (p.584 (*2b*); p.587, 3 images; p.588, 4 images). **Tom Murray**: *Antistea brunnea* (p.92); *Cheiracanthium inclusum* (p.426 (*12b*);
p.435; p.436, 2 images); *Hahnia cinerea* (p.96); *Xysticus elegans* (p.323, 2 images). **Eric Neubauer**: *Zora pumila* lower female (p.449). **Tony Palmer**: *Homalonychus selenopoides* (p.328 (*4a*); p.358; p.359, 5 images); **Jay Paredes**: *Leucauge argyrobapta* (p.195). **Laura Paxson**: *Dolichognatha pentagona* (p.192,
2 images; p.193, 2 images). **Arlo Pelegrin**: *Zodarion rubidum* female and male (p.422, 2 images). **Kevin Pfeiffer**: *Hololena curta* (p.77). **Erin Powell**: *Deinopis spinosa* (p.298, web in anterior legs; p.299, lower male, female with web, and female with web waiting for prey, 3 images). **Lisa Powers**: *Acacesia hamata* female (p.128); *Araneus bicentenarius* (p.131, 2 images); *Araneus juniperi* (p.137); *Misumenoides formosipes* male (p.313); *Peucetia viridans* male on flowers (p.457). **Darius A. Przygoda**: *Zygiella x-notata* (p.190, 2 images). **Michael Rexman**: *Rabidosa carrana* (p.384). **Jesse Rorabaugh**: *Metaltella simoni* (p.63 (*7a*); p.89; p.90, 2 images; p.91, top 2 images). **Jenn Rose**: *Phidippus mystaceus* juvenile
(p.541). **Danielle Siddle**: *Zora pumila* (p.427 (*13a*); p.449, upper female). **John Sloan**: *Agroeca ornata*
(p.361, 2 images); *Amaurobius borealis* (p.80); *Araneus marmoreus* keyhole variant (p.138); *Drassodes neglectus* (p.344, 2 images). **Natalie Stalick**: *Araneus gemmoides* (p.135; p.136, 2 images). **Aaron Stoll**: *Maymena ambita* (p.232; p.233, 2 images). **David Timmons**: *Scoloderus nigriceps* web (p.186).
Leonard Vincent: *Alopecosa kochi* (p.364, 5 images); *Argiope argentata* male and female (p.149, 2 images); *Heteropoda venatoria* female (p.23 (*40b*); p.576; p.577; p.578, 3 images); *Messua limbata* male (p.517, 2 images); *Scotophaeus blackwalli* (p.352, 4 images); *Steatoda nobilis* (p.278, 2 images; p.279,
2 images); *Trachelas pacificus* (p.418, 3 images). **Gerald Wegner**: *Acacesia hamata* male (p.128);
Eriophora ravilla left female, and juvenile (p.158, 2 images); *Nesticodes rufipes* (p.265; p.266, 2 images).
Kevin Wiener: *Argiope aurantia* (p.28 (*45*)); *Kukulcania hibernalis* (p.40 (*7a*); p.41 (*8b*) 2 images; p.54;
p.55, 2 images; p.56, 6 images); *Mecynogea lemniscata* egg sac (p.174); *Phidippus regius* (p.549, 6 images; p.550, 2 images); *Spintharus flavidus* (p.273, 3 images); *Tigrosa georgicola* female with young
(p.397). **Justin Williams**: *Anasaitis canosa* (p.484, 2 images); *Araneus miniatus* female with egg sac
(p.139); *Colonus puerperus* male (p.492); *Lupettiana mordax* (p.433); *Wamba crispulus* (p.293,
2 images). **Susan Wise-Eagle**: *Araneus saevus* (p.144, 2 images); *Cybaeopsis wabritaska* (p.84).
Ian Wright: *Phidippus mystaceus* female (p.541).

space webs, orb webs, and sheet webs highlighted by dew on a foggy morning

THE WONDERFUL WORLD OF SPIDERING

"If the terrestrial world is a stage, then any predator as abundant and ubiquitous as the spider must be a major character in the ensuing ecological and evolutionary dramas" (Wise 1993).

Spiders can be found in almost every terrestrial ecosystem and have been reported from every continent, even Antarctica (Forster 1971). The World Spider Catalog currently states that over 49,000 species of spiders are known worldwide, and the number increases every year. North America (north of Mexico) is home to over 4,000 known species of spiders, but there is still a lot to learn; there are many species out there that have not been described. Spiders are a highly diverse group but often overlooked, especially as most of them are quite small.

Spiders tend to evoke strong emotions in many people. Some truly love these critters, some are apathetic to them, and some are completely terrified of them. Arachnophobia (the fear of spiders) is one of the most reported phobias in Western cultures, but the good news is that phobias can often be overcome. Exposure and education are two key aspects to overcoming fear. Trained therapists can also help people overcome their phobias, and there is no shame in seeking professional help if you need it. Some therapists are showing great success using virtual reality (Carlin, Hoffman, and Weghorst 1997), and as technology advances, the options to help people overcome their fears will also expand. If you have bought this field guide to help you overcome your phobia, I commend you; you have taken a brave first step to appreciating these wonderful creatures.

This field guide will describe what a spider is and give you some of the basic identification knowledge to be able to narrow down which spiders you find. This is not meant to be a way to identify all spiders to species. In many cases, you need to examine the genitalia of a mature species to confirm a species-level identification, and at times, this requires dissection of the specimen. There is no way that a field guide can cover all the species found in North America, so this guide will cover some of the more commonly encountered species and provide tips for identifying them to at least family level.

Identification is often achieved utilizing dichotomous keys, which present a series of choices based on physical structures at each step. Starting at the first couplet, you select the option that best matches the animal you are observing. This will direct you either to another couplet or to the identification.

For a simple example, let's consider the following candies: M&M's®, Hershey's Kisses, starlight mints, peanut butter cups, Skittles®, and gummi bears. The dichotomous key could be as follows:

1a.	Covered in a hard shell	2
1b.	Not covered in a hard shell	3
2a.	Chocolate flavored	**M&M's®**
2b.	Fruit flavored	**Skittles®**
3a.	Hard and mint flavored	**Starlight mints**
3b.	Neither hard nor mint flavored	4
4a.	Chewy, somewhat soft, fruit flavored	**Gummi bears**
4b.	Not chewy, firm but not soft, not fruit flavored	5
5a.	Solid chocolate, usually wrapped in foil	**Hershey's Kisses**
5b.	Peanut butter core covered with chocolate	**Peanut butter cups**

So with this key, if you had Skittles®, you would start at number 1. As Skittles® are covered in a hard shell, you would go to number 2. They are fruit flavored, not chocolate, leading you to the identification of Skittles®. Be sure to read the couplets carefully, as in some cases, combinations of traits (not just one) are required to make the right choice.

In this field guide, you will find dichotomous keys for the class Arachnida and for family-level identification at the start of each of the guild chapters. Usually, the keys are most easily followed when you have a specimen, but clear photographs will often allow you to see the needed characteristics.

Spider identification can be tricky. Sometimes lots of subtle traits need to be observed to make a conclusive identification, and sometimes a family-level identification will be the best you can do. Please don't let this discourage you.

WHAT IS A SPIDER?

Spiders are in the order Araneae and part of the class Arachnida. Many people are easily confused about the class, as arachnophobia usually refers to an irrational fear of spiders, but other animals than spiders are in this class.

A Quick Biology Reminder

All living things are categorized, and all the organisms in a category share some similar traits that distinguish them from other living things. One simple way to remember the order of classification is using mnemonics. For example: "Divine kings play chess on fancy glass stools" or "Do kindly place candy out for good students." By using the first letter of each word, one can remember the taxonomic ranks:

DOMAIN	Divine	Do
KINGDOM	Kings	Kindly
PHYLUM	Play	Place
CLASS	Chess	Candy
ORDER	On	Out
FAMILY	Fancy	For
GENUS	Glass	Good
SPECIES	Stools	Students

All spiders (and all other animals) are in the domain Eukarota and the kingdom Animalia. The phylum Arthropoda (which roughly translates to "jointed foot") includes animals with an exoskeleton, a segmented body, and paired jointed appendages. Members of the class Arachnida have eight walking legs (although these can be modified in some cases). They have chelicerae—appendages on the anterior of the prosoma, each of which is composed of a basal segment and fang—and a pair of pedipalps, small leglike appendages near the front of the animal. Their bodies are divided into two tagmata, or body regions: the prosoma or head, and the opisthosoma or abdomen, although these can be fused and difficult to see in some cases. (In contrast, insects have three body regions and three pairs of walking legs.)

KEY TO THE EXTANT (NOT EXTINCT) ARACHNIDS

1a.	Taillike appendage at end of opisthosoma (abdomen)	**2**
1b.	No taillike appendage at end of opisthosoma (abdomen)	**6**
2a.	Segmented tail with a modified telson (terminal segment) containing a venom bulb and stinger	**Scorpiones (Scorpions)**
2b.	Tail not segmented and lacks venom bulb and stinger	**3**
3a.	Tail short, sometimes uniquely shaped	**Schizomida (Shorttailed whipscorpions)**
3b.	Tail usually long and thin	**4**
4a.	Marine animal with hard shells over prosoma (cephalothorax) and opisthosoma (abdomen)	**Xiphosura (Horseshoe crabs)**
4b.	Not as above	**5**

5a. Very small (less than 5 mm), pedipalps used in locomotion -- **Palpigradi (Microwhip scorpions)**
5b. Larger (usually 25–85 mm), pincerlike pedipalps, sprays acetic or octanoic
acid when disturbed -- **Thelyphonida (Whipscorpions)**

6a. Pedipalps appear clawlike or raptorial -- **7**
6b. Pedipalps neither clawlike nor raptorial--- **8**

7a. Pedipalps look like scorpion claws, usually small (2–8 mm)------------------- **Pseudoscorpions**
7b. Pedipalps raptorial, larger (5–70 mm) ------------------ **Amblypygi (Tailless whipscorpions)**

8a. Body regions fused, making them appear to have only one body section ----------------------- **9**
8b. Body regions not fused, with clear delineation between prosoma and opisthosoma --------- **10**

9a. Legs usually fairly stout --- **Acari (Mites and ticks)**
9b. Legs usually thin and long-- **Opiliones (Harvestmen)**

10a. Cucullus or hood covering anterior of prosoma (cephalothorax) -- **Ricinulei (Hooded tickspiders)**
10b. No cucullus present --- **11**

11a. Chelicerae modified and pincerlike---------------------------------- **Solifugae (Wind scorpion)**
11b. Chelicerae not modified to be pincerlike --------------------------------- **Araneae (Spiders)**

Scorpiones: Scorpions (fig. 1). These animals are best known for a modified telson (terminal tail segment) that has a venom bulb and stinger. The pedipalps are modified into large, grasping, pincerlike claws. Many species are known to fluoresce under black light.

Schizomida: Shorttailed whipscorpions (fig. 2). These are small animals (less than 5 mm) that usually occupy leaf litter or caves. The first pair of legs is antenniform. The tail or flagellum has only one to four segments and may be uniquely shaped.

Xiphosura: Horseshoe crabs (fig. 3). These are marine animals that often live in shallow coastal waters. Both the opisthosoma and prosoma are covered with thick hard shells. They have a long telson that is used to help the animal move and right itself if it is upside down. According to a recent study (Ballesteros and Sharma 2019), horseshoe crabs are arachnids and sister taxa (most closely related) to Ricinulei (hooded tickspiders).

Palpigradi: Microwhip scorpions (fig. 4). These animals are extremely small (less than 5 mm) and tend to live in wet soils in warm habitats. They are colorless. Their pedipalps are used for walking. They have a long taillike flagellum.

Thelyphonida: Whipscorpions, sometimes called vinegaroons (fig. 5). These animals are known to spray an acetic or octanoic acid from their thin whiplike tail as a defense. The first pair of legs is antenniform, and the pedipalps are heavily armored.

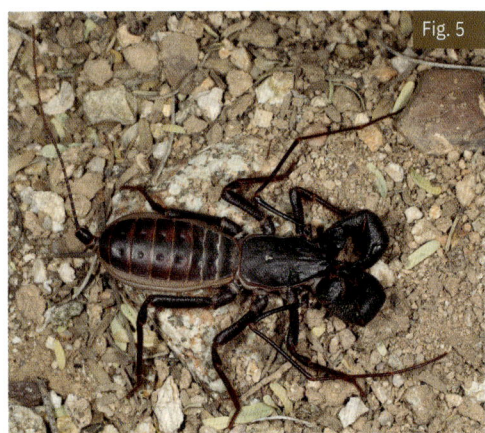

Fig. 4

Fig. 5

Pseudoscorpions: These are small animals (2–8 mm) and often overlooked (fig. 6a). They do not have a stinging tail but otherwise look like miniature scorpions. They produce silk from their chelicerae, and venom is produced in their modified clawlike pedipalps (although they are too small to be of any threat to humans). Some species are known to travel using a technique referred to as phoresy, where they cling onto a larger, usually flying, animal (fig. 6b).

Fig. 6a

Fig. 6b pseudoscorpion traveling using phoresy

Amblypygi: Tailless whipscorpions, sometimes called whipspiders (fig. 7a). These are fierce-looking animals (although harmless to humans) with raptorial pedipalps (fig. 7b). The first pair of legs is extremely long, whiplike, and antenniform (used for sensory purposes). They do not produce silk or venom.

Fig. 7a

Fig. 7b amblypygi showing raptorial palps

Acari: This is a subclass and includes the mites and ticks. It is divided into two superorders: Acariformes or Actinotrichida, which tend to be herbivorous (fig. 8), and the Parasitiformes or Anactinotrichida, which includes many species that are considered parasites (fig. 9).

Fig. 8

Fig. 9

Opiliones: Harvestmen. These animals are sometimes referred to as "daddy-longlegs," which causes much confusion, as other animals are also commonly referred to by this name (fig. 10). Opiliones have fused body regions, so they often look like a football with legs. Oftentimes, they have very long thin legs, although there are some with shorter legs. Opiliones lack venom, but some can give off a noxious chemical defense. They are capable of eating solid pieces, unlike their spider cousins, which must liquefy their meal before ingesting it.

Fig. 10

Ricinulei: Hooded tickspiders (fig. 11). These unique animals hide their face behind a cucullus or hoodlike structure, giving them an unusual appearance. They are eyeless. The third leg is modified in the males for sperm transfer, and the genital opening is located on the pedicel.

Solifugae: Windscorpions, sometimes also called "sun spiders" or "camel spiders" (fig. 12). These fast-running arachnids have large intimidating chelicerae. The long pedipalps cause many people to think, mistakenly, they are ten-legged. They are the subject of many urban legends that exaggerate their size, speed, and toxicity. (They have neither venom glands nor a venom-delivery method.)

Fig. 11

Fig. 12

Araneae: Spiders (fig. 13). So what makes a spider a spider? Like all arachnids, they have eight legs, pedipalps, and chelicerae. Unlike insects, they do not have antennae or wings. Unique to spiders are the spinnerets, used to produce silk. All spiders produce silk, and it has many uses. Some build elaborate capture webs with it, some use it as drag lines, and some make little silken retreats that are used for resting or molting. Silk can be used to wrap up prey, and it is used to cover the egg sacs. Spiders can produce up to seven different kinds of silks. Different silks have different uses.

Fig. 13

Spiders are separated into two suborders: Mesothelae and Opisthothelae. Mesothelae (fig. 14) are considered the most ancestral of the extant spiders. They have a clearly segmented abdomen, and the spinnerets are between the two pairs of book lungs. These spiders are found in Asia, and none is found in North America. They will not be covered in this guide. Opisthothelae have their spinnerets at the end of the opisthosoma. Opisthothelae is separated into two infraorders: the Mygalomorphae and the Araneomorphae. Mygalomorphs include some of the largest spiders and include such creatures as tarantulas. They also tend to be long-lived animals. Most live in burrows or cavities. Araneomorphs, sometimes called the true spiders, are most of the spiders we

Fig. 14

encounter on a day-to-day basis and include such animals as orbweavers, wolf spiders, and cobweb spiders. Araneomorphs can be divided into two groups; the entelegynes (those with more complex genitalia; the female having three external openings—not easily seen—and a sclerotized epigynum; the males with complex palp structures) and the haplogynes (those with more simple genitalia; the female having only one external opening, and the male palp with a simple bulb).

HOW TO SEPARATE THE MYGALOMORPHAE FROM THE ARANEOMORPHAE

To determine if a spider is a Mygalomorphae or an Araneomorphae, look for the following combinations of traits:

Mygalomorphae: Their chelicerae are paraxial, or parallel to the central axis of the body (fig. 15a). When they attack, they raise up their prosoma (or cephalothorax) to unfold the fangs (fig. 15b) and strike downward onto the prey. They have two pairs of book lungs. They have eight eyes that are usually clustered together (fig. 15c). Their legs are usually robust.

Araneomorphae: The chelicerae are diaxial, or at an angle to the central axis of the body (fig. 16a). When they attack, the fangs are used in a pinching motion. They usually have only one pair of book lungs. If they have two pairs of book lungs, they also possess a cribellum (a specialized plate that produces cribellate silk). Most have eight eyes, but some have six, four, two, or none. There are many different eye configurations, but often they are in two rows across the front of the prosoma (fig. 16b).

Fig. 15a

Fig. 15b

Fig. 16a

Fig. 15c

Fig. 16b

Fig. 15a ventral view of *Myrmekiaphila foliata* showing paraxial chelicerae

Fig. 15b *Sphodros* sp. with fangs unfolded

Fig. 15c eye cluster of *Antrodiaetus unicolor*

Fig. 16a ventral view of *Elaver excepta* showing diaxial chelicerae

Fig. 16b eyes in two rows across front of prosoma of *Elaver excepta*

SPIDER ANATOMY

Some basic terminology:

Dorsal is the top side, and *ventral* is the underside. *Anterior* means "toward the front," while *posterior* means "toward the rear." *Lateral* means "to the side"; *median* means "toward the middle." *Proximal* is close to the body core, and *distal* is away from the body.

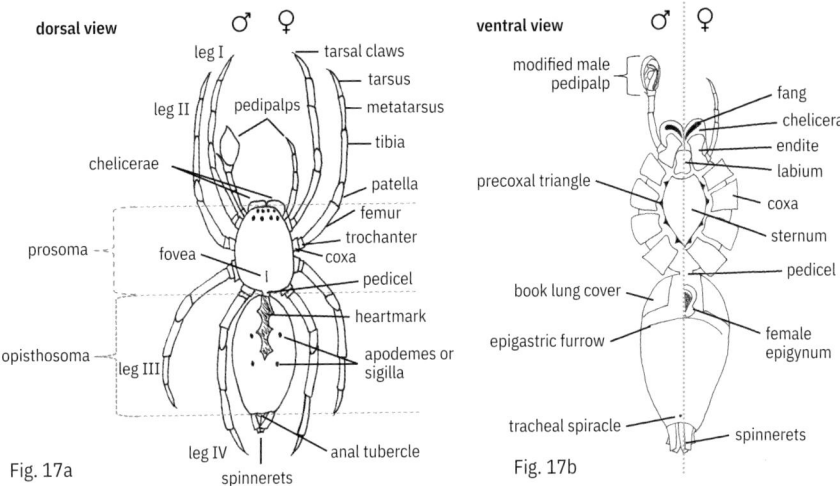

Fig. 17a

Fig. 17b

Fig. 17c

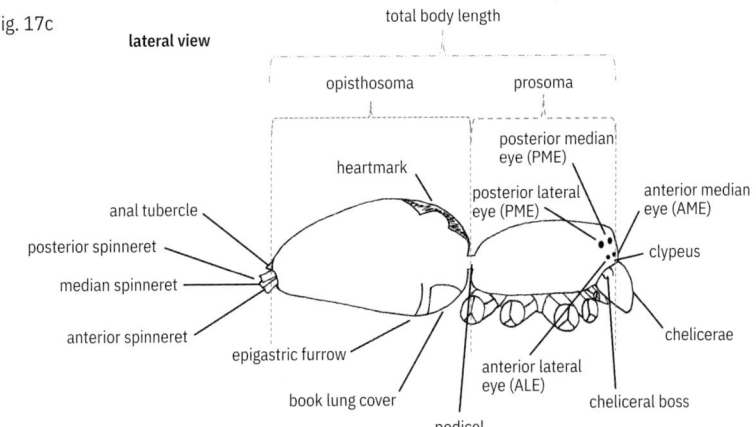

To learn how to identify spiders, one must also have an understanding of the basic anatomy (fig. 17a–c). Spiders have two main body regions or tagmata: the *opisthosoma* (some people refer to this as the abdomen) and the *prosoma* (some people call this the cephalothorax, or head). These two tagmata are connected by a narrow structure called a *pedicel*. The spinnerets are on the opisthosoma, and the mouthparts and legs are on the prosoma. (This is one mistake many make about spiders, as many "spiders" are depicted with their legs coming from the opisthosoma.)

Spiders have four pairs of walking legs. These are numbered starting near the front of the spider and using Roman numerals: first is leg I, then leg II, leg III, and leg IV. The legs are made up of the coxa at the base, the trochanter, femur, patella, tibia, metatarsus, tarsus, and tarsal claws. Spiders also have a pair of *pedipalps*; these often look like small walking legs and are near the chelicerae. Pedipalps have fewer segments (lacking the metatarsi) than the walking legs and are used for food manipulation and chemosensory purposes. Mature male spiders have modified pedipalps (additional structures that often make them look like they are wearing boxing gloves) for sperm transfer. At the base of the pedipalps are the *endites*; some spiders have endites that are lined with tiny teeth, forming what is called a *serrula*.

At the front of the spider is a pair of *chelicerae*; the fangs are attached to these at the distal end. The fangs are usually held folded into a furrow in the chelicerae, which is sometimes lined with teeth. The actual mouth of the spider is beneath the chelicerae. Spiders liquefy their food before consuming it. At the entrance to the mouth are rows of hairs that filter any solid particles out of the food being ingested. The area above the chelicerae but below the eyes is referred to as the *clypeus*.

On the top of the prosoma is often a visible mark called a *fovea*. This is where the muscles for the stomach attach. The ventral side of the prosoma has the sternum. Here you can also see the *coxae*, which are the base segments for the legs. The opisthosoma often has a visible *heartmark* on the dorsal surface. This is directly over where the heart of the spider is, and in some species, you can easily watch the heart beating. The dorsal opisthosoma may be smooth, or there may be visible depressions called *sigilia*; these are points where muscles attach. The *spinnerets* are usually located at the posterior end of the spider, and their number and placement are also useful in spider identification. In some spiders the anterior spinnerets are a modified plate, called a cribellum (and those spiders are referred to as cribellate). The ventral opisthosoma has a crease running transversely across it called the *epigastric furrow*. Mature females will have their *epigynum* or *gonopore* located just anterior to this. A pair of *book lungs* is anterior of the epigastric furrow (usually but not always one pair). The covers of the book lungs are sometimes colored or textured. Most spiders also have a *tracheal spiracle* (or more than one in some cases). This is usually located near the spinnerets. Its placement and number can also be helpful in identification.

All information given has been provided under the assumption that you will have an intact specimen that is not deformed in any way. Deformities can happen. Figure 18a shows a *Dolomedes vittatus* that either hatched with a deformity or suffered some great injury, causing it to have fewer eyes than usual. Injuries can happen, too; loss of legs is common for spiders. Figure 18b shows a *Schizocosa* species (family Lycosidae) with only five legs, and all three missing legs were on the same side. This did not seem to slow her down at all. These spiders seem to fare quite well in their natural environment, but these differences can sometimes make identification tricky if the trait that has been injured or is missing is diagnostic.

Fig. 18a

Fig. 18b

Fig. 18a *Dolomedes vittatus* missing eyes

Fig. 18b *Schizocosa* sp. missing three legs on left side

In the scientific literature, spiders are measured from the front of the prosoma to the end of the opisthosoma. This is referred to as the *total body length*. This is recorded in millimeters (metric is used for scientific literature) but can easily be converted to inches if you remember that 25.4 mm equals 1 in. A ruler marked in inches and millimeters is provided on page 612 to assist you with this.

There can be many variations in the markings of spiders (fig. 19). Marginal bands or stripes will be on the very edge, usually of the prosoma. Median bands or stripes run longitudinally down the middle. Submarginal bands are in the area between the median and marginal bands. An anterior band on the opisthosoma runs transversely across the anterior edge. If a band or stripe is referred to as broken, it is discontinuous, like a dashed line. Paired spots are usually on either side of the medial line. Chevron markings are usually on the opisthosoma. Leg banding creates rings on the legs, whereas leg striping runs longitudinally. These markings can be darker than the background coloration of the spider, or they can be lighter.

One useful trait that may be used to help narrow down the family-level identification is the eye configuration or eye placement. Most spiders have eight eyes, some have six, some have four, a few have two, and some are eyeless. Eyes are often in two rows and are referred to by their placement. The anterior eyes are toward the front of the spider, while the posterior eyes are toward the rear of the head. The median eyes are toward the midline, and the lateral eyes are toward the edge. As such, we end up with the anterior median eyes, the anterior lateral eyes, the posterior median eyes, and the posterior lateral eyes (fig. 20). The posterior and anterior eye rows can be straight or curved. If they are curved so the lateral eyes are closer to the anterior than the median eyes, this is referred to as *procurved*. Conversely, if they are curved so the lateral eyes are closer to the posterior than the median eyes, this is referred to as *recurved*.

Lastly, we should address coloration. Some species can come in several different color forms. In addition, some spiders' coloration can be affected by prey consumption, spiders' coloration will often vary immediately before and after molting, and some spiders are able to shift their color based on their environment. Therefore, coloration may not always be the most useful trait when trying to identify spiders (fig. 21).

How can you look at the ventral side of a spider? Looking at the anatomy and reading the descriptions of many species, you will see that you need to look not only at the dorsal surface of the spider but also at the ventral surface. This can be fairly easy with

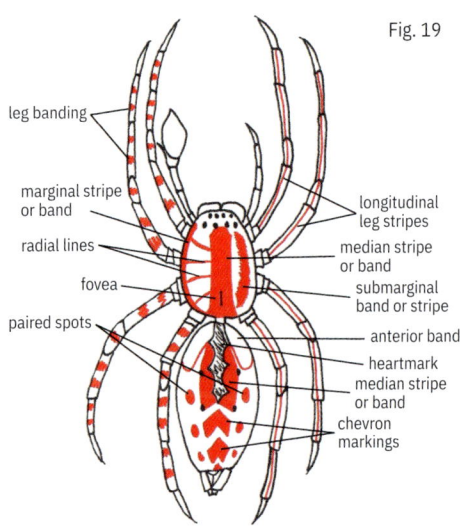

Fig. 19

leg banding

marginal stripe or band

radial lines

fovea

paired spots

longitudinal leg stripes

median stripe or band

submarginal band or stripe

anterior band

heartmark

median stripe or band

chevron markings

Fig. 20

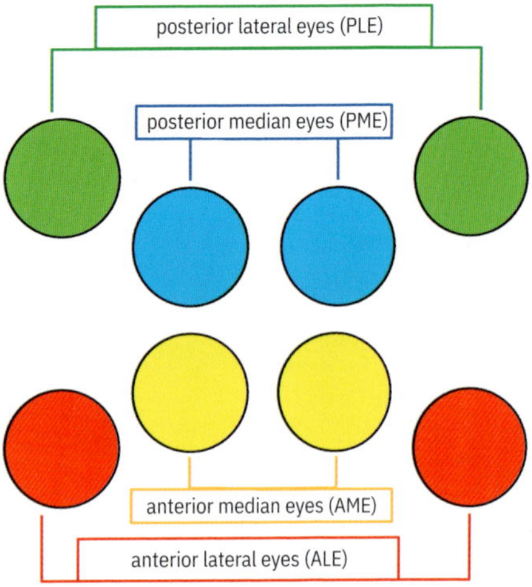

posterior lateral eyes (PLE)

posterior median eyes (PME)

anterior median eyes (AME)

anterior lateral eyes (ALE)

Fig. 21a *Pholcus* that is pink from consuming ladybugs

Fig. 21b *Parasteatoda tepidariorum* with green prey causing color shift

Fig. 21c *Parasteatoda tepidariorum* with an overall green coloration from prey consumption

Fig. 21d the same *Parasteatoda tepidariorum* in the previous photo a few days later

some of the web-building spiders, which sit suspended within their web, but tricky with ground-dwelling spiders. If you are able to catch the spider, there are a couple of ways to get a good view of the ventral side.

First, spiders can be temporarily anesthetized using carbon dioxide (CO_2). A bicycle tire inflator (fig. 22) or keyboard duster that uses CO_2 can be used. Catch the spider in a container that has a lid. Carefully fill the container with CO_2 and place the lid on. (Since you can hurt or kill a spider if you fill the container with CO_2 too rapidly, you will want to try this without a spider in the container the first time.) After a few moments, the spider will become motionless. At this point, you can turn the spider out of the container to obtain a ventral view. Be careful moving spiders, as they are very delicate. Be warned, not all spiders will give you any indication that they are waking

Fig. 22

up, so place the spider where it will not fall or cause harm; it may be a little disoriented upon waking. Also, do not place the spider in your hand, as it can be startling when it awakens.

Fig. 23a spi-pot with the sections separated

Fig. 23b spi-pot with the sections together

Fig. 23c ventral view of a spider in the spi-pot

Fig. 24 spider being held in plastic bag in an embroidery hoop

Second, if you cannot (or don't want to) anesthetize the spider, you can use a spi-pot (fig. 23). Spi-pots are described in the Collins field guide *Spiders of Britain and Northern Europe* (Roberts 1995) and can be made in different sizes to accommodate spiders of different sizes. Lots of online tutorials show how to make them. Remember, spiders are delicate, so you will need to be careful not to hurt the spider when placing it in your spi-pot. Do not keep the spider in the cup for a long time, just long enough to take a closer look or snap a couple photos.

Third, you can trap the spider into a clear sandwich bag and, using an embroidery hoop (fig. 24), hold the spider immobilized while you look at the ventral side.

WHERE DO SPIDERS LIVE?

Since spiders are highly diverse, they can live in just about any terrestrial ecosystem. Species specialize in many different habitats. There are even spiders that are well adapted to living in our homes and are considered synanthropic (from the Greek for "together" and "man"). There is also a diversity of hunting strategies and life histories. So you can find spiders under rocks, on trees, in large webs, running on the ground, hiding in tall grasses, hunting on the surface of water, hiding in the leaf litter, even inhabiting your home.

OBSERVING SPIDERS

Observing spiders in the wild can be difficult but also rewarding. Spiders occupy many different habitats and can be observed almost anywhere. The key is finding them. One of the things to remember is that all spiders produce silk, and that finding their silk can help you track down the spiders.

Techniques for Finding and Collecting Spiders

First and foremost, one needs to be aware of local laws and restrictions on collecting spiders and altering the habitat. Please be aware of local laws and restrictions before trying to collect any animals from the wild. You must also be aware of any threatened or endangered species in your area, so you can be sure that your actions will not harm them in any way. In general, if you do not own the land from which you will be collecting, you would at least need to get permission from the landowner before attempting to collect any animals.

VISUAL SEARCHING

Web-based spiders can be found by following the silk. Webs built for prey capture are highlighted well on foggy mornings, making them great times to hunt for spiders. If you can see the web, careful examination will often lead you to the spider. If the spider is not out in the open on their web, remember to look at the edges, in the vegetation, or on the structure to which the web is attached. Funnel-weaving spiders often hide in the back of their funnel retreat and will rush down at the slightest disturbance. Some of these spiders can be coaxed into the open with a tuning fork or sonic toothbrush (fig. 25), as the vibrations mimic prey. Spiders also like to hide under things. Rolling logs is a good way to find spiders, but please always try to leave things as you found them; replace the log into its original position when you are done searching. Walls, fences, bridges, and cellars are other great places to look for spiders. It is important when searching to take care not to harm the environment or the plants and other animals within it.

Fig. 25a sonic toothbrush luring Agelenidae from its retreat

Fig. 25b sonic toothbrush luring *Amaurobius* sp. from its retreat

SWEEP NETTING

This technique is great for grassy habitats. A canvas or mesh net is swept through the vegetation several times (fig. 26). Be sure to avoid areas with thorny plants, as these will tear or snag your net. Also, if you have plant allergies (for example, to poison ivy), you should avoid sweeping those plants, as the oils and residues from the plants can transfer to your net. Oftentimes, a sheet is laid flat on the ground, and the contents of the net are placed onto it for examination. Many creatures

Fig. 26a sweep netting

Fig. 26b sorting the sweep net sample

Fig. 27 vegetation beating

will make a quick dash to get away, so be prepared to capture as soon as the contents of the net are emptied. Some spiders will play dead (the scientific term is *thanatosis*), so look carefully through the debris from the net for those critters. This technique should be done only in dry conditions, as moisture will make the insects/spiders stick to the debris and can easily cause injury or death.

VEGETATION BEATING
This method is used mainly in areas of woody plants. A sheet or tarp (or even an open umbrella) is placed or held below the branches of vegetation. Then the branches are struck with a pole or branch, or they are grasped and vigorously shaken, dislodging spiders that are within that structure and causing them to drop onto the sheet beneath (fig. 27). This is often easier to do with more than one person, as the dislodged spiders will often run frantically off the sheet. It can be helpful to have one person who is prepared to capture those spiders while another shakes or beats the vegetation. Also, watch for spiders suspended on a line of silk. They will often quickly retreat to the branch on their silk line when the shaking stops. Once again, this technique is best done in dry conditions. If the vegetation is wet, the spiders will stick to the moisture and can be hurt or killed.

SNIFFING FOR SPIDERS
Sniffing for spiders does not use odors or the sense of smell but is a technique for observing spiders at night. Many spiders are more active during the night. This technique requires the use of a bright flashlight or headlamp and is best done during dry weather conditions (as condensation and raindrops will create reflections similar to the eyes of a spider). Many spider eyes have a reflective tapetum, which will reflect the light from a flashlight. The key to this method is the angle of the light. One way to obtain a good angle is to hold the flashlight or place your headlamp below your nose (which is where the name "sniffing" came from). If you sweep the light back and forth across grassy areas, you will see twinkling greenish-blue reflections that likely belong to spiders. Other animals' eyes also have these reflective properties. (Think about wildlife eyes you have seen highlighted by car headlights.)

PITFALL TRAPPING
Pitfall trapping is a great way to capture ground-running spiders, and also a good technique to capture night-active spiders without having to be out at night, as the trap can be set during the day and left overnight before the contents are examined. For pitfall traps, a catch container is placed in the ground so the edge of the container is level with the ground's surface (fig. 28). Some traps can have a funnel top, and you can also provide a roof over the trap to help prevent debris and rain from entering. If you are interested in just looking at the spiders caught, you can leave the trap dry. (Some people even put some moss or leaves in the bottom.) Most scientists using this collection method put in a few inches of chemical that kills and preserves the spiders until

the contents are collected. (Propylene glycol is a good choice, as it is less toxic than other chemicals.) A couple drops of dish soap will help reduce the surface tension, so that the spiders will fall into, rather than sit on top of, the solution. If you are using any chemical solution, your trap should also be covered with a wire screen (chicken wire or hog wire) to help reduce the chances of larger wildlife disturbing your trap and accessing the chemicals. If you are dry trapping, you should check your traps daily and be aware that spiders and other predacious animals confined in the trap will readily eat one another. Traps with chemicals can be left for longer periods of time, but the specimens will degrade over time. Be prepared to have lots of specimens that will need to be transferred to 70 percent ethanol in a timely manner.

Fig. 28a digging the hole for a pitfall trap

Fig. 28b installing the outer container

Fig. 28c checking that the fit is snug and level with the ground surface

Fig. 28d installing the catch container

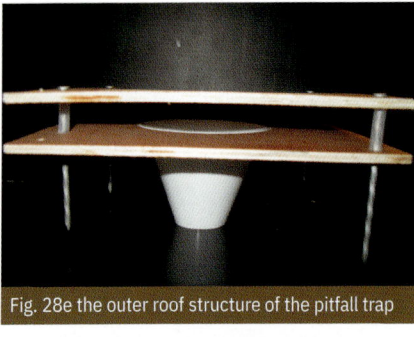

Fig. 28e the outer roof structure of the pitfall trap

Fig. 28g trap roof covered with chicken wire and ready to catch invertebrates

Fig. 28f putting the roof on the pitfall trap

LEAF LITTER SAMPLING

Many spiders live, hide, and hunt in the leaf litter. One easy way to sample them is to scoop samples of leaf litter into a large bag or Berlese funnel (fig. 29a). Wear leather gloves when scooping up the leaf litter to help prevent anything in the litter from causing scratches or wounds. The bag of litter can be spread out on a sheet or tarp similar to that employed in the sweep-net method, or you can use a Berlese funnel. The leaf litter is placed in the top area of the funnel, which has a mesh at the bottom large enough for spiders to go through but fine enough to hold the litter. A catch container is placed below the funnel, and a light or heat source is placed at the top. The spiders will try to avoid the heat and work their way through the funnel and into the catch container. This

Fig. 29a Berlese funnel made from a flower pot

Fig. 29b *Pirata sedentarius* in the catch container soon after setting up Berlese funnel

container can be filled with 70 percent ethanol to kill and preserve the spiders, or it can be left dry if you want to look at live specimens (but you will need to check this frequently). Within minutes of setting up the funnel, you may have spiders in the catch container like the young *Pirata sedentarius* shown in figure 29b; it was collected moments after setting up the funnel. Be careful not to place your light or heat source so close that it ignites your leaf litter. Leaf litter can be collected at any time of year. During winter, many spiders will be inactive, in a state called torpor, so bringing leaf litter samples inside to a warm area will enable you to watch as they "wake up."

POOTERING SPIDERS

A device commonly called a "pooter," or an aspirator, can be used to suction up spiders into a tube or catch container (fig. 30a). There are lots of different designs. Pooters can be made from common household items, or they can be purchased from a variety of outlets. You can also purchase an attachable filter that will prevent you from inhaling things you would not want to introduce into your lungs. (The one shown was purchased from BioQuip fig. 30b.) Pooters are great for collecting small or extremely fast-running animals that could be harmed in the collection process.

Fig. 30a pootering

Fig. 30b pooter from BioQuip with attached filter

Keeping Live Spiders Captive

People have mixed feelings about capturing and keeping wild spiders. In general, if the spider is not an endangered or threatened species, little harm to the overall population would be caused by keeping a solitary spider. Once again, you would need to check your local laws and ordinances to verify that you can collect spiders from a location legally. Additionally, you would need to know that you could provide a suitable habitat and food supply for the animal. Observing spiders in captivity is a great way to learn about these fascinating creatures. Salticidae (jumping spiders), Theraphosidae (tarantulas), and other spiders can be found in the pet trade and purchased from breeders, but please make sure you buy only from reputable sellers. A spider's behavior in captivity may be different from its natural, "wild" behavior, but keeping a spider gives a chance for some up-close observations that can be very educational. Many people have overcome their arachnophobic tendencies by learning about spiders and then keeping one as a pet. Be sure to research the species of spider you plan to acquire as a pet so you know you can provide an adequate home for it.

Any wild-collected and preserved spider should have two labels associated with it to be useful to the scientific community (fig. 31). The first label should list the location where the spider was originally collected and the collection date (if the spider was captive reared, you would also want to note that here), the sampling technique, and the collector's name. The second label should include the identification of the specimen and the person who identified it.

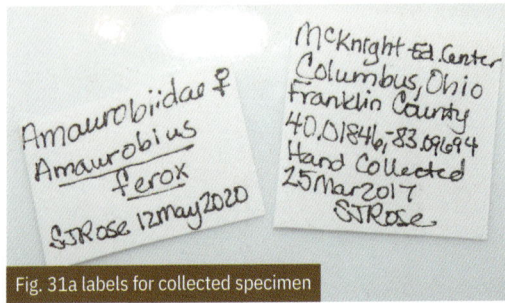

Fig. 31a labels for collected specimen

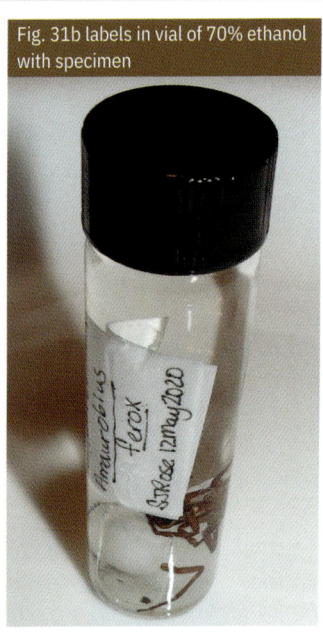

Fig. 31b labels in vial of 70% ethanol with specimen

The labels, usually written on linen in ink (that does not dissolve in ethanol) or pencil, should be placed in the vial with the specimen when it is preserved. It is also a good idea to keep a field notebook if you are collecting multiple animals. In this notebook, you can record the details of obtaining permission to collect as well as permit numbers, notes on the habitat and time of day the collection was made, and any other details about the time and conditions when the collection was made. These data can be extremely valuable for scientists in their research.

Fig. 32 the shed exoskeletons of a *Dolomedes albineus* from when it emerges from the egg sac to adulthood

Fig. 33a immature *Bassaniana versicolor* missing two legs

Fig. 33b *Bassaniana versicolor* freshly molted with regenerated legs

Fig. 33c *Bassaniana versicolor* with regenerated legs

Fig. 34a *Dolomedes albineus* exoskeleton stomach

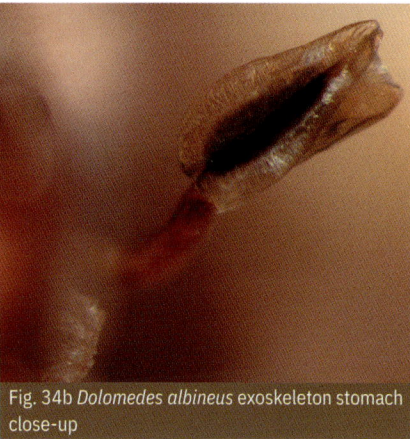

Fig. 34b *Dolomedes albineus* exoskeleton stomach close-up

SPIDER LIFE HISTORY

Development and Growth

Spiders start their life as an egg laid by the female. Some mothers guard their egg sacs, others carry their egg sacs with them, and some build the egg sac and leave their offspring to fend for themselves. In some cases, the young emerge months after the mother has died. When the spiders emerge from the egg, they look like a spider, although they may not look exactly like the adult form of their species. They will then go through a series of molts, or *ecdysis* (shedding of the exoskeleton), to grow (fig. 32). All spiders molt, and many of them can regenerate lost limbs during this process. Legs, chelicerae, pedipalps, or spinnerets lost due to injury, predator attacks, or a previous bad molt can all be regenerated. Figure 33 shows a *Bassaniana versicolor* that was collected missing two legs. He molted into maturity and regenerated both legs. You can see that, when he was freshly molted and he tried to stretch the two new legs

Fig. 34c *Dolomedes albineus* exoskeleton stomach

as much as possible, they were pale. Once his legs hardened, they had the same coloration as the other legs, although they were slightly shorter. In this case, as the spider is now a mature male, he will not molt again. If he were immature and able to molt again, we would likely not be able to tell that any of his legs had regenerated. One interesting fact about spiders is that ecdysis is not limited to the external cuticle we see on the spider. Figure 34 shows a few photos of the stomach and esophagus that were shed during the molting of a *Dolomedes albineus* fishing spider.

How Spiders Disperse

Spiders are known to be one of the first colonizers of recently disturbed habitats, such as areas devastated by fire, flooding, or even human behaviors. For example, during a May 1884 expedition after the eruption of Krakatoa, which cleared the landscape of all animal life, the first (and only) living creature found was a spider on the south side of Rakata (Thornton 1997). How do spiders get to these habitats? They fly using a technique called ballooning (fig. 35). Spiders can balloon at any time of year, and some spiders can balloon at any life stage, but most often ballooning occurs when young spiders disperse from where they emerged. During the spring and fall/early winter, it is not uncommon to have mass ballooning events, where the landscape will be covered by the gossamer silk strands of thousands of spiders taking flight. The photo in figure 36 was taken on December 24, 2015, in central Ohio. You can see the gossamer silk covering the grassy areas of the park. To balloon, spiders usually climb up any structures that can provide them with height. They then stand on their tiptoes and start releasing silk. When enough silk has been released to catch the wind (or electromagnetic field), they take off. Often, they make many failed attempts to balloon and release the long strands of silk before takeoff.

Fig. 35a *Pardosa* sp. wolf spider ballooning

Fig. 35b *Tmarus* sp. crab spider ballooning

Fig. 35c *Philodromus* sp. ballooning

Fig. 35d *Steatoda grossa* ballooning

Fig. 36 field at a park in Central Ohio covered in gossamer silk

Mating and Courtship

Spiders have some unique mating and courtship rituals (fig. 37). Since spiders are so diverse, there are many different strategies, but many of them include some way for the male to be sure the female is the correct species and receptive to mating (as the female can and will readily eat male suitors). A simple internet search will yield numerous videos demonstrating a vast variety of courtship displays. Many a male web-based spider plucks the strands of the female's web to announce his arrival. Male jumping spiders dance and sing in elaborate displays. Some wolf spiders purr and use percussion and stridulation to attract a mate. Although Australian peacock jumping spiders are not found in North America, videos of them are really fun to watch, as the males display their abdomens in their mating dances.

There is a myth that all spider females eat the male once copulation is complete. The notorious black widows have this reputation erroneously, as there are not many quantified cases where the male succumbs to the fangs of the female, especially if the female is in good condition (Johnson et al. 2011). In contrast, studies have shown that the male *Dolomedes tenebrosus* (dark fishing spider) spontaneously dies after mating and then is consumed by the female (Schwartz, Wagner, and Hebets 2014). Some male spiders are known to offer the female a "nuptial gift" of a snack prior to attempting to mate—in the hopes, one would assume, that she would be too busy eating her gift to consider eating the male.

Fig. 37a *Agelenopsis* sp. mating

Fig. 37b *Dolomedes albineus* (in captivity) mating

Fig. 37c *Tetragnatha elongata* mating

Fig. 37d *Tetragnatha elongata* mating showing the locked jaws

Fig. 37e *Micrathena gracilis* mating

Fig. 37f *Phidippus audax* male courting female

Fig. 37g *Phrurolithus goodnighti* mating

Fig. 37h *Trachelas tranquillus* (in captivity) male courting female

Fig. 37i Lycosidae mating

Fig. 37j *Gea heptagon* male courting female

Fig. 37k *Neriene radiata* mating

One unique thing about spider mating is the method of sperm transfer. Male spiders have modified pedipalps, and upon maturing, they build a sperm web onto which they deposit their semen. The pedipalps are charged by dipping into the semen, collecting it. When copulation between spiders occurs, the embolus of the palp is inserted into the female's epigynum or gonopore, and the sperm is delivered to the female (fig. 38). This sperm can be stored by the female for months to years and used to create multiple egg sacs. A captive *Steatoda grossa* (false black widow) produced thirteen viable egg sacs over eighteen months from a single mating encounter (fig. 39). Egg sacs come in many different shapes and sizes. Some are carried by the females, some are guarded by the females, and some are left to fend for themselves (fig. 40).

Fig. 38 male *Steatoda grossa* with palp inserted

Fig. 39 *Steatoda grossa* making an egg sac

Fig. 40a *Bassaniana* sp. guarding egg sac with young emerging

Fig. 40b *Heteropoda ventoria* with egg sac

Fig. 40c *Parasteatoda tabulata* with egg sac

Fig. 40d *Trachelas tranquillus* egg sac

Fig. 40e *Pholcus manueli* with egg sac

Fig. 40f *Ero* sp. egg sac

Fig. 40g *Tigrosa helluo* with egg sac

Fig. 40h *Uloborus glomsus* with egg sac

Fig. 40i *Neospintharus trigonum* with egg sac

Fig. 40j *Dolomedes albineus* with egg sac

Fig. 40k *Euryopis* sp. egg sac

Fig. 40l *Cyclosa turbinata* with egg sac in trashline

Fig. 40m *Mermessus tridentatus* egg sac

Fig. 40n *Mastophora hutchinsoni* egg sac

Fig. 40o *Enoplognatha marmorata* with egg sac

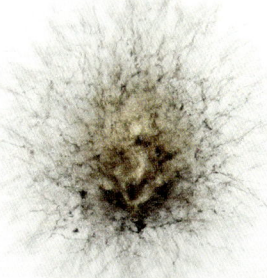

Fig. 40p *Tetragnatha laboriosa* egg sac

Fig. 40q *Theridiosoma gemmosum* egg sac

Fig. 40r *Theridion* sp. guarding egg sac with young emerging

Fig. 40s *Rhomphaea fictilium* egg sac

Fig. 40t *Oxyopes scalaris* guarding egg sac with young emerging

Fig. 40u Gnaphosidae guarding egg sac

Fig. 40v *Phrurtimpus* sp. egg sac

WHAT ARE THE BENEFITS OF HAVING SPIDERS AROUND?

Spiders often elicit strong responses in people. Some people love them; many people strongly dislike them. Whether you like them or not, there are lots of benefits to having spiders around.

Spiders are one of the top predators of the invertebrate world. There is only one documented "vegetarian" spider: *Bagheera kiplingi*, which is native to Central America. It has forgone the predatory lifestyle. The other spiders utilize prey for at least some of their food intake. The fact that spiders are predators is highly beneficial to humans. It has been calculated that the mass of insects that all spiders consume each year totals more than the mass of the entire human population (Nyffeler and Birkhofer 2017). Without spiders, we would be overrun with many insects that people don't like to have around (as they eat us, our food, or our homes). Figure 41 shows just a few examples of spiders with their invertebrate prey. Most spiders are generalist predators and can even act as our first line of defense against potentially invasive insects, as you can see in figure 42, which shows a *Neoscona* orbweaver feasting on a spotted lanternfly (*Lycorma delicatula*); in figure 43, which shows a *Phidippus audax* feeding on a brown marmorated stink bug (*Halyomorpha halys*); and in figure 44, which shows an *Araneus marmoreus* preying upon a German yellowjacket (*Vespula germanica*). All of the insects are non-native, and the spotted laternfly and brown marmorated stink bug are considered invasive species. Spiders are not usually considered of economic importance,

Fig. 41a *Euryopis* sp. with ant

Fig. 41b *Trachelas tranquillus* with Indianmeal moth

Fig. 41c *Pholcus manueli* with spider prey

Fig. 41d *Dolomedes tenebrosus* with roach

Fig. 41e *Leucauge venulsta* with mosquito

Fig. 41f *Neoscona crucifera* with crane fly

Fig. 41h *Leucauge venusta* with midge

Fig. 41g *Synema parvulum* with ant

Fig. 41i *Oecobius* sp. with ant

Fig. 41j *Micrathena gracilis* with horsefly

Fig. 41k *Theridion frondeum* with Opiliones

Fig. 41l *Neriene clathara* with fruit fly

Fig. 42 *Neoscona* sp. feeding on a spotted lanternfly (*Lycorma delicatula*)

Fig. 43 *Phidippus audax* feeding on a brown marmorated stink bug (*Halyomorpha halys*)

Fig. 45 *Argiope aurantia* with a five-lined skink (*Plestiodon fasciatus*)

Fig. 44 *Araneus marmoreus* with German yellowjacket (*Vespula germanica*)

but the role they play in pest control is extremely valuable. Every so often, you will find that spiders are able to capture and overpower vertebrates, too. Figure 45 shows an *Argiope aurantia* that has caught a five-lined skink (*Plestiodon fasciatus*).

Spiders also serve as prey to many animals, including but not limited to birds (fig. 46), lizards, snakes, small mammals, fish, insects, even other spiders. A study has even shown that baby birds that were fed spiders have greater brain development (Arnold et al. 2007), and many birds also utilize spider webs in nest building (fig. 47). Additionally,

Fig. 46 Grackle with spider prey

Fig. 47a Hummingbird and nest

Fig. 47b Acadian flycatcher nest

hummingbirds have been observed picking prey insects out of a *Micrathena gracilis* web. The spider was not consumed in this case. It may be that the spines on the spider were a deterrent to it serving as a prey item for the hummingbird. To have birds, we need spiders.

Spiders are also hosts for a number of parasites and parasitoids. Quite a few insects will prey upon spiders in various life stages. There are wasps that specialize on spiders, paralyzing them for their larvae to feed upon (fig. 48). Some of these wasps remove the spider's legs before securing the body in a burrow with an egg laid on it (fig. 49). There are wasps that provide multiple spiders for each larva, and that pack mud tubes full of spiders (fig. 50). Small-headed flies are internal parasites that consume the spider from the inside until they are ready to pupate into flies (fig. 51). There are also wasps that lay their egg on a spider, and the larva will feed on the spider while it goes about its normal life (fig. 52). When the wasp is ready to pupate, it can use mind control to make the spider build an appropriate web for its cocoon. Then it will finish off the spider before pupating.

Fig. 48 Wasp with paralyzed spider

Fig. 49 Wasp with spider with legs removed

Fig. 50 *Trypoxylon* wasp nest with paralyzed spiders

Fig. 51a *Philodromus* sp. harboring a smallheaded fly larva

Fig. 51b Smallheaded fly larva emerging from host *Philodromus* sp.

Fig. 51c Smallheaded fly larva emerging from host *Philodromus* sp.

Fig. 51d Smallheaded fly larva

Fig. 51e Smallheaded fly larva

Fig. 51f smallheaded fly

Mantidflies will ride on a spider as a larva (usually near the pedicel), and the larva will then consume the spider's eggs before transforming into an adult (fig. 53). There are worms that enter a spider and feed on it internally (fig. 54). There are mites that are ectoparasites; they feed externally on spiders (fig. 55). There are also fungi that can turn spiders into zombies, eating them from the inside out. The life of a spider is not an easy one (fig. 56).

Spiders are pioneer colonizers, one of the first animals to take up residence at recently disturbed sites (Marx, Guhmann, and Decker 2012). One study demonstrated that the placement of their webs may contribute to the spatial patterns of developing plant and animal communities (Hodkinson et al. 2001). So, if spiders are the first to colonize an area, what do they eat? Well, as generalist predators, spiders often readily eat other spiders; many spiders have been shown to feed on nectar and pollen of plants; and some spiders have also been shown to scavenge for food.

Fig. 52a wasp larva on small Theridiidae

Fig. 52b wasp that developed from larva in previous photo

Fig. 53a *Wulfila* with mantidfly larva

Fig. 53b mantidfly

Fig. 54a Lycosidae with nematode

Fig. 54b Lycosidae with nematode

Fig. 54c nematode from Lycosidae

There are lots of benefits to humans from spiders beyond their ecological and pest-control services. Spider silk is a great resource, and it has been collected and used for a variety of purposes over the years. In years gone by, spider silk was used to stop bleeding and promote the healing of wounds. There is a species (*Coras medicinalis*, Hentz, 1821) that was so named because its web was used for medicinal purposes. Spider silk was traditionally used to create the crosshairs in rifle scopes; there are companies that manufacture cloth from spider silk; and there are even violin strings made from spider silk. Spider venom may also be of great use to humans, as lots of research is being performed into the properties of various venoms and potential human uses for them.

Fig. 55a mite on *Ocrepeira* sp.

Fig. 55b mite on prosoma of *Allocosa funera*

Fig. 56a *Colonus sylvanus* consumed by fungus

Fig. 56b *Pholcus* consumed by fungus

Fig. 56c Anyphaenidae consumed by fungus

INTRODUCED SPECIES

Throughout this field guide, you will see it noted that many species are considered introduced. This means that they are not native to the region and were likely transported into the area by human activity. The terms "introduced species" or "non-native species" are often misinterpreted to mean "invasive," but these are two completely different concepts. To be considered an invasive species, a species must be shown to cause a negative impact. These negative impacts could be to other living organisms in the ecosystem, to human health, or to human resources. There are even cases where native species can become invasive. The term is not restricted to those species that have been introduced into a new region.

ARE SPIDERS VENOMOUS OR POISONOUS?

Nelsen et al. (2014) tried to reduce the ambiguity around the terms used to describe various biological toxins. They defined each toxin based on the delivery method. Poisons are ingested, inhaled, or absorbed and are delivered passively. Venoms are delivered to internal tissue via mechanical trauma that creates a wound. Toxungens are intentionally delivered without an accompanying wound. Some examples from Nelsen: Golden dart frogs produce a toxin on their skin that is harmful if ingested or if contact is made with the skin and the toxin is absorbed, making this species poisonous. Fire salamanders spray potential predators with a toxin, and they can aim their spray, making this species toxungenous. The Brazilian casque-headed tree frog has spicules on its head; when threatened, the tree frog uses the spicules to cause wounds, into which a toxin is introduced. This species is considered venomous. To put this in spider terms, most spiders are venomous, as their toxin is introduced into a wound made by their fangs. (There are spiders that lack venom, but they are the exception, not the rule.) No documented spider species is considered poisonous to humans. Scytodidae (spitting spiders) could be considered toxungenous, as they intentionally spray a toxin silk concoction onto their prey. Note that this is a fairly recent attempt at clarifying these terms, and even recent publications will often still refer to spiders as poisonous.

SPIDER BITES AND POTENTIALLY DANGEROUS VENOM

The vast majority of spiders are not considered medically significant (or dangerous) to humans, but according to the Centers for Disease Control, two groups of spiders in the United States have venom that is considered medically significant to humans: *Latrodectus* species, the widows (fig. 57), and *Loxosceles* species, the recluse spiders (fig. 58). If you are bitten by one of these

Fig. 57a *Latrodectus mactans*

Fig. 58a
*Loxosceles
reclusa*

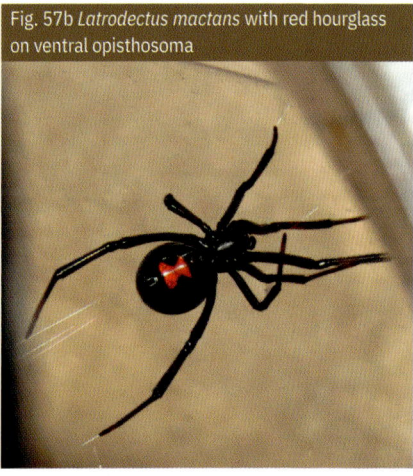

Fig. 57b *Latrodectus mactans* with red hourglass on ventral opisthosoma

Fig. 58b *Loxosceles reclusa* close up showing "violin" on prosoma

spiders, you should seek medical attention (and take the spider, even if it is squished, with you, so that the medical professional can seek out an arachnologist to confirm the identification, if needed). Although these spider bites are considered medically significant, deaths from bites are extremely rare. According to Nentwig (2011), there has not been a confirmed death from any *Latrodectus* species in over forty years. There have been only a couple cases in the United States in the last 40 years where a *Loxosceles* species has been implicated in a death, and according to Rick Vetter (personal communication), these were "most often due to hemolytic anemia and renal issues in children, however, it is an extremely rare outcome."

There is a long-standing myth that any wound with two holes is likely a spider bite. This not only promotes the idea that spider bites are common (which they are not) but also implies a diagnosis can be made by the appearance of the wound (it cannot). Many arthropods like to feed on humans and will actively seek us out to feed (and this does not include any species of spider), and it is quite possible for those arthropods that are feeding on humans to bite more than once, leaving two holes. In contrast, spiders usually bite only in self-defense, as a last resort, often when they are being squished against the skin. Also, many other skin conditions can cause welts, marks, blisters, or even necrosis that are completely unrelated to spiders. If a spider was not observed causing the wound (or if a dead, squished spider is not found close to the wound), the odds are, it is not a spider bite. Even many medical professionals seem unaware of this, which is why there are papers in the medical literature about misdiagnoses of spider bites, especially of bites by the necrosis-causing brown recluse. Please see Stoecker, Vetter, and Dyer (2017), an informative paper that presents a mnemonic device to help doctors avoid those misdiagnoses.

ONLINE RESOURCES

There are many resources online, and most people are very familiar with googling things. One source that is extremely useful when trying to identify spiders is the World Spider Catalog, which offers free membership. This site keeps an up-to-date listing of all the known spiders in the world. Diligent staff members constantly update the catalog as new papers are published. The odds are that, by now, at least one species listed in this guide has had either a name change or a change in family assignment. This is due not to any real error, but to our ever-expanding knowledge. If you search the World Spider Catalog by scientific name, the website will let you know if the name is still valid and, if it is not, inform you of the new name. The site also provides access to much of the scientific literature, if you were looking for figures of the genitalia, say, or wanted other information about a species.

There is a plethora of other sites; it would be impossible to name them all here. One word of caution, though: some sites use a majority-rules mentality. This means that if you post a photo, members of the site could easily "outvote" the expert who provides you with the correct identification. Many sites do not have any way to recognize who the experts are. There are also apps that will make a "guess" based on your photo. Most of these are fairly new and have yet to build up a good database of accurately identified animals. Every day, these sites and their AI are learning and getting better, but many cases of misidentification can still be made. Please use these sites and apps as a jumping-off point for your own research, not as a means to an absolute answer.

A QUICK WORD ABOUT COMMON NAMES

One frequently asked question is "What is the common name?" or "English please!" Since there are over 49,000 known species of spider, it would not be feasible to give each one a meaningful common name. On the other hand, each species has been given a unique binomial name, which is the most accurate way to refer to that organism and is consistent worldwide. In the Americas, the American Arachnological Society (AAS) maintains a list of "approved" common names for arachnids. This guide will refer to those names as the common name, but the write-ups for some spiders will note other colloquial names. Many of the colloquial names will vary from region to region. A name used to refer to a specific spider in one area may refer to a completely different species of spider in another area. Some colloquial names are even used for vastly different taxa.

I will give one example: "daddy-longlegs." Think about what "daddy-longlegs" means to you. Is it a Pholcidae (cellar spider), is it an Opiliones (harvestman), or is it a Tipulidae (cranefly)? All three animals are often referred to as "daddy-longlegs." So when someone asks if a "daddy-longlegs" is venomous, you have to inquire further to know what they are referring to or risk making the wrong assumption. And as only one of the "daddy-longlegs" taxa listed is venomous, if you didn't seek clarification, you could be providing inaccurate information. On the other hand, using the scientific name removes most of the confusion.

Spider Ranges

Many ranges of known species are not well documented. This guide will provide range maps, but in most cases, these will be best guesses. These guesses are based on the following: the list maintained by the AAS, which documents not only the species known from North America (north of Mexico) but also the states and provinces in which the species have been documented according to the scientific literature; specimens from several museums cataloged by the Symbiota Collections of Arthropods Network (SCAN); specimens collected or obtained by the author; and additional locations recorded in the scientific literature but not included in the AAS listings. Locations that were reported on BugGuide.net or iNaturalist (or other social media sites or apps) were usually not included, as many of those were not verified with a specimen and errors remain to be resolved on these sites. Based on these known locations for a species, a range map was created to best fit the data. Some records were not included in the maps, as they were single specimens found great distances from other records and may have been isolated individual introductions.

HOW TO IDENTIFY SPIDERS

Identifying spiders is usually a multistep process that first involves observing characteristics that can be used to categorize the spider into a guild, and then narrowing that to family. Below are the guild descriptions. Once you have determined the guild to which you feel your spider belongs, turn to the chapter on that guild. Each guild chapter starts with a dichotomous key to help you identify a specimen to the family level.

The Spider Guilds: How Spiders Catch Their Prey

Cardoso et al. (2011) proposed a guild structure for the spider families, and this guide will follow that outline, although it has been slightly modified. Spiders can be separated into eight guilds based on how they hunt for and capture their prey. One word of caution: when you find a spider that is not in a web, you cannot rule out the web-building species, especially in the case of wandering males.

Four guilds use webs for prey capture: the Sensing Web Weavers, Sheet Web Weavers, Orb Web Weavers, and Space Web Weavers. Those that don't use webs are the Ambush Hunters, Ground Active Hunters, and Other Active Hunters. Lastly, the Spider Hunters specialize on hunting and feeding on other spiders.

Sensing Web Weavers: These spiders tend to live in a burrow or retreat that has silk lines that radiate outward and are used to detect prey. The silk is not used to ensnare the prey, just to detect it. Then the spider rushes out to seize the prey item (fig. 59).

Fig. 59 sensing web of *Ariadna bicolor*

Sheet Web Weavers: These webs are a platform of horizontal silk that is sometimes associated with a funnel retreat also made of silk (fig. 60).

Orb Web Weavers: This is what most people imagine when visualizing a spider web: the wheel-shaped web that is composed of radial lines and a sticky spiral. Some spiders make modified versions of this web: either just a portion of the wheel or a web that is pulled into a cone shape (fig. 61).

Space Web Weavers: These webs take up three-dimensional space and do not seem to have any sort of organized structure. They include those webs commonly referred to as cobwebs or tangle webs (fig. 62).

Fig. 60 sheet web of an *Agelenopsis* sp.

Fig. 61 orb web of an Araneidae

Fig. 62 space web of a Theridiidae

Ambush Hunters: These are sit-and-wait predators that do not rely on silk for prey capture. They often resemble the environment on which they sit. Unsuspecting prey will walk right up to them (fig. 63).

Ground Active Hunters: These spiders primarily pursue prey running on the ground, and they do not rely on any web structures for prey capture (fig. 64).

Other Active Hunters: These spiders also do not use a web for prey capture. They hunt on various surfaces (plants, walls, etc.) and chase or pounce on their prey (fig. 65).

Spider Hunters: These spiders usually do not build a web of their own; rather, they prey upon other spiders, sometimes invading the webs of their host (fig. 66).

Fig. 63 ambush hunter *Misumenoides formosipes*

Fig. 64 ground active hunter Lycosidae (with young)

Fig. 65 other active hunter *Colonus* sp.

Fig. 66 spider hunter *Mimetus puritanus* feeding on *Parasteatoda* sp.

The spiders in the Sensing Web Weaver Guild use a series of silk lines to detect prey; these "trip lines" are often associated with a burrow-type retreat and sometimes radiate from the burrow. One can think of the trip lines as a prey-detection system; they do not trap or ensnare prey but alert the spider to the prey's location, so that the spider can quickly grab it. In North America, the Sensing Web Weaver Guild includes both Mygalomorphae and Araneomorphae. The Mygalomorphae families in this guild are Antrodiaetidae, Atypidae, Cyrtaucheniidae, Euctenizidae, Halonoproctidae, Nemesiidae, and Theraphosidae. The Araneomorphae families in the guild are Filistatidae, Hersiliidae, Oecobiidae, and Segestriidae.

KEY TO FAMILY

1a. Mygalomorphae: Chelicerae are paraxial (parallel to central axis of body); when spider attacks, it raises its prosoma up to unfold fangs and strikes downward onto prey; two pairs of book lungs; eight eyes that are usually clustered together; usually with robust legs ------- **2**

1b. Araneomorphae: Chelicerae diaxial (at an angle to central axis of body); when spider attacks, fangs are used in a pinching motion; one pair of book lungs ---------------------------- **7**

2a. With one of more tergites (sclerotized areas of cuticle) on opisthosoma ----------------------- **3**

2b. Without tergites on opisthosoma -- **4**

a

b

Antrodiaetus unicolor

Myrmekiaphila foliata

3a. Short endites (less than half the width of sternum), short chelicerae (less than half the length of carapace), collapsible door at entrance to burrow, 6.0–25.0 mm total body length ----------------------------- **Antrodiaetidae** (page 41)

3b. Long endites (greater than half the width of sternum), long chelicerae (over half the length of carapace), silk tube extending vertically or horizontally from burrow entrance, 8.0–27.0 mm total body length -------- **Atypidae** (page 44)

a

b

Antrodiaetus unicolor

Sphodros rufipes

4a. Burrow with door at entrance --- **5**

4b. Burrow without door at entrance -- **6**

a

b

Bothriocyrtum californicum

Aphonopelma sp.

5a. Digging spines (short thornlike spines) on pedipalps and tarsi, metatarsi, and tibiae of legs I and II, burrow door thin and wafer-like or thick and cork-like, 10.0 mm to >30.0 mm total body length -------- **Halonproctidae** (page 49)

5b. Digging spines neither on pedipalps nor on leg I and II, dense scopulae present, burrow with turret or thin flexible door, 7.0–32.0 mm total body length --- **Euctenizidae** (page 46)

a

Ummidia audouini

b

Aptostichus simus

6a. Claw tufts on all tarsi, urticating hairs on posterior opisthosoma, 9.0–75.0 mm total body length --- **Theraphosidae** (page 53)

6b. Lacks claw tufts, lacks urticating hairs, velvety appearance, rodlike setae in a row on upper inner chelicerae (best seen with a microscope), 16.0–30.0 mm total body length --- **Nemesiidae** (page 52)

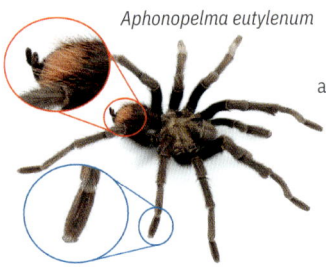

Aphonopelma eutylenum

a

Calisoga longitarsis

b

7a. With cribellum (cribellate) --- **8**

7b. Without cribellum (ecribellate) --- **9**

a

Kukulcania hibernalis

b

Ariadna bicolor

8a. Large anal tubercle, starburst legs, eyes clustered on semicircular prosoma, typically creates mat of silk on walls or other surfaces, 0.9–3.2 mm total body length --- **Oecobiidae** (page 58)

8b. Lacks anal tubercle, short calamistrum, eyes are clustered on central mound, 1.5–18.0 mm total body length, typically hides in crevices lined with cribellate silk --- **Filistatidae** (page 54)

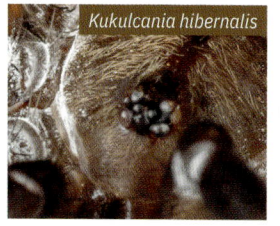

Oecobius navus

Kukulcania hibernalis

a

b

9a. Six eyes (anterior median eyes missing), first three pairs of legs directed forward, two tracheal spiracles near book lungs, lives in silk tubular retreats, 4.0–16.0 mm total body length ---------------------------------- **Segestriidae** (page 59)

9b. Eight eyes on a raised ocular area, posterior spinnerets extremely elongated (as long as or longer than abdomen), very long slender legs, 6.0–8.0 mm total body length ------------------------- **Hersiliidae** (page 57)

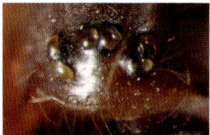

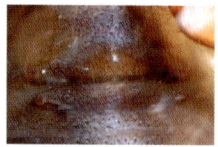

a

Ariadna bicolor

* The World Spider Catalog lists the family Cyrtaucheniidae as present in North America. This family is represented in North America by only one species: *Cyrtauchenius talpa*, which was described in 1891 from a specimen collected in California. There do not seem to be any additional records of this species or other members of this family in North America. They are similar to Halonoproctidae but lack the thornlike setae on the tarsi and metatarsi. This family is not represented in this guide.

b

Neotama mexicana

ANTRODIAETIDAE ▶ FOLDINGDOOR SPIDERS

Other Colloquial Names Collardoor spiders, foldingdoor trapdoor spiders, turret spiders, trapdoor spiders

Genera in This Family *Aliatypus* (14), *Antrodiaetus* (16), *Atypoides* (3), and *Hexura* (2)

Total Body Length 6.0–25.0 mm

Description These mygalomorphs have one or more tergites (sclerotized cuticle) on the dorsal opisthosoma. These spiders are usually tan to brown, although they can be dark enough to appear black in low-light situations. The endites are short. (At most, their length is half the width of the sternum.) The anterior lateral eyes are the largest. These spiders have only two pairs of spinnerets. The distal segments of the posterior lateral spinnerets are about three times as long

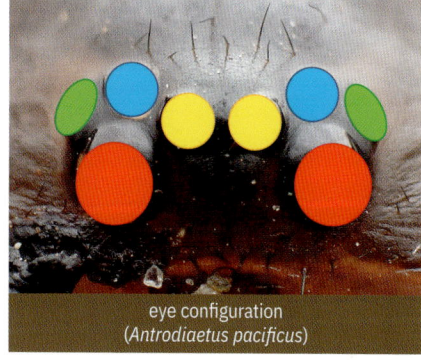

eye configuration
(*Antrodiaetus pacificus*)

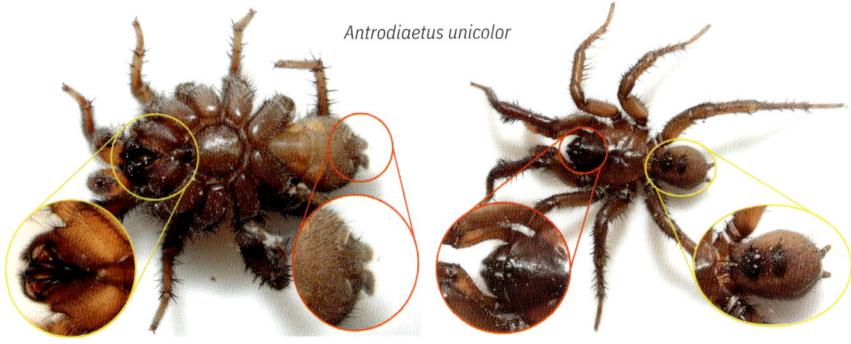

Antrodiaetus unicolor

short endites   two pairs of spinnerets   short chelicerae   tergites

as they are wide. The pattern of the leg spination and the shape of tibia I of the males can aid in species-level identification; lateral views are the most helpful. Foldingdoor spiders excavate a burrow lined with silk. The burrow entrance has a turret or collapsible door, or a thin hinged door. They are usually most active at night. They wait at the entrance to the burrow for unsuspecting prey to trigger their sensing web lines. They are difficult to locate when they are in their burrows during the day, as their burrow entrances are well camouflaged in the habitat. Most often, males are encountered when they wander looking for females.

Similar Families Atypidae, Euctenizidae, Euagridae, Halonoproctidae, Hexurellidae, Megahexuridae, Microhexuridae

How to distinguish Antrodiaetidae from other similar families:

Antrodiaetids can be distinguished from euctenizids, euagrids, halonoproctids, and microhexurids by the presence of the tergites on the dorsal opisthosoma, as euctenizids, euagrids, halonoproctids, and microhexurids lack this trait. Antrodiaetids can be distinguished from atypids by their endites: atypids have long endites (about three-fourths the width of the sternum), whereas antrodiaetids have short endites (less than half the width of the sternum). Additionally, atypids create a purseweb, whereas antrodiaetids have a burrow with a turret entrance. Most antrodiaetids can be distinguished from hexurellids and megahexurids by the stouter distal segment of the posterior lateral spinnerets. On most antrodiaetids, the distal segment is only about three times as long as it is wide; in hexurellids and megahexurids, it is at least five times as long as it is wide.

Antrodiaetus montanus
(Chamberlin & Ivie, 1935)

female

Common Name This species does not have a common name.

Total Body Length Female 14.0 mm, Male 11.0 mm

Description Coloration can vary from pale yellow to light reddish brown. The ends of the chelicerae are covered with scattered golden hairs. The females have robust chelicerae, while the chelicerae of males tend to be somewhat skinnier. The females have a single tergite on the dorsal opisthosoma; the males have two or three tergites on the top of the opisthosoma. The prolateral surface of tibia I of the males has an area of dense spines that are helpful in determining species-level identification. The male palpal tibia is swollen. These are long-lived spiders, and most likely take a few years to reach sexual maturity. The presence of the hairs on the chelicerae helps distinguish *A. montanus* from other *Antrodiaetus* species.

Habitat These spiders live in a burrow with a collapsible collar at the entrance. When the burrow door is pulled shut, they are nearly impossible to detect. They seem to be generalist predators and will feed on any prey that triggers their trip lines and they can overpower. Specimens have been collected at elevations up to 7,200 ft. The males are often found wandering in the fall. The entrance to their burrow is often under debris.

Known records from ID, NV, OR, UT, WY

Antrodiaetus pacificus (Simon, 1884)

Common Name This species does not have a common name.

Total Body Length Female 13.0 mm, Male 11.0 mm

Description Coloration can vary from pale yellow to brown. The fovea is longitudinal. The females have robust chelicerae; the males' chelicerae tend to be somewhat skinnier. Two to three tergites are on the top of the opisthosoma. The prolateral surface of tibia I of the males has a brushlike area of spines. The male palpal tibia is swollen. These are long-lived spiders and most likely take a few years to reach sexual maturity.

female

Habitat These spiders live in a burrow with a collapsible collar at the entrance. When the burrow door is pulled shut, they are nearly impossible to detect. They seem to be generalist predators and will feed on any prey that triggers their trip lines and they can overpower. The males are often found wandering in late summer (June–early November). They have been found at elevations up to 7,500 ft. They tend to live in humid areas with loamy or sandy soils and occasionally have been found to create burrows in rotting logs. They build their burrows in sheltered areas that are not susceptible to flooding.

male

Known records from BC (Canada); AK, CA, ID, OR, WA (United States)

Antrodiaetus unicolor (Hentz, 1842)

Common Name This species does not have a common name.

Total Body Length Female 20.0 mm, Male 17.0 mm

Description Coloration can vary from pale yellow to brown. The fovea is longitudinal. The females have robust chelicerae; the males' chelicerae tend to be somewhat skinnier. The females have a single tergite on the dorsal opisthosoma, while the males have three tergites on the top of the opisthosoma. The prolateral surface of tibia I of the males has an area of dense spines that are helpful in determining species-level identification. The male palpal tibia is swollen. These are long-lived spiders and most likely take a few years to reach sexual maturity. *A. unicolor* is very similar in appearance to *A. robustus*, which has been found in MD, OH, PA, VA, and WV.

female

male

female at burrow

male inside burrow

Habitat These spiders live in a burrow with a collapsible collar at the entrance. When the burrow door is pulled shut, they are nearly impossible to detect. They seem to be generalist predators and will feed on any prey that triggers their trip lines and they can overpower. They are usually found in dense forested habitats with sandy loam soils. In areas with favorable conditions, you might find multiple individuals in close proximity. Males can be found wandering in search of a mate from July through December. They have been found at elevations up to 6,600 ft.

Known records from AL, AR, GA, IL, IN, KY, MD, MI, NC, NJ, NY, OH, PA, SC, TN, VA, WV

ATYPIDAE › PURSEWEB SPIDERS

Other Colloquial Names Atypical tarantulas

Genera in This Family *Atypus* (1) and *Sphodros* (7)

Total Body Length 8.0–27.0 mm

Description These mygalomorphs have one or more tergites (sclerotized cuticle) on the dorsal opisthosoma. The thoracic furrow is quadrangular. The labium and sternum are fused, and the chelicerae are usually long and robust (greater than half the length of the prosoma).

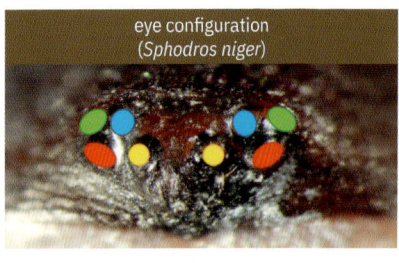
eye configuration
(*Sphodros niger*)

Purseweb spiders have three pairs of spinnerets. They live in a burrow with a long silk tube that can extend vertically or horizontally from the burrow entrance. When prey is detected walking on the silk tubes, the spider attacks by piercing through the tube with its long fangs to secure the prey. These spiders can live seven (or more) years. The name "purseweb spider" comes from the tube's similarity to the elongated silken purses carried by women of years gone by.

Sphodros rufipes
long chelicerae

Sphodros rufipes purseweb

Similar Families Antrodiaetidae, Euctenizidae, Halonoproctidae, Hexurellidae, Megahexuridae

How to distinguish Atypidae from other similar families:

Atypids can be distinguished from euctenizids and halonoproctids by the presence of the tergites on the dorsal opisthosoma, as euctenizids and halonoproctids lack this trait. Atypids can be distinguished from antrodiaetids, hexurellids, and megahexurids by their endites. Atypids have long endites (about three-fourths the width of the sternum), whereas the others have short endites (less than half the width of the sternum). Also, hexurellids are small animals (2.5–5.0 mm total body length), and atypids usually are larger (8.0–27.0 mm total body length). In addition, atypids are the only spiders known to make the purseweb that extends from their burrow.

Sphodros niger (Hentz, 1842)

Common Name This species does not have a common name.

Other Colloquial Names Black purseweb spider

Total Body Length Female 22.0 mm, Male 10.5 mm

Description Females are overall red brown; males are overall dark red brown to black. They have robust porrect (forward-pointing) chelicerae.

female

male

Habitat This species is known for building horizontal webs along the ground and through leaf litter, especially in areas with a thick duff layer or accumulations of pine needles. This is the most northerly recorded of the *Sphodros* species.

Known records from ON (Canada); CT, IL, IN, KS, MA, MI, MO, NC, NJ, NY, OH, PA, TN, VA, WI, WV (United States)

Sphodros rufipes (Latreille, 1829)

Common Name Redlegged purseweb spider

Total Body Length Female 24.0–25.0 mm, Male 14.5 mm

Description The female is usually overall dark brown; the male has a very dark brown to black opisthosoma and prosoma, but the legs are bright orange red. *Sphodros fitchi* and *Sphodros atlanticus* also have red legs, but note that in *S. rufipes* the entire femur and distal segments of the legs are red.

Habitat The purseweb is usually built vertically up the trunk of a small tree. The web is often adorned with debris, helping to camouflage it. The species is often found in forested areas with sandy loam soils.

Known records from AL, DC, FL, GA, IL, IN, KS, LA, MA, MD, MS, NC, NY, OH, PA, RI, TN, TX, VA

male

purseweb

EUCTENIZIDAE > WAFERLID TRAPDOOR SPIDERS

Genera in This Family *Apomastus* (2), *Aptostichus* (40), *Crypoctenzia* (1), *Entychides* (1), *Euctenizia* (2), *Myrmekiaphila* (12), *Neoapachella* (1), and *Promyrmekiaphila* (2)

Total Body Length 7.0–32.0 mm

Description These mygalomorphs do not have tergites on the dorsal opisthosoma. The thoracic furrow is procurved to straight. The posterior lateral spinnerets

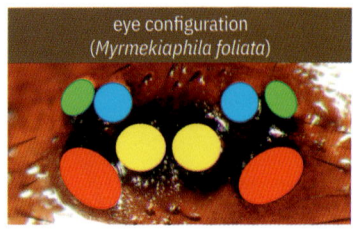

eye configuration (*Myrmekiaphila foliata*)

Myrmekiaphila foliata

Aptostichus simus

lacks digging spines and has dense scopulae

lacks tergites

are fairly short, and the distal segment is less than half the length of the medial segment. Waferlid trapdoor spiders live in silk-lined burrows with a thin flexible trapdoor.

Similar Families Antrodiaetidae, Atypidae, Halonoproctidae

How to distinguish Euctenizidae from other similar families:

Euctenizids can be distinguished from antrodiaetids and atypids by the lack of tergites on the dorsal opisthosoma that both those families have. Some halonoproctids have a truncated opisthosoma with

Apostichus burrow door, closed

Apostichus burrow door, open

longitudinal grooves and a sclerotized disc at the posterior end; this is unlike any euctenizid. Euctenizids can be distinguished from the other halonoproctids by the lack of the "digging spines" that are seen on the lateral surface of the pedipalps and legs I and II of female halonoproctids.

Aptostichus simus Chamberlin, 1917

Common Name This species does not have a common name.

Other Colloquial Names Southern coastal dune trapdoor spider

Total Body Length Female 12.0–18.0 mm, Male 9.0–12.0 mm

Description The coloration of the females ranges from brown to reddish brown. The anterior portion of the prosoma is elevated in the females. The female opisthosoma has a subtle chevron pattern. The males are tan to light brown.

Habitat This species is known only from southern California, often building its burrows in sandy dunes.

Known records from CA

female

Myrmekiaphila species

Description The members of this genus look very similar and are difficult to identify to species without looking at the genitalia, usually of a preserved specimen. The males of this genus are easily separated from other genera by the swollen distal portion of metatarsus I. Coloration can range from yellow to dark brown but is often orange to reddish brown.

Habitat These spiders live in burrows that include side chambers that can be shut off from the rest of the burrow by a second trapdoor. They have been shown to be active at all times of year. Interestingly, the temperature at the bottom of the burrows was never below freezing, even when surface temperatures were below freezing.

Myrmekiaphila comstocki Bishop & Crosby, 1926

Common Name This species does not have a common name.

Other Colloquial Names Comstock's wafer trapdoor spider

Total Body Length Female 23.5–25.0 mm, Male 16.5 mm

Known records from AR, LA, OK, TX

male

male

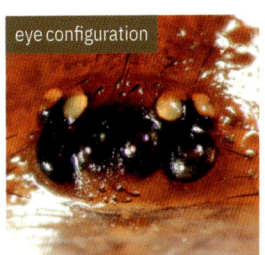

eye configuration

Myrmekiaphila foliata Atkinson, 1886

Common Name This species does not have a common name.

Total Body Length Female 22.0–26.0 mm, Male 14.0 mm

Known records from AL, GA, KY, NC, OH, TN, VA, WV

male

male

male

male

Myrmekiaphila tigris Bond & Ray, 2012

Common Name This species does not have a common name.

Other Colloquial Names Auburn tiger trapdoor spider

Total Body Length Female 21.0–23.0 mm, Male 15.0 mm

Known records from AL, GA

male

female

HALONOPROCTIDAE ⟩ CORKLID TRAPDOOR SPIDERS

(Previously **CTENIZIDAE**)

Genera in This Family *Bothriocyrtum* (1), *Conothele* (1), *Cyclocosmia* (2), *Hebestatis* (1), and *Ummidia* (18)

Total Body Length 10 mm to >30 mm

Description These mygalomorphs lack tergites on the dorsal opisthosoma. The thoracic furrow is strongly procurved. The tarsi and metatarsi have short thornlike spines commonly referred to as "digging spines." They live in silk-lined burrows, with a door made of silk and packed substrate often referred to as a "corkdoor." The burrows are often very cryptic. The spider hides below a slightly open door waiting to detect prey, which it lunges at and pulls back to its burrow.

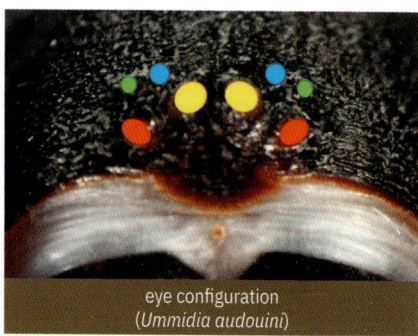

eye configuration
(*Ummidia audouini*)

Similar Families Antrodiaetidae, Atypidae, Euctenizidae

Ummidia audouini

lacks tergites

thornlike spines

How to distinguish Halonoproctidae from other similar families:

Halonoproctids can be distinguished from antrodiaetids and atypids by the lack of tergites on the dorsal opisthosomas that both those families have. Unlike any euctenizid, some halonoproctids have a truncated opisthosoma with longitudinal grooves and a sclerotized disc at the posterior end. Otherwise, halonoproctids can be distinguished from euctenizids by the presence of digging spines on the lateral surface of the pedipalps and legs I and II. Digging spines are absent in euctenizids.

Bothriocyrtum californicum
(O. Pickard-Cambridge, 1874)

Common Name California trapdoor spider

Total Body Length
Female 25.0–33.0 mm,
Male 15.0–24.0 mm

Description The prosoma is brown and shiny. The fovea is U shaped, and the open end points anteriorly. The opisthosoma is gray brown, sometimes yellow orange with a slightly visible darker heartmark, and covered with hairs. These are stocky spiders.

female

Habitat The inner side of the door to the burrow has holes where the spider, if it feels threatened, can use its fangs to hold the door shut. According to *Guinness World Records 2010,* this is the world's strongest spider, as "it is able to resist a force 38 times its own weight when defending its trapdoor. ... this equated to a man trying to hold a door closed while it is being pulled on the other side by a small jet plane." The burrows are often built on sunny slopes.

Known records from CA

Bothriocyrtum californicum
burrow entrance

Cyclocosmia truncata (Hentz, 1841)

Common Name This species does not have a common name.

Other Colloquial Names Ravine trapdoor spider

Total Body Length Female 33.0 mm, Male 19.0 mm

Description The prosoma is brown to black. The fovea is U shaped, and the open end points anteriorly. The opisthosoma is very distinctive: it has a series of ridges that end in a flat sclerotized disc that looks like it could be used to stamp a wax seal on an envelope. The females are mahogany brown, while the males are matt black.

Habitat The burrows have a side chamber in which the spider can hide and where it uses its unusual abdomen to block the entrance. The spider tends to build its burrows in areas with sandy clay soils with limited leaf litter.

Known records from AL, FL, GA, LA, TN

female

Ummidia audouini
(Lucas, 1835)

Common Name This species does not have a common name.

Other Colloquial Names
Audouin's trapdoor spider

Total Body Length Female 28.0 mm, Male 15.0 mm

male

male

male

saddle-like depression
on leg III tibia

male

Description The prosoma is dark brown to black. The fovea is U shaped, and the open ends point anteriorly. The opisthosoma is dark brown to black. The males are overall black but sometimes have a pale opisthosoma. The tibia of leg III has a saddlelike depression that is diagnostic for this genus. Species identification is extremely difficult with this genus.

Habitat The thick doors to their burrows are very effectively camouflaged and very difficult to detect. Burrows are often built on banks and hillsides in forested habitats.

Known records from FL, LA, OH, TX

NEMESIIDAE ▷ VELVETEEN TARANTULAS

Other Colloquial Names Wishbone spiders, false tarantulas

Genera in This Family *Calisoga* (5)

Total Body Length 16.0–30.0 mm

Description These are large hairy mygalomorphs. They have rodlike setae in a row on the upper inner chelicerae, although this trait is best seen with a microscope. The opisthosoma lacks tergites. They lack claw tufts and have three tarsal claws. They live in burrows or crevices lined with silk.

Similar Families Theraphosidae

How to distinguish Nemesiidae from similar families:

Nemesiids can be distinguished from theraphosids by the rodlike setae on the upper inner chelicerae. They also lack claw tufts and urticating hairs; both traits are seen in theraphosids.

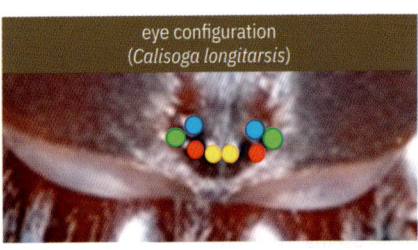

eye configuration
(*Calisoga longitarsis*)

lacks claw tufts

Calisoga longitarsis

lacks tergites, velvet-like appearance

Calisoga longitarsis (Simon, 1891)

Common Name This species does not have a common name.

Other Colloquial Names False tarantula

Total Body Length Female 18.0–46.0 mm, Male 15.0–18.0 mm

Description The prosoma is bluish brown to silver. The opisthosoma is dark brown to dark gray. The males are usually darker than the females. These spiders resemble the theraphosids (tarantulas) but have shorter hairs, giving them a more velvety look. They also have longer posterior lateral spinnerets than theraphosids. The males have a curved metatarsus I. When disturbed, these spiders will readily rear up, exposing their fangs in a defensive posture.

female

Habitat These spiders often create their own burrows (which resemble a wishbone shape), but they have also been documented taking over rodent burrows. Their burrows have no door or turret, but they do have silk that extends outward from the burrow entrance. They can be found on open hillsides, fields, and orchards. Mature males can be found wandering in the fall.

Known records from CA, ID, NV, TX (It is speculated that the ID and TX records may be in error.)

THERAPHOSIDAE > TARANTULAS

Other Colloquial Names Baboon spiders, earth tigers

Genera in This Family *Aphonopelma* (29) and *Tliltocatl* (1)

Total Body Length 9.0–75.0 mm

Description These are large hairy mygalomorphs. They lack tergites on the opisthosoma. They have claw tufts on all tarsi. An urticating hair patch is on the posterior dorsal opisthosoma. They live in subterranean burrows that lack any door-like closure, and the perimeter is lined with silk. They are active at night, when they sit at the burrow entrance waiting for prey. Species identification is notoriously difficult.

Similar Families Nemesiidae

How to distinguish Theraphosidae from other similar families:

Theraphosids can be distinguished from nemesiids by the presence of claw tufts (and only two tarsal claws), the presence of an urticating hair patch on the posterior dorsal opisthosoma, and the lack of rodlike setae on the upper inner chelicerae.

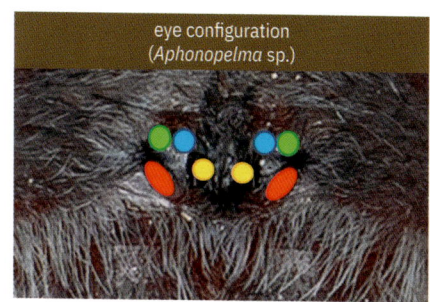

eye configuration
(*Aphonopelma* sp.)

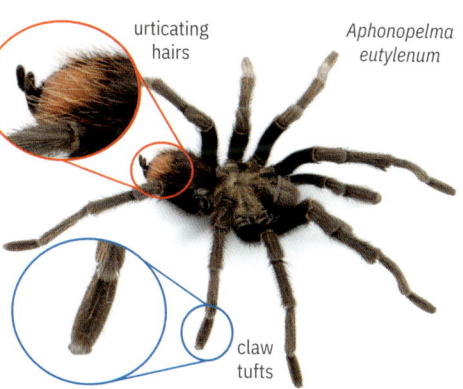

urticating hairs

Aphonopelma eutylenum

claw tufts

Aphonopelma eutylenum Chamberlin, 1940

Common Name California ebony tarantula

Total Body Length Female 43.0–57.0 mm, Male 40.0–42.0 mm

Description The prosoma is tan to brown, with golden highlights. The chelicerae are dark brown. The opisthosoma is dark brown and covered with long hairs. The femurs are dark brown, and the distal leg segments are brown to tan. The female is often paler than the male. The prosoma is lighter in color than the opisthosoma.

female

male

Habitat These spiders are usually found in bunchgrass or scrub oak habitats in coastal areas. Males can often be found wandering in search of females in the fall (September–November).

Known records from CA

Aphonopelma hentzi (Girard, 1852)

Common Name Texas brown tarantula

Other Colloquial Names Oklahoma brown tarantula

Total Body Length Female 44.0–58.0 mm, Male 38.0–52.0 mm

Description The prosoma is tan to light brown. The chelicerae are dark brown. The opisthosoma is brown to dark brown and covered with reddish hair. The legs are brown to dark brown.

Habitat This is the most widely distributed species of *Aphonopelma* in the United States and can be found in a wide variety of habitats. The males can be found wandering in the spring/summer (April–August) or in the fall (September–October).

Known records from AR, AZ, CO, KS, LA, MO, NM, OK, TX

female   male

FILISTATIDAE ▷ CREVICE WEAVERS

Genera in This Family *Filistatinella* (6), *Filistatoides* (1), and *Kukulcania* (5)

Total Body Length 1.5–18.0 mm

Description These are long-lived haplogyne spiders, taking several years to reach maturity. The females are known to continue molting after maturity, which is rare for an Araneomorphae. They are cribellate and have a short calamistrum. Their eyes are clustered on a central mound. Leg autospasy occurs at the patella-tibia joint. They have a pair of book lungs and a single wide tracheal spiracle (or two small spiracles) located between the spinnerets and the epigastric furrow. They typically hide in cribellate silk-lined crevices from which trip lines extend.

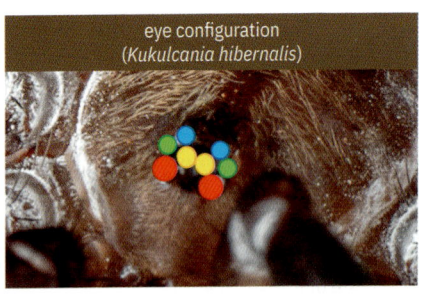

eye configuration
(*Kukulcania hibernalis*)

Similar Families Amaurobiidae, Plectreuridae, Zoropsidae

How to distinguish Filistatidae from other similar families:

Filistatids can be distinguished from other similar families as their eyes are clustered on a central mound. The eyes of amaurobiids, plectreurids, and zoropsids all appear in two rows occupying over half of the width of the prosoma. Additionally, autospasy at the patella-tibia joint and the anteriorly placed tracheal spiracle (or spiracles) distinguish filistatids from amaurobiids, plectreurids, and zoropsids.

Kukulcania hibernallis

Kukulcania hibernallis

cribellum, anterior tracheal spiracles   short calamistrum

Filistatoides insignis (O. Pickard-Cambridge, 1896)

Common Name This species does not have a common name.

Total Body Length Female 4.5 mm, Male 3.0 mm

Description The prosoma is tan to yellow. There is a dark median stripe with irregular borders. There are stripes from the eye region to the chelicerae. The opisthosoma is tan to light brown, with a series of five to six dark brown chevrons; sometimes these appear broken. The legs are tan to yellow, with some dusky banding.

female

Habitat These spiders are often found in cracks and crevices in rocks and walls. They occur in dry habitats and are sometimes associated with buildings.

Known records from AZ, TX

Kukulcania hibernalis (Hentz, 1842)

Common Name Southern house spider

Total Body Length Female 13.0–19.0 mm, Male 9.0–10.0 mm

Description These spiders have a very deep fovea. The females have a dark gray to jet-black prosoma, opisthosoma, and legs; while the males have tan to brown prosoma, opisthosoma, and legs. The males have very long pedipalps, which are often held in a folded position. Due to the velvety appearance of females, they can be confused with Theraphosidae and Nemesidae. They can be distinguished from the Mygalomorphae by the orientation of the fangs (diaxial) and the short spinnerets. The males are often confused with *Loxosceles reclusa* (brown recluse) due to their coloration and carapace markings. They can be distinguished from *L. reclusa* by the number of eyes (eight clustered eyes, in contrast to *Loxosceles* species, which have six eyes in pairs), the presence of leg spines, and the long pedipalps. These spiders are very difficult to identify to species without collecting a specimen. Additional species of *Kukulcania* occur in the southwestern United States.

Habitat These spiders are well adapted to human structures, as their common name implies.

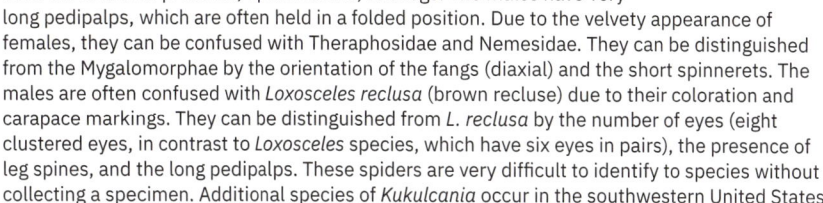

female

male

female

male

female with spiderlings

spiderlings

They will often build their webs in cracks and crevices within buildings, bridges, or barns. They can occasionally be found under bark.

Known records from AL, AR, CA, FL, GA, IL, LA, MS, NM, NC, NY, OK, SC, TN, TX, VA

HERSILIIDAE ▶ LONGSPINNERET SPIDERS

Other Colloquial Names Two-tailed spiders, tree trunk spiders

Genera in This Family *Neotama* (1), *Murrica* (1), and *Yabisi* (1)

Total Body Length 6.0–12.0 mm. (Note: this does not include the spinnerets.)

Description These spiders are entelegyne and ecribellate. They have a raised ocular area but are otherwise somewhat flattened in appearance. They have eight eyes in two strongly recurved rows. They have a pair of book lungs and a single tracheal spiracle located near the spinnerets. The posterior spinnerets are extremely elongated (as long as or longer than the abdomen). They have very long slender legs. They create silken mats that serve as a prey-detection system.

Similar Families Oecobiidae, Philodromidae

How to distinguish Hersiliidae from other similar families:

Both hersiliids and oecobiids have a similar prey-capture technique, but the extremely long posterior spinnerets distinguish hersiliids from oecobiids and philodromids, as neither of those families has elongated spinnerets.

eye configuration
(*Neotama mexicana*)

long spinnerets

Neotama mexicana

raised ocular area

Neotama mexicana (O. Pickard-Cambridge, 1893)

Common Name This species does not have a common name.

Other Colloquial Names Long-spinneret spider, Mexican two-tailed spider

Total Body Length Female 6.5–11.8 mm, Male 6.1–8.2 mm

Description The prosoma is usually mottled with grays, orange, and greens and slightly elevated in the eye region. There is a darker border around the edge. The opisthosoma is mottled with grays, greens, and browns. The heartmark is usually darker, and there is often a dark band around the anterior. The spinnerets are very long, often longer than the opisthosoma. When capturing prey, the spider holds the spinnerets over the prey item and attaches the prey to the surface with silk. The legs are very long and usually banded in tan and brown.

female

Habitat These spiders are often found on the trunks and branches of trees. Their coloration allows them to camouflage effectively on tree trunks, especially those covered in lichen.

Known records from TX

OECOBIIDAE ▸ FLATMESH WEAVERS

Other Colloquial Names Discweaver spiders, disc weavers, hackled band weaver spider, spiraalspinnen (spiral spider), wall spiders, scheibennetzspinnen (disk net spider)

Genera in This Family *Oecobius* (7) and *Platoecobius* (1)

Total Body Length 0.9–3.2 mm

Description These small entelegyne spiders are cribellate. Their eight eyes are clustered on their semicircular prosoma. They have a large anal tubercle fringed with long hairs, which is used to comb out silk from the large posterior spinnerets (in a similar way that most cribellate spiders comb silk from the cribellum with the calamistrum). The legs are positioned in a diagnostic starburst pattern. These spiders have a pair of book lungs and a single inconspicuous tracheal spiracle near the spinnerets. The spiders are often found under rocks or on walls. They are commonly found near human-built structures. When prey is detected, they quickly run circles around it, laying swaths of silk. They will readily attack and prey upon ants.

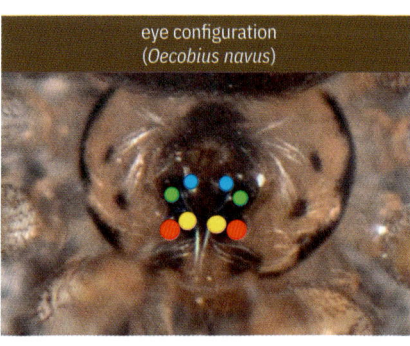

eye configuration
(*Oecobius navus*)

Oecobius navus

starburst legs, eyes clustered

Similar Families Hersiliidae

Both hersiliids and oecobiids have a similar prey-capture technique, but the extremely long posterior spinnerets seen in hersiliids distinguish them from oecobiids.

Oecobius navus Blackwall, 1859

Common Name This species does not have a common name.

Total Body Length Female 2.5–2.9 mm, Male 2.0–2.6 mm

Description This species is most easily recognized by the pattern on the prosoma: a dark central area and three dashes or dots on each side. The prosoma is semicircular and comes to a point at the front. The eight eyes are clustered near the middle of the prosoma. The opisthosoma is grayish and mottled with white; there is a dark unmottled heartmark. The legs are banded, although this might be quite faint. The males and females look similar, although the males tend to be skinnier and have obviously modified pedipalps. Egg sacs can contain three to ten eggs and receive no parental care once created.

Habitat This species is considered synanthropic and is often found on and in human structures.

Known records from QC (Canada); AL, AZ, CA, FL, IL, LA, MA, ME, MS, NC, NM, NY, OH, OR, TN, TX, WA (United States). This is an introduced species.

female

female

SEGESTRIIDAE ▸ TUNNEL SPIDERS

Other Colloquial Names Tubeweb spiders

Genera in This Family *Ariadna* (3), *Citharoceps* (2), and *Segestria* (2)

Total Body Length 4.0–16.0 mm

Description These spiders are entelegyne and ecribellate. They have six eyes (the anterior median eyes are missing); the eyes are in two rows occupying over half the prosomal width. They have a pair of book lungs and two tracheal spiracles near the epigastric furrow. They rest with their first three pairs of legs directed anteriorly. They tend to be nocturnal hunters that live in tubular silk retreats with trip lines radiating from the entrance.

Similar Families Dysderidae

Segestriids can be distinguished from dysderids by their eye configuration. Both families have only six eyes (the anterior median eyes are missing for both), but in dysderids, the eyes are clustered and occupy less than a third of the prosomal width. Segestriids' eyes are in two rows occupying over half the prosomal width. Also, segestriids rest with the first three pairs of legs directed anteriorly; dysderids usually sit with only the first two pairs of legs directed anteriorly.

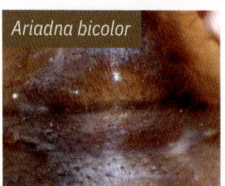

eye configuration
(*Ariadna bicolor*)

Ariadna bicolor

LEFT: pair of tracheal spiracles near book lungs

BELOW: first three pairs of legs directed forward

Ariadna bicolor

Ariadna bicolor (Hentz, 1842)

Common Name Tubeweb spider

Total Body Length Female 6.1–15.0 mm, Male 5.4–7.2 mm

Description The prosoma is usually brown, with the eye region notably darker. There are six eyes. The opisthosoma is lighter in color with a pale stripe running its length. The legs are brown to pale tan; the first pair is usually the darkest. The males have modified macrosetae on the distal tibia of the first pair of legs. They sit with the first three pairs of legs directed anteriorly, which is characteristic of this family.

female

Habitat These spiders build a silken tube retreat within rotting logs, in cracks in buildings and rocks, or in tree bark. Running a small paintbrush gently over the trip lines that radiate from the tube web will oftentimes tease the spider into revealing itself.

Known records from ON (Canada); AL, AR, AZ, CA, CO, CT, DC, FL, GA, IL, IN, KY, LA, MA, MD, ME, MO, MS, NC, NJ, NM, NY, OH, OK, PA, TN, TX, UT, VA, WV (United States)

male

female

juvenile

in web

Segestria pacifica Banks, 1891

Common Name Western tubeweb spider

Total Body Length Female 5.8–16.0 mm, Male 7.0–10.0 mm

Description The prosoma is brown, with a darker eye region; indistinct radial lines are just visible. There are six eyes. The opisthosoma is pale brown with a dark brown-purple median line with four to five diamond or square lobes. There is often dark mottling on the lateral edges. The legs are tan to brown, with dusky banding. The spiders sit with the first three pairs of legs directed anteriorly, which is characteristic of this family.

female

Habitat These spiders build a silken tube retreat within rotting logs, in cracks in buildings and rocks, or in tree bark. They do not balloon, even as juveniles, so there often will be large congregations of individuals in a localized area.

Known records from BC (Canada); CA, OR, WA (United States)

The spiders in the Sheet Web Weaver Guild construct a nonsticky platform of horizontal silk that may or may not be associated with a funnel-like retreat that is also made of silk. This guild contains both Mygalomorphae and Araneomorphae families and includes the following families: Euagridae, Hexurellidae, Megahexuridae, and Microhexuridae (Mygalomorphae), and Agelenidae, Amaurobiidae, Cybaeidae, Desidae, Hahniidae, Linyphiidae, Ochyroceratidae, Pimoidae, Telemidae, and Zoropsidae (Araneomorphae).

KEY TO FAMILY

1a. Mygalomorphae: Chelicerae are paraxial (parallel to central axis of body); when spider attacks, it raises its prosoma up to unfold fangs and strikes downward onto prey; two pairs of book lungs; eight eyes that are usually clustered together; usually with robust legs -------- **2**

1b. Araneomorphae: Chelicerae diaxial (at an angle to central axis of body); when spider attacks, fangs are used in a pinching motion; usually only one pair of book lungs; if spider has two pairs of book lungs, it also possesses a cribellum ----------------------------- **5**

2a. Without tergites on opisthosoma -- **3**
2b. With one of more tergites on opisthosoma --- **4**

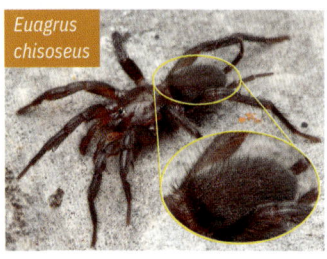

3a. Thoracic furrow is longitudinal, males with mid-ventral spine on tibia I, 2.5–6.0 mm total body length --------------------------------- **Microhexuridae** (page 68)
3b. Thoracic furrow is shallow and rounded or a deep depression, males with mid-ventral spine or spines on tibia II, 3.5–17.0 mm total body length ------ **Euagridae** (page 65)

4a. Two tergites (sometimes only one visible), pleurite extension on carapace, 13.0–18.0 mm total body length ---------------------------- **Megahexuridae** (page 67)

4b. Two tergites, lacks pleurite extensions on carapace, small, 2.5–5.0 mm total body length --------------------------------------- **Hexurellidae** (page 66)

Megahexura fulva

a

Hexurella

b

5a. With cribellum (cribellate) --- **6**

5b. Without cribellum (ecribellate) --- **8**

Amaurobius ferox

a

Cybaeus signifer

b

6a. Long calamistrum --- **Desidae (*Badumna*)** (page 90)

6b. Short calamistrum --- **7**

a

Badumna longinqua

b

Amaurobius ferox

7a. Three pairs of ventral spines on anterior tibiae, orange prosoma that darkens in eye region, lateral eyes larger than median eyes, five or more teeth on both margins of chelicerae furrow (difficult to see in life specimens), compact male pedipalps, 6.0–10.0 mm total body length, introduced species known to be established in AR, CA, FL, GA, LA, MS, NC, and TX, usually synanthropic (found in and around human structures) --------------------------------- **Desidae (*Metaltella*)** (page 90)

7b. None to variable number of ventral spines on anterior tibia, prosoma variable in color, fewer than five teeth on margins of chelicerae furrow (difficult to see in live specimens), eyes of similar size or anterior median eyes reduced or absent, 1.3–20.0 mm total body length, widespread distribution, often found in natural habitats (occasionally found on buildings) ------------ **Amaurobiidae** (page 79)

Metaltella simoni

a

b

Amaurobius ferox

8a. Six eyes in three diads --- **9**
8b. Eight eyes -- **10**

Usofila pacifica

a

Eratigena duellica

b

9a. Book lungs present, 1.0–2.0 mm total body length, North American records are from FL only, reproduces parthenogenetically, no males recorded, usually orange to orange brown -------------------------- **Ochyroceratidae** (page 120)
9b. Lacks book lungs, 1.0–1.7 mm total body length, living specimens have a brown prosoma and purplish-brown opisthosoma, records from AB, BC (Canada); AK, CA, OR, UT, WA, and WY (United States) ------ **Telemidae** (page 122)

Theotima minutissima

a

b

Usofila pacifica

10a. Spinnerets in transverse row, 1.2–5.1 mm total body length,
tracheal spiracle placed anteriorly from spinnerets ---------------- **Some Hahniidae** (page 91)
10b. Spinnerets not in a transverse row, tracheal spiracle near spinnerets ------------------------- **11**

Neoantisea magna

a

b

Cybaeus signifer

11a. Posterior spinnerets distinctly elongated, 4.0–18.0 mm total body length
-- **Agelenidae** (page 69)
11b. Posterior spinnerets not distinctly elongated --- **12**

*Agelenopsis
pennsylvanica*

a

b

Cybaeus signifer

12a. One to five pairs of ventral spines on anterior tibiae --- **13**
12b. Lacks ventral spines on anterior tibiae --- **14**

Islandiana flaveola

a

b

Cicurina arculata

13a. Anterior eye row usually straight, usually dark, 1.4–14.0 mm total
body length --- **Cybaeidae** (page 84)
13b. Anterior eye row usually procurved, usually pale, 2.0–13.0 mm total
body length --- **Some Hahniidae (*Cicurina*)** (page 91)

Cybaeus signifer

Cicurina arculata

a

b

14a. Fovea conspicuous and ovate, 1.4–12.0 mm total body length ----------- **Pimoidae** (page 121)
14b. Fovea variable, sometimes lacking, not usually conspicuous or ovate,
 <1.0–7.0 mm total body length -- **Linyphiidae** (page 97)

Pimoa altioculata

Ceraticelus similis

a

b

EUAGRIDAE ⟩ AMERICAN FUNNELWEB SPIDERS

Other Colloquial Names Curtainweb spiders, funnelweb spiders

Genera in This Family *Euagrus* (3)

Total Body Length 6.5–17.0 mm

Description These mygalomorphs lack tergites on the dorsal opisthosoma. The thoracic furrow is shallow and rounded or a deep depression. The males have a large spine or spines on the mid-ventral surface of tibia II. They have eight eyes in a compact quadrangle twice as wide as it is long. (There are some eyeless troglobites known from Mexico.)

Similar Families Antrodiaetidae, Hexurellidae, Megahexuridae, Microhexuridae

How to distinguish Euagridae from other similar families:

Euagrids can be distinguished from antrodiaetids, hexurellids, and megahexurids by their lack of tergites on the dorsal opisthosoma; tergites are seen in those families. They can be distinguished from the microhexurids by their overall size. Euagrids are 6.5–17.0 mm total body length, whereas microhexurids are small animals, only 2.5–6.0 mm total body length. Additionally, male microhexurids have a modified tibia I with one or more large spines; euagrids have a modified tibia II with one large spine.

eye configuration
(*Euagrus chisoseus*)

eyes in a compact quad

Euagrus chisoseus

spine, mid-ventral
surface of tibia II

Euagrus chisoseus Gertsch, 1939

Common Name This species does not have a common name.

Total Body Length Female 8.5–10.3 mm, Male 7.5–8.0 mm

Description The female prosoma is yellow to light orange brown; the legs, pedipalps, and chelicerae are usually slightly darker than the prosoma.

male

female

The opisthosoma is grayish yellow to brownish purple. The male prosoma is tan to light brown; the legs are a similar color. The opisthosoma is usually tan to brown. This species looks very much like the other *Euagrus* species. The spines on the male tibia II and metatarsus II are helpful in determining species identification. *Euagrus comstocki* is almost indistinguishable without looking at the genitalia but is usually found only at the southernmost point of Texas, whereas *E. chisoseus* is found in more central parts of Texas and west into New Mexico and Arizona.

Habitat Webs are built under rocks in areas where at least some leaf litter is present.

Known records from AZ, NM, TX

HEXURELLIDAE ▶ DWARF FUNNELWEB SPIDERS

Other Colloquial Names Dwarf tarantula

Genera in This Family *Hexurella* (3)

Total Body Length 2.5–5.0 mm

Description These mygalomorphs have two tergites on the dorsal opisthosoma. They have six spinnerets; the posterior lateral spinnerets have four segments. These and Microhexuridae are among the smallest Mygalomorphae of North America.

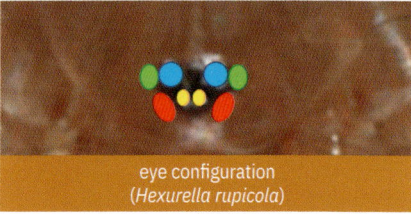

eye configuration
(*Hexurella rupicola*)

Similar Families Antrodiaetidae, Atypidae, Euagridae, Megahexuridae, Microhexuridae

How to distinguish Hexurellidae from other similar families:

Hexurellids can be distinguished from most antrodiaetids by the longer distal segment of the posterior lateral spinnerets. In most antrodiaetids, the distal segment is only about three times as long as it is wide; in hexurellids, it is at least five times as long as it is wide. Hexurellids can be distinguished from atypids

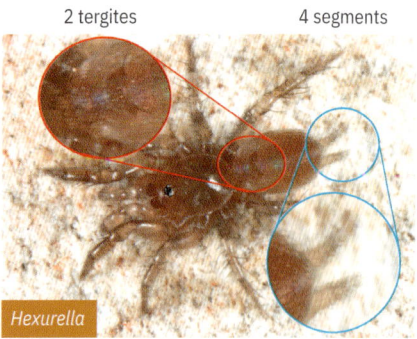

2 tergites   4 segments

Hexurella

by their endites: atypids have long endites (about three-fourths the width of the sternum), while the endites are short in hexurelliids. Hexurellids can be distinguished from euagrids and microhexurids by the presence of the dorsal tergites on the opisthosoma; euagrids and microhexurids lack these. Hexurellids can be distinguished from megahexurids by the animal's overall size: hexurellids are only 2.5–5.0 mm total body length, whereas megahexurids are 10.0–18.0 mm total body length.

Hexurella rupicola Gertsch & Platnick, 1979

Common Name This species does not have a common name.

Total Body Length Female 4.5 mm, Male 2.3 mm

Description The prosoma is orange brown and has some darker, but faint, radiating lines. The posterior eye row is slightly procurved. The opisthosoma is orange brown, sometimes with a slight iridescent sheen to it. The legs are orange brown, with copious spines. The males of this species lack clasping spines on leg I.

Habitat These spiders are often found in chaparral.

Known records from CA

MEGAHEXURIDAE > RAVINE FUNNELWEB SPIDERS

Other Colloquial Names Funnelweb tarantulas

Genera in This Family *Megahexura* (1)

Total Body Length 13.0–18.0 mm

Description These mygalomorphs have two tergites on the dorsal opisthosoma, although at times only one may be visible. They tend to be brown to dark brown. They have pleurite extensions on the posterolateral prosoma.

Similar Families Antrodiaetidae, Atypidae, Euagridae, Hexurellidae, Microhexuridae

How to distinguish Megahexuridae from other similar families:

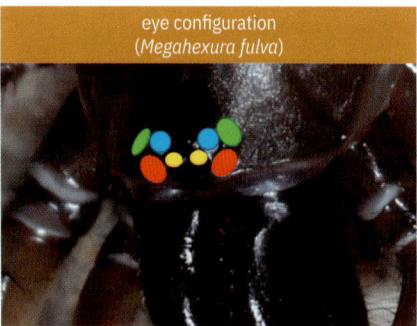

eye configuration (*Megahexura fulva*)

Megahexurids can be distinguished from euagrids and microhexurids by the presence of tergites on the opisthosoma; tergites are lacking in those two families. Megahexurids can be distinguished from hexurellids by the animal's overall size: hexurellids are only 2.5–5.0 mm total body length, whereas megahexurids are 10.0–18.0 mm total body length. Megahexurids can be distinguished from most antrodiaetids by the long distal segment of the posterior lateral spinnerets.

pleurite extensions

2 tergites

Megahexura fulva

In most antrodiaetids, the distal segment is only about three times as long as it is wide; in megahexurids, it is at least five times as long as it is wide. Megahexurids can be distinguished from atypids by their endites: atypids have long endites (about three-fourths the width of the sternum), whereas megahexurids have short endites (less than half the width of the sternum).

Megahexura fulva (Chamberlin, 1919)

Common Name This species does not have a common name.

Other Colloquial Names Tawny dwarf tarantula

Total Body Length Female 13.0–18.0 mm, Male 8.0–13.0 mm

Description The prosoma is brown to dark brown. The opisthosoma is reddish brown to pink, with a darker heartmark. These spiders have two tergites on the dorsal opisthosoma, but sometimes only one of the two tergites is visible. The legs are brown to pale brown.

Habitat The sheet webs are often built in areas with deep leaf litter, under rocks, or under logs.

Known records from CA

female

MICROHEXURIDAE ▸ TINY FUNNELWEB SPIDERS

Other Colloquial Names Funnelweb spiders

Genera in This Family *Microhexura* (2)

Total Body Length 2.5–6.0 mm

Description These mygalomorphs lack tergites on the dorsal opisthosoma. The thoracic furrow is longitudinal. The tibia I of males has a large spine on the mid-ventral surface. These, along with Hexurellidae, are among the smallest Mygalomorphae in North America.

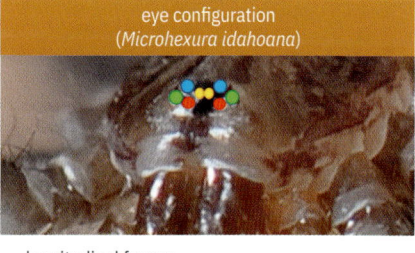

eye configuration
(*Microhexura idahoana*)

longitudinal furrow

Microhexura montivaga

Microhexura idahoana

Similar Families Antrodiaetidae, Euagridae, Hexurellidae, Megahexuridae

How to distinguish Microhexuridae from other similar families:

Microhexurids can be distinguished from antrodiaetids, hexurellids, and megahexurids by the lack of the tergites on the dorsal opisthosoma that is seen in those families. Microhexurids can be distinguished from euagrids by their overall size: euagrids are 6.5–17.0 mm total body length, whereas microhexurids are small animals, only 2.5–6.0 mm total body length. Additionally, male microhexurids have a modified tibia I with one or more large spines; euagrids have a modified tibia II with one large spine.

Microhexura Idahoana Chamberlin & Ivie, 1945

Common Name This species does not have a common name.

Total Body Length Female 2.9–5.6 mm, Male 3.1–4.5 mm

Description The prosoma is shiny brown with a sculpted appearance, darker in the eye region. The opisthosoma is brown and velvety. The legs are brown to pale brown.

Habitat The web is often built in a depression in the duff layer.

Known records from BC (Canada); ID, MT, WA (United States)

female in web

female with egg sac

female with egg sac

AGELENIDAE ▶ FUNNEL WEAVER

Other Colloquial Names Funnel web spiders, grass spiders

Genera in This Family *Agelenopsis* (14), *Barronopsis* (5), *Calilena* (20), *Callidalena* (1), *Coras* (14), *Eratigena* (4), *Hololena* (30), *Melpomene* (1), *Novalena* (5), *Rualena* (9), *Tegenaria* (3), *Tortolena* (1), and *Wadotes* (11)

Total Body Length 4.0–18.0 mm

Description These medium-sized to large spiders are entelegyne and ecribellate. They have elongated posterior lateral spinnerets

eye configuration
(*Eratigena duellica*)

that are widely separated. They have eight eyes. They build a web made up of an exposed sheet web with a funnel-like retreat. The spiders are very reactive to anything touching the web, as the silk is nonsticky and prey can escape easily. They can often be coaxed out of their retreat (if they

elongated
posterior lateral
spinnerets

*Agelenopsis
pennsylvanica*

Tegenaria domestica

funnel-shaped web

are hungry) by touching a turned-on sonic toothbrush to the edge of the web. They are extremely fast.

Similar Families Amaurobiidae, Cybaeidae, Dictynidae, Hahniidae, Lycosidae, Miturgidae

How to distinguish Agelenidae from other similar families:

Agelenids can be distinguished from amaurobiids, as they are ecribellate and amaurobiids are cribellate. Some of the dictynids are cribellate, which would distinguish them from the agelenids. Those that are ecribellate are found only in desert salt crust habitats of the Southwest and intertidal zones in Florida, which would not be ideal habitat for most of the agelenids. They can be distinguished from cybaeids, dictynids, hahniids, and miturgids by their spinnerets, as agelenids have elongated posterior lateral spinnerets. In contrast, in cybaeids the anterior lateral spinnerets tend to be only slightly longer than the posterior lateral spinnerets; in dictynids and miturgids, the distal segment of the posterior lateral spinnerets is very short; and in many of the hahniids, the spinnerets are in a transverse row, and those hahniids whose spinnerets are not in a transverse row do not have elongated spinnerets. Agelenids can be distinguished from lycosids by their eye configuration. Lycosids have enlarged posterior median eyes, and the posterior lateral eyes are placed back on the sides of the prosoma, whereas agelenids have fairly equally sized eyes in two rows.

Agelenopsis pennsylvanica (C. L. Koch, 1843)

Common Name The genus has the common name "grass spider"; this species does not have a common name.

Total Body Length Female 10.0–17.0 mm, Male 9.0–12.0 mm

Description The prosoma is tan, and two brown bands run on either side of the central line. The opisthosoma has a light brown central area bordered

female

female

male

male

spiderlings

by a tan stripe that breaks up toward the posterior end; the opisthosoma is darker on the lateral edges. *Agelenopsis* species have an elongated distal segment on the posterior lateral spinnerets (twice as long as the basal segment), creating very visible spinnerets at the end of the opisthosoma. The palps of a mature male have an easily seen coiled embolus that is diagnostic of this genus. This species looks very similar to other species of *Agelenopsis*. To conclusively identify it, you would need to look at the genitalia using a microscope.

Habitat These spiders' sheet webs with a funnel retreat are often built in grassy areas, but these spiders also readily build on human structures.

Known records from ON (Canada); CO, CT, ID, IL, IN, KS, LA, MA, MD, MI, MN, MO, ND, NM, NY, OH, OK, OR, PA, TN, UT, VA, WA, WI, WV (United States)

Agelenopsis utahana (Chamberlin & Ivie, 1933)

Common Name The genus has the common name "grass spider"; this species does not have a common name.

Total Body Length Female 9.0–12.0 mm, Male 8.0–10.0 mm

Description The prosoma is pale yellow, and two brown bands radiate out each side of the central line. The opisthosoma is tan to light brown and mottled with darker brown. The legs and pedipalps are pale yellow and faintly banded. *Agelenopsis* species have an elongated distal segment on the posterior lateral spinnerets

female

female

female ventral

juvenile

(twice as long as the basal segment), creating very visible spinnerets at the end of the opisthosoma. The palps of a mature male have an easily seen coiled embolus that is diagnostic of this genus. This species looks very similar to other species of *Agelenopsis*. To conclusively identify it, you would need to look at the genitalia using a microscope.

Habitat These spiders' sheet webs with a funnel retreat are often built in woody areas.

Known records from AB, BC, MB, NB, NL, NS, NT, ON, PE, QC, SK, YT (Canada); AK, AR, CO, IL, KY, MA, ME, MI, MN, MT, ND, NH, NM, NY, OH, PA, UT, VA, VT, WI, WY (United States)

Barronopsis texana (Gertsch, 1934)

Common Name This species does not have a common name.

Total Body Length Female 6.8–10.0 mm, Male 6.9–10.0 mm

Description The prosoma is pale tan, and two brown bands radiate out each side of the central line. The opisthosoma has a light reddish-brown

female

female ventral

female

juvenile

central area bordered by a broken tan stripe; the opisthosoma is darker on the lateral edges. The legs are boldly banded. This species looks very similar to other species of *Barronopsis*. To conclusively identify it, you would need to look at the genitalia using a microscope.

Habitat These spiders' sheet webs with a funnel retreat are often built in the corners of human-built structures and can occur in high densities. The stacked webs look complex, almost apartment-like. The spiders tend to build their webs up off the ground.

Known records from AL, FL, GA, LA, MS, NC, SC, TN, TX, VA

Calilena arizonica Chamberlin & Ivie, 1941

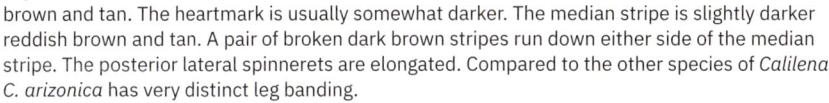

Common Name This species does not have a common name.

Total Body Length Female 7.3–11.3 mm, Male 9.7 mm

Description The prosoma is tan to gray tan. There is a dark brown submarginal band. The eyes are in two strongly procurved rows. The eye region is somewhat elevated. The opisthosoma is mottled with reddish brown and tan. The heartmark is usually somewhat darker. The median stripe is slightly darker reddish brown and tan. A pair of broken dark brown stripes run down either side of the median stripe. The posterior lateral spinnerets are elongated. Compared to the other species of *Calilena*, *C. arizonica* has very distinct leg banding.

Habitat These spiders often build their webs fairly close to the ground in loose rocks, piles of debris, or fences.

Known records from AZ, NM

male

male

Coras lamellosus (Keyserling, 1887)

Common Name This species does not have a common name.

Total Body Length Female 8.1–13.2 mm, Male 8.5–12.3 mm

Description The prosoma is longer than it is wide, narrower in the eye region, and widest at legs III and IV. The prosoma is elevated in the eye region, and the sides are nearly parallel in the front third of its length. It is light reddish brown, with two dark submarginal markings. The fovea is short and deep and located at the start of a mark shaped like a tuning fork that leads to the lateral eyes. Both eye rows are slightly procurved. The robust chelicerae have a prominent lateral boss. The opisthosoma is tan to brown, and a dark heartmark with pale chevron markings tapers down the opisthosoma. The jointed spinnerets are usually very visible. The legs are tan with brown banding. *Coras* species can look very similar, and all can vary somewhat in their coloration and markings. To confirm a species-level identification, you would need good views of the genitalia.

female

female

female ventral

male

male

egg sac and female

Habitat These spiders often build their webs in cracks and crevices in hollow trees or logs.

Known records from ON, NS, QC (Canada); AR, CA, DC, IA, IL, IN, KS, KY, LA, MD, ME, MN, MO, MS, NE, OH, OK, OR, PA, TX, WI (United States)

Coras medicinalis (Hentz, 1821)

Common Name This species does not have a common name.

Total Body Length Female 9.5–13.3 mm, Male 9.0–12.6 mm

Description The prosoma is longer than it is wide, narrower in the eye region, and widest at legs III and IV. The prosoma is elevated in the eye region, and the sides are nearly parallel in the front third of its length. It is light reddish brown, with two dark, quite broad submarginal markings. The fovea is short and deep and located at the start of a broad mark shaped like a tuning fork that leads to the lateral eyes. Both eye rows are slightly procurved. The robust chelicerae have a prominent lateral boss. The opisthosoma is dark brown, with a dark heartmark, and pale tan to yellow chevron markings taper down the opisthosoma. The jointed spinnerets are usually very visible. The legs are reddish brown with dark brown banding. *Coras* species can look very similar, and all can vary somewhat in their coloration and markings. To confirm a species-level identification, you would need good views of the genitalia. This species gets its name from the medicinal uses of the webs in years gone by.

Habitat These spiders can be found in and around human structures, under loose bark, or in hollow trees.

Known records from NS, ON, QC (Canada); AL, CO, CT, DC, FL, GA, IL, MD, ME, MI, MN, MS, NJ, NY, NC, OH, OK, TN, TX, VA (United States)

Eratigena agrestis (Walckenaer, 1802)

Common Name Hobo spider

Total Body Length Female 10.0–15.0 mm, Male 7.0–10.0 mm

Description The prosoma is light tan, with indistinct darker longitudinal bands. The eyes are in two slightly procurved rows. The sternum is brown with a pale midline. The lack of pale spots near the coxa helps distinguish this species from *Eratigena duellica*. The opisthosoma is mottled with tan and brown; the heartmark is usually light. A series of chevrons runs down the midline of the dorsal opisthosoma and sometimes resembles a string of hearts or triangles. *Eratigena* can be distinguished from the very similar *Tegenaria* by the lack of banding on the legs. Contrary to popular belief, these spiders are not medically significant to humans.

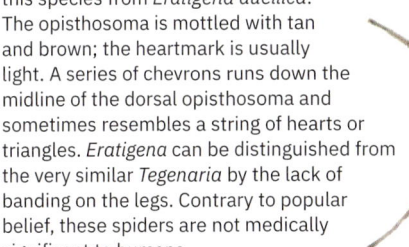

female

male

female

Habitat The name comes from the Latin *agrestis*, which means "field," as these spiders often dwell in fields and grassy areas, but they will readily build their webs on human structures.

Known records from AB, BC, ON (Canada); CO, ID, MT, OR, UT, WA, WY (United States)

This species was introduced to North America. It has been present in the Pacific Northwest for some time and was introduced to Ontario, Canada, recently.

male ventral

Eratigena duellica (Simon, 1875)

Common Name Giant house spider

Total Body Length Female 11.0–16.0 mm, Male 10.0–14.0 mm

Description The prosoma is light tan to tan, with two dark tan to brown longitudinal stripes. The eyes are in two slightly procurved rows. The sternum is dark with a pale midline and three to four pairs of light spots near the leg coxa. Each coxa has paired dark spots on it. The opisthosoma is tan with brown mottling; the heartmark is usually light. A series of chevrons runs down the midline of the dorsal opisthosoma and sometimes resembles a string of hearts or triangles. *Eratigena* can be distinguished from the very similar *Tegenaria* by the lack of banding on the legs.

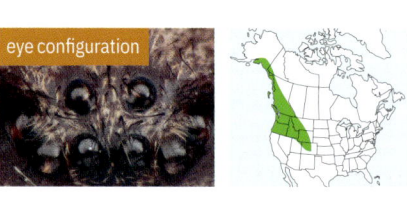

eye configuration

female

female ventral

male

sternum and coxae

Habitat This species is often associated with human structures, and they tend to build their webs in sheltered, dark areas.

Known records from AB, BC (Canada); AK, CO, OR, WA, WY (United States) This is an introduced species.

Hololena curta (McCook, 1894)

Common Name Corner funnel weaver

Total Body Length Female 10.0–12.0 mm, Male 9.0–10.0 mm

Description The prosoma is tan to light yellowish brown, with a dark border and two broad, distinct, dark longitudinal stripes. The eyes are in two strongly procurved rows. The opisthosoma has a dark heartmark within a pale gray midline stripe with a faint chevron pattern. This midline is bordered in pale tan, which becomes broken toward the posterior end. The lateral edges are darker brown. The legs are tan with dark bands. *Hololena* species are very difficult to identify to species without a preserved specimen. They are also difficult to distinguish from *Novalena* and *Rualena* species. As such, you will see some online sources refer to this group of spiders as "Western Ageleninae."

female

Habitat These spiders tend to build their webs in shrubs, bushes, small trees, fences, or buildings.

Known records from CA

Tegenaria domestica
female ventral

Tegenaria domestica
(Clerck, 1757)

Common Name Barn funnel weaver

Other Colloquial Names Common house spider, drain spider, lesser European house spider

Total Body Length Female 6.3–11.5 mm, Male 6.0–9.0 mm

Description The prosoma is tan to light brown, with two dark submarginal bands and dark edges. The opisthosoma has a grayish heartmark and a light chevron pattern that

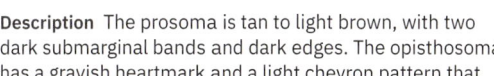

female

female

male

male

juvenile

web

narrows posteriorly. The legs are pale tan with brown banding. (The banding can be faint in some specimens.) The spinnerets are not always visible from a dorsal view.

Habitat These spiders tend to build their webs in or around human structures and seem to like areas with cracks and crevices that can serve as retreats.

Known records from AB, BC, LB, MB, NB, NL, NS, ON, QC, SK (Canada); AK, AL, CA, CO, FL, ID, IL, IN, KE, KS, MA, MD, ME, MI, MN, MT, NC, ND, NE, NJ, NM, NY, OH, OR, PA, SD, TX, UT, VA, WA, WI, WY (United States)

This species was introduced from Europe and is now widespread throughout North America.

Wadotes calcaratus (Keyserling, 1887)

Common Name This species does not have a common name.

Total Body Length Female 5.9–11.0 mm, Male 5.9–8.7 mm

Description The prosoma is dark brown and darker in the eye region. The eye region is elevated; the sides are almost parallel for the front third of the prosoma. The eyes are in two slightly procurved rows. The opisthosoma is dark brown, and a faint chevron pattern is sometimes visible. The legs are brown to reddish brown, and faint banding is sometimes visible. *Wadotes* species are difficult to identify to species without looking at genitalia.

Habitat These spiders like to build their webs in forested areas, often webbing near the ground and under rocks.

Known records from NB, NS, ON, QC (Canada); AL, CT, DE, IL, IN, KY, MA, MD, ME, MI, NC, NH, NJ, NY, OH, PA, RI, SC, TN, VA, VT, WI, WV (United States)

female

female

female

male

AMAUROBIIDAE ▷ HACKLEDMESH WEAVERS

Other Colloquial Names Lace weavers

Genera in This Family *Amaurobius* (25), *Arctobius* (1), *Callobius* (25), *Cavernocymbium* (2), *Cybaeopsis* (9), *Parazanomys* (1), *Pimus* (10), and *Zanomys* (8)

Total Body Length 1.3–20.0 mm

Description These entelegyne spiders are cribellate with a divided cribellum and a short calamistrum. They usually have eight eyes, although some have reduced or absent anterior median eyes. They have a pair of book lungs and a single tracheal spiracle located near the spinnerets. Their webs are often built on the forest floor or under logs, bark, rocks, and leaf litter. In sunlight, the silk can appear to have a bluish coloration.

eye configuration
(*Amaurobius ferox*)

Similar Families Agelenidae, Cybeidae, Desidae, Dictynidae, Filistatidae, Hahniidae, Titanoecidae, Zodariidae, Zoropsidae

79

fewer than 5 teeth on the cheliceral margins

short calamistrum

Amaurobius ferox

divided cribellum

How to distinguish Amaurobiidae from other similar families:

Amaurobiidae are cribellate, which distinguishes them from agelenids, cybaeids, hahniids, and zodariids, which are ecribellate. They can be distinguished from *Badumna longinqua* (Desidae) by the lack of hairs on the prosoma, as *B. longinqua* tends to have a thick covering of silvery hairs, and the anterior median eyes are 1.4 times larger than the anterior lateral eyes, which is unlike any of the amaurobiids. They can be distinguished from *Metaltella simoni* (Desidae) by the compact structures of the male pedipalps in *M. simoni*, and shorter prosoma seen in amaurobiids. Also, if a specimen has been collected, *M. simoni* can easily be distinguished by the presence of five or more teeth on both the prolateral and retrolateral margins of the chelicerae. The cribellum is entire (or at least appears entire) in dictynids, whereas it is divided in amaurobiids. They can be distinguished from filistatids by their eye configuration: amaurobiids' eyes are in two rows that occupy more than half the width of the prosoma, whereas filistatids' eyes are clustered on a central mound. They can be distinguished from titanoecids by the short calamistrum: in titanoecids, the calamistrum is over 70 percent of the length of metatarsus IV, whereas in amaurobiids it is less than 50 percent of the length.

Amaurobius borealis
Emerton, 1909

male

Common Name
This species does not have a common name.

Total Body Length
Female 4.3–6.0 mm, Male 3.5–5.0 mm

Description The prosoma is dark yellow to pale brown. The opisthosoma is reddish brown, with a pale median stripe in the form of broken chevrons; these can be so faint that they are not visible on some specimens. The femurs are paler than the prosoma, and the legs darken distally.

Habitat These spiders are often found in leaf litter. They have been documented feeding on black flies.

Known records from AB, BC, LB, MB, NB, NL, NS, ON, QC, SK (Canada); MA, ME, MI, MN, NH, NY, PA, VT, WI (United States)

Amaurobius ferox (Walckenaer, 1830)

Common Name This species does not have a common name.

Other Colloquial Names Black lace weaver

Total Body Length Female 8.5–14.0 mm, Male 8.0–12.5 mm

Description The prosoma is dark brown, almost black in the eye region. These spiders have robust chelicerae. The opisthosoma is dark brown, with a pale heartmark, and pale chevrons are sometimes visible. The legs are brown, and light banding can occasionally be detected. The male pedipalps have a large pale tegulum, which makes it look like this spider is carrying snowballs while wearing mittens. Matriphagy occurs with this species; the female guards the young, and they will then eat her before dispersing.

eye configuration

female ventral

female

female

female

juvenile

juvenile

juvenile

male

male

male

Habitat These spiders are often found in and around human structures.

Known records from NS, ON, QC (Canada); CT, IL, MA, MD, ME, MI, NJ, NY, OH, PA, RI, VA, WA, WV (United States)

This species was introduced from Europe.

Callobius bennetti (Blackwall, 1846)

Common Name This species does not have a common name.

Other Colloquial Names Bennett's laceweaver

Total Body Length Female 5.0–12.0 mm, Male 5.0–9.0 mm

Description The prosoma is dark orange to reddish brown, darker in the eye region; sometimes some radiating lines are visible. The opisthosoma is dark reddish brown, and a pale chevron pattern runs down its length; the heartmark is dark. The legs are uniformly reddish brown. Several species of *Callobius* look similar. To confirm an identification, you may need to look at the genitalia.

Habitat These spiders often build their webs under debris and rocks.

Known records from AB, LB, MB, NB, NL, NS, ON, QC (Canada); CT, IA, IL, IN, MA, MD, ME, MI, MN, MT, NC, NH, NJ, NY, OH, PA, TN, VA, VT, WI, WV (United States)

female

female

female

female

male

male

male

Cybaeopsis wabritaska
(Leech, 1972)

male

Common Name This species does not have a common name.

Total Body Length Female 4.5–6.5 mm, Male 4.0–5.0 mm

Description The prosoma is light to medium brown, and the eye region is slightly darker; there may be a faint radiating line. The opisthosoma is golden brown, with a slightly paler median stripe made up of lobed and broken chevron shapes. The legs are medium brown, with indistinct dusky banding.

Habitat These spiders are most often found in forested habitats, where they build their webs under litter and debris.

Known records from AB, BC (Canada); AK, OR, WA (United States)

CYBAEIDAE ❯ SOFT SPIDERS

Other Colloquial Name Water spiders

Genera in This Family *Allocybaeina* (1), *Blabomma* (10), *Calymmaria* (29), *Cryphoeca* (2), *Cybaeina* (3), *Cybaeota* (4), *Cybaeozyga* (1), *Cybaeus* (48), *Dirksia* (1), *Ethobuella* (2), *Neocryphoeca* (2), *Willisus* (1), and *Yorima* (5)

Total Body Length 1.4–14.0 mm

Description These small to medium-sized spiders are entelegyne and ecribellate. Most cybaeids have eight eyes, but all eyes may be reduced or missing in cave-dwelling species. The anterior lateral spinnerets are slightly longer than the posterior lateral spinnerets. Cybaeids tend to live in moist habitats, such as the litter of the forest floor or caves.

Similar Families Agelenidae, Amaurobiidae, Dictynidae, Hahniidae, Pimoidae

How to distinguish Cybaeidae from other similar families:

Cybaeids can be distinguished from agelenids by their spinnerets: cybaeids' anterior lateral spinnerets tend to be slightly longer than the posterior lateral spinnerets, whereas agelenids have elongated posterior lateral spinnerets. Cybaeids can be distinguished from amaurobiids, as they are ecribellate and

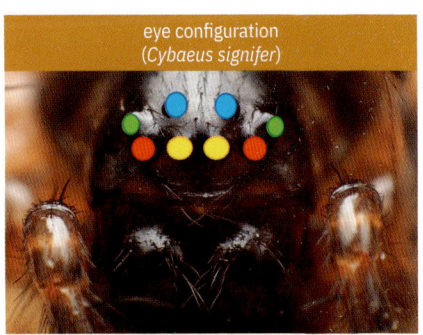

eye configuration
(*Cybaeus signifer*)

anterior lateral spinnerets longer than posterior lateral spinnerets

Cybaeus signifer

amaurobiids are cribellate. Some of the dictynids are cribellate, which would distinguish them from the cybaeids. The dictynids that are ecribellate are found only in desert salt crust habitats of the Southwest and intertidal zones in Florida, which would not be ideal habitat for any of the cybaeids. They can be distinguished from hahniids (genus *Cicurina*) by location, as the cybaeids that look similar to *Cicurina* species are found only on the West Coast, and most of the *Cicurina* species are seen in more eastern locations. They can be distinguished from pimoids, as autospasy occurs at the patella-tibia joint in pimoids.

Blabomma californicum (Simon, 1895)

Common Name This species does not have a common name.

Total Body Length Female 7.0–8.0 mm, Male 5.0–7.4 mm

Description The prosoma is reddish brown, with faint radiating lines. A dark marking starts at the fovea, splits, and goes below the eyes. The eye region is somewhat darker. The anterior median eyes are very small and could be overlooked easily. The opisthosoma is yellow to pale brown. A pair of dark markings are on either side of the heartmark. A series of chevrons runs down the midline, and there are mottled dark markings laterally. The legs are yellow brown with indistinct banding.

Habitat These spiders often build their webs in woody debris, including woodpiles.

Known records from BC (Canada); CA, OR, WA (United States)

female

female ventral

male eyes

male

Calymmaria persica (Hentz, 1847)

Common Name This species does not have a common name.

Other Colloquial Names Basket-web weaver

Total Body Length Female 4.0–9.7 mm, Male 6.1–7.3 mm

Description The prosoma is pale yellow, with a mark shaped like a tuning fork on the anterior; there is a series of dark spots around the edge, and it is bordered with a dark rim. The opisthosoma has a dark heartmark, and a series of dark chevrons runs down the pale midline; the edges are mottled with brown. The legs are banded in tan and brown.

juvenile

juvenile

This species resembles *Tegenaria domestica*, but the carapace markings are distinctly different, with the spots on the lateral edge.

Habitat These spiders are often found in caves or on rock faces. The web is a sheet web within a cone- or bowl-shaped web.

Known records from AL, GA, IN, KY, MS, NC, OH, SC, VA, WV

female with prey in web

web

female with prey in web

Cybaeus giganteus Banks, 1892

Common Name This species does not have a common name.

Total Body Length Female 9.0–12.0 mm, Male 8.0–10.0 mm

Description The prosoma is dark brown. The opisthosoma is very dark, usually obscuring any visible pattern. The legs are usually unbanded or at most have indistinct banding on the proximal segments only.

male

Habitat These spiders build their webs in moist area in woodlands.

Known records from KY, NC, NY, OH, PA, TN, VA, WV

Cybaeus reticulatus Simon, 1886

Common Name This species does not have a common name.

Other Colloquial Names Common west coast woodland spider

Total Body Length Female 8.7–9.2 mm, Male 8.0 mm

Description The prosoma is orange to light brown. There is a dusky submarginal band. The eye region is notably darker. The opisthosoma is dark gray. The heartmark is brown, and on either side, a series of pale diamonds continues down the opisthosoma to the spinnerets. The legs are orange with broad bands.

male

Habitat These spiders are often found in temperate rainforests, where they build their webs under rocks and other debris. The female is known to decorate the egg sac with debris.

Known records from BC (Canada); AK, CA, OR, WA (United States)

male ventral

Cybaeus signifer Simon, 1886

Common Name This species does not have a common name.

Other Colloquial Names Night-hunting woodland spider

Total Body Length Female 8.5–10.0 mm, Male 7.0–7.7 mm

Description The prosoma is brown, with a darker eye region and dark markings. All eyes are approximately the same size. The opisthosoma has a dark heartmark, and a series of pale chevrons runs down its length; the lateral edges are mottled with brown. This is the only *Cybaeus* species in North America on which the anterior median eyes are the same size as the other eyes.

Habitat These spiders are often found in forests with Douglas fir, cedar, or hemlocks. Their webs are often under bark or in rotting logs or moss. They have been observed hunting at night on Douglas fir trees, and it is thought that this is possible due to their well-developed anterior median eyes.

Known records from BC (Canada); AK, CA, OR, WA (United States)

female

female ventral

male

eye configuration

Yorima angelica Roth, 1956

Common Name This species does not have a common name.

Total Body Length Female 3.9 mm, Male 2.9 mm

Description The prosoma is dark yellow to light brown, usually slightly darker in the eye region. They have only six eyes; the anterior median eyes are absent. There may be faint radial lines visible. The opisthosoma is dark gray with a slightly lighter heartmark. A pair of slightly lighter markings are usually on either side of the heartmark, and there is a series of slightly lighter chevrons toward the posterior end. The legs are golden brown. There is a dark ring around the spinnerets.

Habitat These spiders tend to be found in areas with high humidity, such as in leaf litter or under rotting logs.

Known records from CA

DESIDAE ▸ SALTWATER SPIDERS

Other Colloquial Names Intertidal spiders

Genera in This Family *Badumna* (1) and *Metaltella* (1)

Total Body Length 5.0–12.0 mm

Description These are entelegyne cribellate spiders with a divided cribellum and a long calamistrum. They have eight eyes. Only two species from this family are in North America, and both are introduced.

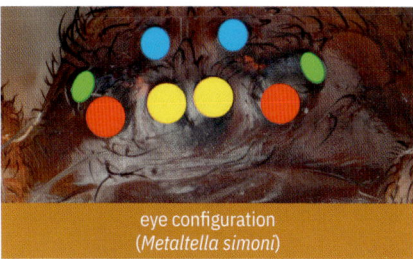

eye configuration
(*Metaltella simoni*)

Similar Families Amaurobiidae, Titanoecidae, Zoropsidae

How to distinguish Desidae from other similar families:

long calamistrum

Badumna longinqua tends to have a thick covering of silvery hairs on the prosoma, which is not seen in other similar families. *Metaltella simoni* can be distinguished from amaurobiids by the compact structures of the male pedipalps in *M. simoni*, and shorter prosoma seen in amaurobiids. *M. simoni* can be distinguished from titanoecids by the shorter calamistrum: in titanoecids, the calamistrum is over 70 percent of the length of metatarsi IV, whereas in *M. simoni,* it is approximately 40 percent of the length. Many of the zoropsids are ecribellate, which easily distinguishes them from the desids. Those zoropsids with a cribellum (*Zoropsis* and *Zorocrates* species) can be distinguished by the fact that the posterior eyes are recurved in *Zoropsis* (they are procurved in Desidae), and fine fuzzy hairs are seen on the prosoma of *Zorocrates* that are not found in desids.

Badumna longinqua

Badumna longinqua (L. Koch, 1867)

Common Name This species does not have a common name.

Other Colloquial Names Gray house spider

Total Body Length Female 10.0–18.0 mm, Male 8.0–10.0 mm

Description The prosoma is brown and covered with gray hairs. The anterior median eyes are 1.4 times larger than the anterior lateral eyes. The opisthosoma is brown and covered in thick gray hairs. The heartmark is dark; there may also be paired dark spots toward the posterior end. The legs usually show some banding, although it may be faint. In the past, it was thought that bites from this species might be medically significant, causing necrotic wounds. This has been shown not to be the case, and the bites are not considered medically significant (Isbister and Gray 2004).

female

Habitat These spiders are often found in and around human structures. They often hide deep in their retreats during the day and are active at night.

Known records from CA, FL, OR, WA
This is an introduced species.

Metaltella simoni (Keyserling, 1878)

Common Name This species does not have a common name.

Other Colloquial Names South American toothed hacklemesh weaver

Total Body Length Female 7.0–10.0 mm, Male 6.0–9.0 mm

Description The prosoma is dark orange, with a darker eye region. The opisthosoma is brownish gray; sometimes the heartmark appears pale; and a series of pale chevrons can be seen down its length. The legs are yellow to orange.

female dorsal

female ventral

female

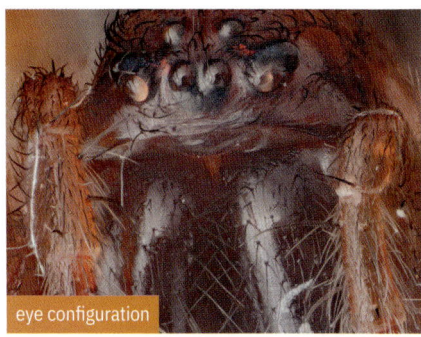

eye configuration

Habitat This is a synanthropic species and is often associated with human structures. The spiders tend to build their webs under debris close to the ground.

Known records from AB (Canada); AR, CA, FL, GA, LA, NC, MS, TX (United States)

This is an introduced species and appears to be expanding its range in North America.

female

HAHNIIDAE ▶ COMBTAIL SPIDERS

Genera in This Family *Antistea* (1), *Cicurina* (108), *Hahnia* (7), and *Neoantistea* (11)

Total Body Length 1.2–13.0 mm

Description These small to medium-sized spiders are entelegyne and ecribellate. They usually have eight eyes, but there are six-eyed or eyeless species in this family. The spinnerets of some species are arranged in a transverse row, in the general appearance of a comb. In species with this spinneret configuration, the tracheal spiracle is located well anterior of the spinnerets. These spiders build small simple sheet webs, sometimes over moss, rocks, or uneven ground, under rocks, or in leaf litter.

Similar Families Agelenidae, Amaurobiidae, Cybaeidae, Gnaphosidae

How to distinguish Hahniidae from other similar families:

In many hahniids, the spinnerets are in a transverse row, which is not seen in any other families. In these species, the placement of the tracheal spiracle anterior of the spinnerets

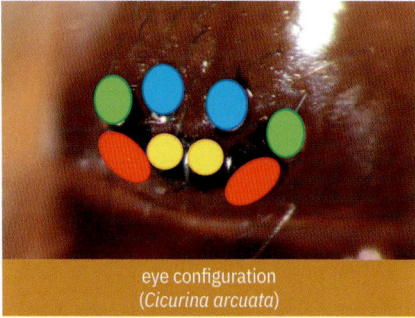

eye configuration
(*Cicurina arcuata*)

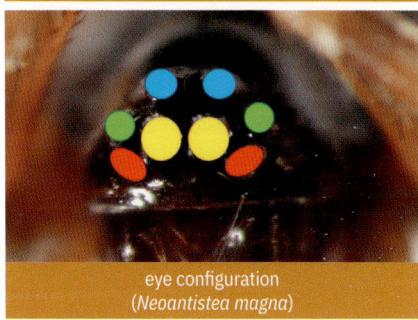

eye configuration
(*Neoantistea magna*)

will also distinguish them from other similar families. Those species in the genus *Cicurina* (which had been moved from several different families before being assigned to Hahniidae) can be distinguished from agelenids by the lack of the elongated posterior lateral spinnerets that are seen in many agelenids. Hahniids can be distinguished from the amaurobiids, as they are ecribellate, and amaurobiids are cribellate. They can be distinguished from most of the cybaeids by location, as the similar-looking cybaeids are found only on the West Coast, and most of the *Cicurina* species are seen in more eastern locations. *Cicurina* can be distinguished from gnaphosids by the lack of the enlarged cylindrical anterior lateral spinnerets that are characteristic of most of the gnaphosids. Those gnaphosids that lack the prominent spinnerets differ from *Cicurina,* as they have iridescent scales, which *Cicurina* lack.

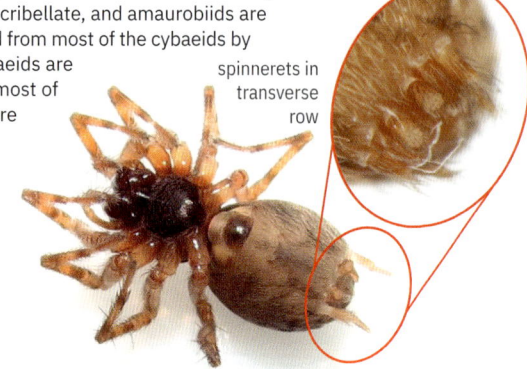

spinnerets in transverse row

Antistea brunnea (Emerton, 1909)

Common Name This species does not have a common name.

Total Body Length Female 2.6–5.1 mm, Male 2.2–2.3 mm

Description The prosoma is light brown and sometimes has faint radial lines. The anterior median eyes are as large as or larger than the anterior lateral eyes. The opisthosoma is dark brown to gray, with five to six chevron shapes (that may be broken) toward the posterior end. The distal segments of the lateral spinnerets are shorter than the proximal segments. The legs are brown. The femur of the male pedipalp has a spur on the ventral side that is nearly as long as the width of the segment.

female

Habitat These spiders often build their small sheet webs in leaf litter.

Known records from AB, BC, MB, NB, NL, NS, ON, QC, SK (Canada); AK, CT, IL, MA, ME, MI, MN, NJ, NY, OH, WI (United States)

Cicurina arcuata Keyserling, 1887

Common Name This species does not have a common name.

Other Colloquial Names Curved meshweaver

Total Body Length Female 4.3–5.7 mm, Male 3.7–7.0 mm

Description The prosoma is yellow brown to pale orange, slightly darker in the eye region. The opisthosoma is gray, with gray markings that give the appearance of indistinct chevron shapes on the posterior half. The ventral opisthosoma is mottled with dark gray markings. The legs are pale reddish brown and darker distally. *C. arcuata* can look very

female

female

similar to other *Cicurina* species. To confirm the identification, you would need to examine a specimen under a microscope.

Habitat These spiders often build their webs under rocks, logs, or other debris. They are active even in the winter months.

Known records from MB, NB, NS, ON, QC, SK (Canada); AL, AR, CO, FL, GA, IL, IN, KS, LA, MA, MD, ME, MI, MN, MO, NC, NE, NJ, NM, NY, OH, OK, PA, TN, TX, VA, WA, WI, WV (United States)

female ventral

Cicurina brevis (Emerton, 1890)

Common Name This species does not have a common name.

Other Colloquial Names Small-eared meshweaver

Total Body Length Female 2.8–5.0 mm, Male 3.2–4.0 mm

Description The prosoma is orange to orange brown, slightly darker in the eye region. The opisthosoma is orange brown. The heartmark is often gray, and posterior to it there are often paired gray spots, although these may be very pale and indistinct in some specimens. The legs are orange to orange brown and notably darker distally.

egg sac under rock

freshly molted male

C. brevis can look very similar to other *Cicurina* species. To confirm the identification, you would need to examine a specimen under a microscope.

Habitat These spiders often build their webs in leaf litter, under rocks, or under other debris. The egg sacs are secured to the web and often adorned with debris.

Known records from MB, NB, NS, ON, PE, QC (Canada); CT, DE, IL, IN, MA, MD, ME, MI, MO, NC, NH, NY, OH, PA, TN, VA, WI, WV (United States)

female

female ventral

Cicurina pallida Keyserling, 1887

Common Name This species does not have a common name.

Other Colloquial Names Pallid funnel-web spider

Total Body Length Female 4.6–6.0 mm, Male 4.4–5.3 mm

female

female

female

Description The prosoma is orange to orange brown, slightly darker in the eye region. The opisthosoma is orange brown and lacking the gray markings often seen in other species of *Cicurina*. The legs are orange brown. *C. pallida* can look very similar to other *Cicurina* species. To confirm the identification, you would need to examine a specimen under a microscope.

Habitat These spiders often build their webs under rocks and other debris, or in leaf litter.

Known records from NB, NS, ON, QC (Canada); CT, DC, IL, IN, MA, MD, ME, MI, NH, NY, OH, PA, WI (United States)

Cicurina robusta Simon, 1886

Common Name This species does not have a common name.

Total Body Length Female 5.8–9.1 mm, Male 5.0–6.7 mm

Description The prosoma is yellowish orange to light brown, slightly darker in the eye region. The legs are uniformly orange. The opisthosoma has an orange-tan background with darker markings throughout, making ambiguous V-shaped patterns. *C. robusta* can look very similar to other *Cicurina* species. To confirm the identification, you would need to examine a specimen under a microscope.

female

female
ventral

female

male

male

Habitat These spiders are often found under rocks or in leaf litter.

Known records from AB, MB, NS, QC, SK (Canada); AR, AZ, CO, CT, DE, IA, IL, IN, MT, NM, OH, PA, TX, UT, WI, WY (United States)

Hahnia cinerea Emerton, 1890

Common Name This species does not have a common name.

Other Colloquial Names Ash hahniid spider

Total Body Length Female 2.0–2.5 mm, Male 1.5–2.0 mm

Description The prosoma varies from light to dark brown and has a dark pattern. The opisthosoma is gray with light markings. The legs are orange, with darker femurs and some faint banding. The spiracle is located slightly more anteriorly from the spinnerets than in most other species in this family. The spinnerets are arranged in a single transverse row and look like the teeth of a comb.

female

Habitat These spiders are often found in leaf litter.

Known records from AB, BC, MB, NB, NL, NS, ON, QC, SK, YT (Canada); AK, AL, AR, AZ, CO, CT, DE, FL, GA, IA, ID, IL, IN, LA, MA, MD, ME, MI, MN, MO, MS, MT, NC, ND, NH, NJ, NY, OH, OK, OR, PA, RI, SC, SD, TN, TX, UT, VA, WA, WI, WV, WY (United States)

Neoantistea magna
(Keyserling, 1887)

Common Name This species does not have a common name.

Total Body Length Female 3.0–4.0 mm, Male 2.7–3.7 mm

eye configuration

Description The prosoma is reddish brown and very shiny. The opisthosoma is tan and brown, with a distinct chevron pattern down the midline. The legs are banded. The anteriorly placed spiracle is located closer to the epigastric furrow than to the spinnerets. The spinnerets are arranged in a single transverse row and look like the teeth of a comb. The lateral spinnerets are usually very visible when the spider is observed from above.

female

female

female

female ventral

male

egg sac with female

Habitat These spiders often build their webs in leaf litter or under other debris. The egg sac is a white sphere suspended in the web, and well secured.

Known records from AB, BC, MB, NB, NL, NL, NS, ON, PE, QC, SK, YT (Canada); AK, CA, CO, DE, FL, ID, IL, IN, MA, MD, ME, MI, MN, MT, NC, ND, NH, NJ, NV, NY, OH, OR, PA, RI, SD, TN, UT, VA, VT, WA, WI, WV, WY (United States)

LINYPHIIDAE ⟩ DWARF SHEETWEB WEAVERS & SHEETWEB WEAVERS

Other Colloquial Names Money spiders

Genera in This Family *Acanthoneta* (2), *Agnyphantes* (1), *Agyneta* (70), *Allomengea* (3), *Anacornia* (2), *Anibontes* (2), *Annapolis* (1), *Anthrobia* (4), *Aphileta* (2), *Arcterigone* (1), *Arcuphantes* (12), *Baryphyma* (2), *Bathyphantes* (24), *Blestia* (1), *Carorita* (1), *Caviphantes* (1), *Centromerita* (1), *Centromerus* (9), *Ceraticelus* (32), *Ceratinella* (14), *Ceratinops* (10), *Ceratinopsidis* (1), *Ceratinopsis* (24), *Cheniseo* (4), *Cnephalocotes* (1), *Collinsia* (14), *Coloncus* (5), *Coreorgonal* (3), *Dactylopisthes* (1), *Dicymbium* (2), *Diplocentria* (5), *Diplocephalus* (4), *Diplostyla* (1), *Disembolus* (24), *Dismodicus* (3), *Drapetisca* (2), *Entelecara* (3), *Epiceraticelus* (2), *Eridantes* (3), *Erigone* (40), *Erigonella* (1), *Eskovia* (1), *Esophyllas* (2), *Estrandia* (1), *Eulaira* (14), *Floricomus* (13), *Florinda* (1), *Frederickus* (2), *Frontinella* (2), *Glyphesis* (2), *Gnathonargus* (1), *Gnathonarium* (2), *Gnathonaroides* (1), *Gonatium* (1), *Goneatara* (4), *Gongylidiellum* (1), *Grammonota* (25), *Graphomoa* (1), *Halorates* (1), *Helophora* (4), *Hilaira* (9), *Horcotes* (3), *Hybauchenidium* (4), *Hypomma* (3), *Hypselistes* (4), *Idionella* (9), *Improphantes* (1), *Incestophantes* (4), *Islandiana* (15), *Ivielum* (1), *Jalapyphantes* (1), *Kaestneria* (2),

eye configuration
(*Pityohyphantes costatus*)

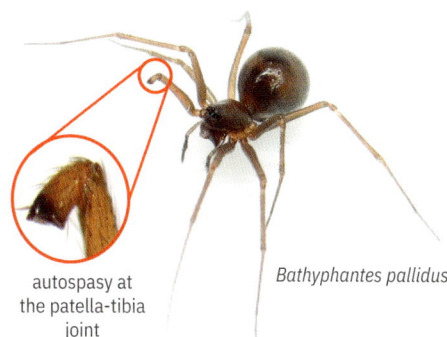

autospasy at the patella-tibia joint

Bathyphantes pallidus

97

Lepthyphantes (13), *Leptorhoptrum* (1), *Lessertia* (1), *Linyphantes* (19), *Linyphia* (9), *Lophomma* (2), *Macrargus* (1), *Maro* (3), *Masikia* (3), *Maso* (4), *Masoncus* (4), *Masonetta* (1), *Mecynargus* (5), *Megalepthyphantes* (1), *Mermessus* (37), *Metopobactrus* (2), *Micrargus* (2), *Microctenonyx* (1), *Microlinyphia* (5), *Microneta* (3), *Montilaira* (1), *Mythoplastoides* (2), *Neodietrichia* (1), *Neriene* (8), *Oaphantes* (3), *Oedothorax* (10), *Oreoneta* (15), *Oreonetides* (7), *Oreophantes* (1), *Origanates* (1), *Oryphantes* (1), *Ostearius* (1), *Pacifiphantes* (1), *Paracornicularia* (1), *Pelecopsidis* (1), *Pelecopsis* (6), *Perregrinus* (1), *Perro* (1), *Phanetta* (1),

stridulatory file on chelicerae

Mermessus fradeorum

Phlattothrata (2), *Pityohyphantes* (16), *Pocadicnemis* (3), *Poeciloneta* (10), *Porrhomma* (6), *Praestigia* (3), *Procerocymbium* (1), *Saaristoa* (1), *Satilatlas* (8), *Savignia* (1), *Sciastes* (6), *Scirites* (2), *Scironis* (2), *Scolopembolus* (1), *Scotinotylus* (33), *Scylaceus* (2), *Scyletria* (1), *Semljicola* (3), *Silometopoides* (2), *Sisicottus* (9), *Sisicus* (3), *Sisis* (2), *Sisyrbe* (1), *Sitalcas* (1), *Smodix* (1), *Soucron* (1), *Souessa* (1), *Souessoula* (1), *Sougambus* (1), *Souidas* (1), *Soulgas* (1), *Sphecozone* (1), *Spirembolus* (41), *Stemonyphantes* (1), *Styloctetor* (2), *Subbekasha* (1), *Symmigma* (1), *Tachygyna* (15), *Tapinocyba* (14), *Tapinopa* (2), *Tapinotorquis* (1), *Taranucnus* (1), *Tarsiphantes* (1), *Tennesseellum* (2), *Tenuiphantes* (7), *Thaleria* (1), *Thyreosthenius* (1), *Tibioplus* (1), *Tiso* (2), *Tmeticus* (2), *Traematosisis* (1), *Tunagyna* (1), *Tusukuru* (1), *Tutaibo* (1), *Typhochrestus* (2), *Vermontia* (1), *Wabasso* (2), *Walckenaeria* (68), *Walckenaerianus* (1), *Wubana* (7), *Zornella* (2), and *Zygottus* (2)

Total Body Length Some are minute: <1.0 mm–7.0 mm

Description These spiders are extremely diverse but often overlooked due to their usually small size. They are entelegyne and ecribellate. Males often have stridulatory files on the sides of their chelicerae, and they use their palps to strum them in courtship displays. These spiders usually have eight eyes in two rows, although some males have highly modified prosomas and eye arrangements. They have a pair of book lungs and a posteriorly placed tracheal spiracle. Autospasy of the legs occurs at patella-tibia joint. These spiders build a sheet web, and these are often overlooked, but foggy mornings are a great time to spot them. Linyphiids can balloon at any life stage. On days with ideal weather conditions, many spiders ballooning at the same time can coat an area in gossamer silk.

In the United Kingdom, these spiders are called "money spiders," as it was thought that when a ballooning linyphiid landed on a person, it indicated that he or she would soon come into money.

The spiders can be extremely difficult to identify, even with microscopic views. For those collecting spiders, there is a great resource available on the website of the Field Museum of Natural History at https://linepig.fieldmuseum.org/

Similar Families Leptonetidae, Mysmenidae, Nesticidae, Pimoidae, Oonopidae, Theridiidae, Tetragnathidae, Theridiosomatidae

Linyphiids can be distinguished from leptonetids and oonopids by their eye configuration: in leptonetids, the posterior medians are usually set back from the other eyes (some are eyeless, and a few have contiguous eyes that occupy less than half the width of the prosoma), and oonopids have six eyes in a cluster, whereas linyphiids usually have their eight eyes in two rows. Linyphiids can be distinguished from mysmenids by the lack of the sclerotized spot on femur I that is present in mysmenids. Linyphiids lack the tarsal comb that is seen in nesticids. Also, linyphiids build a sheet web, rather than the tangle web built by nesticids. Pimoids are most easily distinguished from linyphiids by the structures of the genitalia: pimoid males have an integral retrolateral paracymbium, and the female epigynum has a tonguelike projection. Theridiids have a tarsal comb on leg IV, which linyphiids lack. Theridiosomatids have pits in the prolateral area of the sternum, which are not seen in linyphiids. Linyphiids build a sheet web. In contrast, mysmenids and theridiids build a tangle web, most tetragnathids build an orb web (or are ground hunters in a few cases), and theridiosomatids build an orb web modified into a cone shape.

Agyneta micaria (Emerton, 1882)

Common Name This species does not have a common name.

Total Body Length Female 1.9 mm, Male 1.5–1.8 mm

Description The prosoma is brown to reddish brown, slightly darker around the edges. This species has four opisthosoma patterns, but each is unique, making this one of the linyphiids that are easy to identify. The first pattern has wide dark bands on the anterior and posterior, with a yellow band in the middle. The second pattern has a yellow-brown opisthosoma with the following markings: two black dots on either side of the midline, about halfway down; two pairs of black lines that start on the lateral edges, converge toward the middle (but do not meet), and point somewhat anteriorly; and black around the spinnerets, with a white dot just anterior to that. The third pattern is very similar to the second one but lacks the two black dots, and the white marking near the spinnerets is smaller. The last pattern is similar to the second and third. It lacks the black dots, and the posterior black markings run parallel to one another. There may be some other variations of these general markings. The legs are yellow and unbanded.

Habitat These spiders are often found in low herbaceous vegetation, under logs, in leaf litter, or under rocks.

Known records from ON (Canada); AL, AR, CT, DC, FL, GA, IL, IN, KS, KY, LA, MA, MD, MO, NC, NY, OH, OK, PA, SC, TN, TX, WI, WV (United States)

female

male

female

Agyneta unimaculata (Banks, 1892)

Common Name This species does not have a common name.

Total Body Length Female 1.8 mm, Male 1.8 mm

Description The prosoma is brown, and faint darker markings are visible. The opisthosoma is uniformly brown. The legs are tan to yellow, without banding. The male pedipalps are robust and box shaped. The female epigynum protrudes from the opisthosoma.

Habitat These spiders often build small sheet webs near the ground and under rocks and logs.

Known records from MB, NB, NS, ON, PE, QC, SK (Canada); FL, GA, IL, IN, KY, MD, ME, MI, MN, MO, NC, ND, NE, NJ, NY, OH, PA, TX, VA, WI, WV (United States)

female

female

male

male

Bathyphantes pallidus (Banks, 1892)

Common Name This species does not have a common name.

Other Colloquial Names Pale sheetweb weaver

Total Body Length Female 2.3–2.8 mm, Male 1.9–2.1 mm

Description The prosoma is orangish brown. The opisthosoma is brown, and pale markings are sometimes visible. The legs are pale yellow and unbanded. The female epigynum has a long scape that points toward the posterior of the spider. The male pedipalps are cylindrical, and the embolus almost makes a complete circle at the distal end.

Habitat These spiders build their webs close to the ground in humid areas.

Known records from AB, BC, MB, NB, NL, NS, NT, ON, QC, SK, YT (Canada); AK, AZ, CO, CT, ID, IL, IN, KY, MA, MD, ME, MI, MN, MT, NC, NH, NJ, NY, OH, PA, RI, SD, VA, VT, WA, WI, WV (United States)

female

female

male

Centromerus cornupalpis
(O. Pickard-Cambridge, 1875)

Common Name This species does not have a common name.

Total Body Length Female 2.0–2.6 mm, Male 2.0–2.8 mm

Description The prosoma is yellow brown. The opisthosoma is dark gray. The legs are yellow brown and unbanded. The male cymbium is drawn out to a point or horn, giving him a distinctive look.

Habitat These spiders are usually found in leaf litter or close to the ground in wet habitats. This species is known to overwinter as adults.

Known records from NL, NS, ON, QC (Canada); CT, IL, IN, MA, ME, MI, MO, NC, NH, NY, OH, TN, UT, VA, WI, WV, WY (United States)

male palps, ventral view

male

Ceraticelus similis (Banks, 1892)

Common Name This species does not have a common name.

Total Body Length Female 1.2–1.6 mm, Male 1.3–1.6 mm

Description The prosoma is red to orange, with a dark mask around the eyes. The opisthosoma is reddish orange, and in the males, the dorsal surface is covered by a large scutum (a hardened plate or disc) that has lots of little depressions in it. The legs are orange and unbanded.

Habitat These spiders are often found in low vegetation.

Known records from AB, MB, NB, NL, NS, ON, PE, QC (Canada); AR, AZ, CO, CT, DC, FL, GA, IL, LA, MA, ME, NC, NH, NY, OH, PA, TX, WI (United States)

female

female

Ceratinopsis laticeps Emerton, 1882

Common Name This species does not have a common name.

Total Body Length Female 1.7 mm, Male 1.5 mm

Description The prosoma is orange to dark orange. A dark brown to black marking starting at the fovea extends to the eye region in the shape of a triangle with the point pointing posteriorly. The opisthosoma is pale orange on the dorsal surface, slightly darker laterally and posteriorly. The spinnerets are black. The legs are yellow to orange and unbanded.

female

Habitat These spiders are often found under woody debris or in low shrubby vegetation.

Known records from ON (Canada); CT, FL, IL, IN, MA, ME, MI, NY, OH, TX (United States)

female

female ventral

male

male ventral

male

Drapetisca alteranda
Chamberlin, 1909

Common Name This species does not have a common name.

Other Colloquial Names Northern long-toothed sheetweaver

Total Body Length Female 4.0–4.5 mm, Male 3.2–3.8 mm

Description The prosoma is tan to yellow, bordered in dark brown. There is a central black spot that is shaped like a tulip; the tips of the petals point toward the posterior lateral eyes. There are radial black spots around the border. The opisthosoma is white and black, mottled with a series of dark chevrons down the posterior half. The legs are boldly banded. The male palp has an obvious paracymbium structure that projects like hooks.

male

Habitat This species is often collected from tree trunks, where its coloration provides it with effective camouflage.

Known records from AB, BC, MB, NB, NL, NS, ON, PE, QC, SK (Canada); AK, CO, CT, MA, ME, MI, NE, NH, NM, PA, VA, WA, WI (United States)

Erigone autumnalis Emerton, 1882

Common Name This species does not have a common name.

Other Colloquial Names Autumn money spider

Total Body Length Female 1.2–1.7 mm, Male 1.2–1.5 mm

Description The prosoma is shiny bright orange to yellow. The opisthosoma is gray to brown. The legs are yellow. There are many other

female

female

female
ventral

male

male

species of *Erigone*, but most are much darker, and the males often have more complex pedipalps than those possessed by this species.

Habitat These spiders are often found in the late fall and early winter, when mass ballooning events happen and they can be found on fences and other suitable structures from which to balloon. Otherwise, they can be found in leaf litter or low grassy vegetation.

Known records from MB, NB, NS, ON, PE, QC (Canada); AR, AZ, CA, CT, DC, FL, IL, IN, KS, KY, LA, MA, MD, ME, MO, MS, NC, NE, NH, NJ, NM, NY, OH, OK, TN, TX, UT, VA, WV (United States)

Erigone dentigera O. Pickard-Cambridge, 1874

Common Name This species does not have a common name.

Other Colloquial Names Teethed dwarf weaver

Total Body Length Female 2.3–2.5 mm, Male 2.1 mm

Description The prosoma is dark brown, sometimes almost black. The male prosoma is highly elevated at the anterior, and there are teeth on the lateral margins. The opisthosoma is dark brown, almost black. The legs are dusky yellow. The male pedipalps are complex, lined with teeth, and, due to their length, usually quite obvious. The female *Erigone* spiders are difficult to identify, even when dissected.

Habitat These spiders often build their webs in low vegetation close to the ground.

Known records from AB, BC, MB, NB, NL, NS, NT, ON, QC, SK, YT (Canada); AK, CO, CT, FL, IL, MA, ME, MI, MT, NC, NJ, NY, OH, TX, WA, WI (United States)

female

female with egg sac

male

male

Florinda coccinea (Hentz, 1850)

Common Name Scarlet sheetweaver

Other Colloquial Names Black-tailed red sheetweaver

Total Body Length Female 3.5 mm, Male 3.0 mm

Description The prosoma and opisthosoma are both brilliant red, although some can be more orange. Above the spinnerets at the posterior end of the opisthosoma is a black tubercle, giving the end of the opisthosoma a very distinctive shape.

female

juvenile

female ventral

There are other linyphiids that are red with black markings posteriorly, but the shape of this species' opisthosoma will help with the identification.

male

male and female

Habitat This species' sheet webs can often be found in lawns and grassy areas.

Known records from AR, FL, GA, IL, KY, LA, MD, NC, NY, OH, TX, VA

web

Frontinella pyramitela (Walckenaer, 1841)

Common Name Bowl and doily weaver

Total Body Length Female 3.0–4.0 mm, Male 3.0–3.3 mm

Description The females have a brown prosoma. The opisthosoma is brown, with white banding on the lateral edges that creates stripes when viewed laterally; the white stripes darken to yellow toward the ventral side. The legs are pale brown. The males have a red prosoma. The opisthosoma is dark orange, sometimes with white banding similar to that on the females (although usually fainter). The legs are orange.

Habitat These spiders will build their webs in a wide variety of vegetation, including low tree branches, grassy habitats, shrubs, or bushes. This is one species that can be identified easily by the structure of its web, which comprises a bowl-shaped web (from which the spider hangs inverted at the bottom) and a lacy sheet web underneath, looking like a bowl sitting on a lace doily. The males and females will often share a web.

male

Known records from AB, BC, MB, NB, NS, NT, ON, PE, SK (Canada); AK, AZ, CA, CO, CT, DC, FL, GA, IL, MA, MD, MI, MN, MO, MT, NM, NV, OH, OK, SC, TX, WI, WY (United States)

female

male and female

male and female

web

shed exoskeleton, female and male

Grammonota inornata Emerton, 1882

Common Name This species does not have a common name.

Total Body Length Female 2.0–2.5 mm, Male 1.9 mm

Description The prosoma is dark brown. The opisthosoma is dark gray. The legs are orange, with dusky bands at the distal ends of the femurs and tibiae.

Habitat These spiders often build their webs in low vegetation close to the ground. They can be found in lawns.

Known records from MB, NS, ON, QC (Canada); AR, CT, IL, MA, ME, MO, NY, OH, RI, TX (United States)

female

male

female

female with egg sac

Hypselistes florens (O. Pickard-Cambridge, 1875)

Common Name Splendid dwarf spider

Other Colloquial Names Peatland sheetweb weaver

Total Body Length Female 2.3–3.0 mm, Male 2.0–2.5 mm

Description The prosoma is bright orange to red. The female prosoma is slightly elevated in the eye region. The male prosoma has a cephalic lobe that is paler, with a pair of pits at the base. The posterior median eyes are located on the lobe. The opisthosoma is dark bluish black. The femurs are red to orange, but the other segments are gray to black.

Habitat These spiders are often found in low vegetation and leaf litter.

Known records from AB, BC, LB, MB, NB, NS, NT, ON, PE, QC, SK, YT (Canada); AK, CT, DC, IL, IN, MA, MD, ME, MI, MT, NH, NY, OH, PA, VA, WA, WI (United States)

female

female

Islandiana flaveola (Banks, 1892)

Common Name This species does not have a common name.

Other Colloquial Names Unequal whiskered money spider

Total Body Length Female 1.4–1.5 mm, Male 1.3–1.4 mm

male on a penny for scale

Description The prosoma is orange to orange brown. A dusky marking is often just anterior of the fovea. Sometimes there are dusky radiating lines. The border is slightly darker. The eyes are ringed in black. The opisthosoma is brown to dark gray and unmarked. The legs are yellow to orange, usually slightly darker at the distal ends of the femurs, otherwise unmarked.

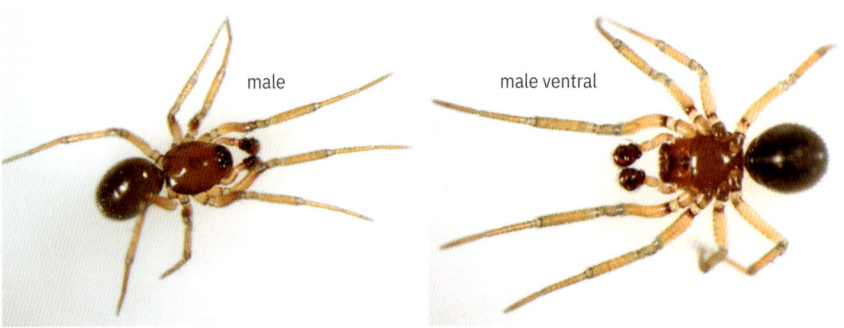

male

male ventral

Habitat These spiders are often found under rocks or other debris near the ground, and they can also be found in caves. They can easily be overlooked due to their small size.

Known records from AB, BC, MB, NB, NL, NS, ON, QC, SK (Canada); AR, AZ, CO, IA, IL, IN, MA, ME, MO, MT, ND, NE, NM, NY, OH, PA, TX, WI, WY (United States)

Lepthyphantes turbatrix (O. Pickard-Cambridge, 1877)

Common Name This species does not have a common name.

Total Body Length Female 2.0–2.8 mm, Male 2.0–2.6 mm

Description The prosoma is dark brown, slightly darker in the eye region. The opisthosoma is dark brown. Mottled cream-white markings on the lateral edges converge toward the midline, creating the appearance of dark brown chevrons or lobed shapes. The legs are yellow with wide dusky bands. The female epigynum has a long posterior-pointing scape.

Habitat These spiders are often found under logs or in leaf litter.

Known records from AB, BC, MB, NB, NL, NS, ON, PE, QC, SK (Canada); AZ, ME, MI, NH, NM, NY, OH, WI (United States)

female

female

female ventral

female with penny for scale

Mermessus fradeorum (Berland, 1932)

Common Name This species does not have a common name.

Total Body Length Female 2.9–3.0 mm, Male 3.0–3.1 mm

Description The prosoma is orange brown and darker in the eye region. Sometimes a dusky band is around the edge, and some have a dusky

female

female

female

female ventral

female with egg sac

male

male pedipalps

male ventral

marking that splits from the fovea to the posterior lateral eyes. The males have pointed denticles and a mastidion on each chelicera. The opisthosoma is yellow to orange with gray chevrons toward the posterior. There may also be a gray midline on the posterior half. The legs are yellow to orange and unbanded.

Habitat These spiders are often found in grasslands, especially near forest edges. They tend to build their webs under debris or on the surface of the soil where there is a depression.

Known records from ON (Canada); AL, AR, FL, GA, IL, KY, LA, MS, NC, NJ, OH, TN, TX (United States)

Mermessus tridentatus (Emerton, 1882)

Common Name This species does not have a common name.

Total Body Length Female 2.3 mm, Males 2.3 mm

Description The prosoma is dark brown, with faint dark radial lines. The opisthosoma is dark brown; sometimes some paler transverse bands are seen on the posterior half. The legs are yellow to orange.

Habitat These spiders are often found in moist habitats: in leaf litter, under logs, under rocks, or on rock faces. The egg sacs are slightly domed and attached to the surface of the substrate near the web. The egg sacs are cream to pale peach. The female sits close to the egg sac, guarding it.

Known records from MB, ON, QC (Canada); AR, CO, CT, DC, FL, GA, IL, IN, KY, LA, MA, MD, MO, NC, NE, NJ, NY, OH, PA, RI, TN, TX, VA, VT, WI, WV (United States)

female

female with egg sac

male

male ventral

Neriene clathrata (Sundevall, 1830)

Common Name This species does not have a common name.

Other Colloquial Names Latticed sheet-web weaver

Total Body Length Female 3.0–5.2 mm, Male 3.6–4.8 mm

female

female

female ventral

female with egg sac

Description The prosoma is brown to reddish brown. The opisthosoma is dark brown, and some males look almost black. There is a white anterior band. The female opisthosoma has a series of pale chevrons that are highlighted by white at the lateral edges. The male often has no markings other than the anterior white band. Some males have a constriction about halfway down the opisthosoma. The females' legs are yellow with dusky bands in the middle of each segment and dark bands at the distal ends of tibiae and femur IV. The males have dark reddish-brown legs.

male

Habitat These spiders like to build their webs between the roots at the base of large trees, under litter, and beneath rocks.

Known records from AB, MB, NB, NL, NS, ON, PE, QC, SK (Canada); CA, CT, DC, IL, IN, MA, ME, MI, MN, MO, NC, NH, NJ, NY, OH, PA, WI, WV (United States)

Neriene litigiosa (Keyserling, 1886)

Common Name Sierra dome spider

Total Body Length Female 5.2–8.5 mm, Male 5.1–6.8 mm

Description The prosoma is tan to light brown, sometimes orange. There is a dark median stripe, and a thin dark band runs around the border. The eye region is slightly elevated. The opisthosoma is white, with a gray median stripe that widens with white markings within it at the posterior end. There are faded gray markings on the lateral edges, and some of the white shifts to yellow. The spinnerets are ringed in gray. The legs are uniformly light brown to gray. The males tend to be somewhat darker than the females, but the overall coloration and pattern are similar.

Habitat These spiders build their web in the shape of a dome. The web is said to be larger than that of *Neriene radiata*. They build their dome-shaped sheet web in coniferous forests, usually between shrubs and trees, and in grassy areas.

Known records from BC (Canada); CA, CO, MT, NV, OR, UT, WA, WY (United States)

female

female

Neriene radiata (Walckenaer, 1841)

Common Name Filmy dome spider

Total Body Length Female 4.0–6.5 mm, Male 3.5–5.3 mm

Description The female has a brown prosoma bordered in lighter orange. The opisthosoma is brown, with a pair of white bands on the lateral edges

female

female ventral

male and female

male

web

and often a series of paired white marks on either side of the midline. The white stripes turn to yellow on the sides of the opisthosoma. The male has an orange prosoma. The opisthosoma is marked similarly to the female, but the males tend to be skinnier.

Habitat The web is often built near the ground in shaded forested areas. These spiders build a dome-shaped sheet web. Lightly misting the web with water from a spray bottle can make the structure more obvious. The spider usually sits on the underside of the dome at its peak. Sometimes, the male and female will occupy the same web. *Neriene litigiosa* builds a larger dome web.

Known records from AB, BC, MB, NB, NL, NS, NT, ON, QC, SK, YT (Canada); AK, AL, AR, AZ, CO, CT, DC, FL, IA, ID, IL, IN, LA, MA, MD, ME, MI, MN, MO, MS, MT, NC, NE, NH, NJ, NM, NY, OH, OK, PA, RI, SD, TX, VA, VT, WA, WI, WY (United States)

Pityohyphantes costatus (Hentz, 1850)

eyes

Common Name Hammock spider

Total Body Length Female 5.0–7.0 mm, Male 4.5–6.0 mm

Description The prosoma is tan to orange brown. A dark mark shaped like a tuning fork runs from the posterior end of the prosoma to the posterior median eyes. The opisthosoma is tan to white, with a series of nested brown triangles that decrease in size toward the posterior down the midline. The legs are tan to yellow with brown bands and spots. The leg spines are quite obvious.

Habitat This species' hammock-style sheet web can be found in forested areas.

Known records from AB, BC, MB, NB, NL, NS, ON, QC, SK (Canada); IA, IN, MA, ME, MI, MO, NC, NH, OH, WI, WV (United States)

female

female

female

male

egg sac

male

male

Soulgas corticarius
(Emerton, 1909)

male palp

Common Name This species does not have a common name.

Total Body Length Female 2.0 mm, Male 2.0 mm

Description The prosoma is reddish brown. Faint radiating lines may be visible, and the edges are slightly dusky. The opisthosoma is gray to brown with short white hairs. The legs are dark yellow to pale orange and unbanded. The end of the embolus of the male pedipalp looks almost like the swirl on the top of soft ice cream.

Habitat These spiders can be found in vegetation close to the ground.

male

Known records from NB, NL, NS, ON, PE, QC (Canada); CT, FL, IL, MA, MI, NC, NY, OH, RI, TX, WI, (United States)

Styloctetor purpurescens
(Keyserling, 1886)

Common Name This species does not have a common name.

Total Body Length Female 1.8– 2.3 mm, Male 1.8 mm

Description The prosoma is bright orange red, and the eye region is darker, almost black. The opisthosoma is also bright orange red, often with a black dot at the posterior end, near the spinnerets. The spinnerets are often dark. The legs are orange red, with no banding.

Habitat These spiders are often found in low vegetation and leaf litter.

female

Known records from ON (Canada); CT, DC, GA, IL, IN, KY, MA, MD, MO, NC, NJ, NY, OH, TX, WI (United States)

female

female

female

female
ventral

Tennesseellum formica
(Emerton, 1882)

Common Name Antlike sheetweaver

Other Colloquial Names Ant sheetweb weaver

Total Body Length Female 1.8–2.4 mm, Male 1.8–2.5 mm

Description The prosoma is dark brown. The opisthosoma is dark brown, with a band around the middle that is often paler. The tracheal spiracle is anteriorly located, which creates a slight depression midway across the opisthosoma; this resembles the body shape of many ants. The legs are pale brown to tan, with darker femurs.

Habitat This species is very common in grassy and agricultural areas.

female

female
ventral

male in vial for scale

web

male

male

Known records from AB, BC, MB, NL, NS, ON, QC, SK, YT (Canada); AK, AZ, CA, CO, CT, DC, FL, GA, IA, IL, IN, LA, ME, MN, MO, NC, ND, NE, NM, NY, PA, RI, SC, TN, TX, UT, WA, WI, WV (United States)

Tenuiphantes tenuis (Blackwall, 1852)

Common Name This species does not have a common name.

Total Body Length Female 2.1–4.3 mm, Male 2.0–3.2 mm

Description The prosoma is reddish brown. The opisthosoma is brown, with dark brown and white markings. The dark brown markings are almost chevron shaped with white scattered throughout; these markings may be difficult to see on the males. The legs are yellow and unbanded.

female

female

juvenile

Habitat These spiders tend to build their webs in low vegetation and leaf litter. The egg sacs are enclosed in white silk and attached to the surface of the area they occupy.

Known records from BC, NL, ON (Canada); AK, CO, IN, MA, OH, OR, WA (United States) This is an introduced species.

male

egg sac

Walckenaeria atrotibialis (O. Pickard-Cambridge, 1878)

Common Name This species does not have a common name.

Total Body Length Female 2.5–3.1 mm, Male 2.0–2.6 mm

Description The female prosoma is reddish brown and darker in the eye region. The opisthosoma is gray brown. The male prosoma is reddish brown and darker in the eye region and is modified with lobes and lateral pits. The posterior median eyes are located on top of the lobe. The legs on both the male and female are orange yellow. The femurs and tibiae on legs I and II are darker, and the tarsi and metatarsi on legs I and II are white.

female

male

male

Habitat These spiders are often found in wooded and grassy areas.

Known records from AB, BC, MB, NB, NL, NS, NT, ON, QC, SD, SK, YT (Canada); AK, ID, IL, MA, ME, MI, NH, NY, OH, VT, WA, WI, WV (United States)

OCHYROCERATIDAE > TINY GROUND WEAVERS

Genera in This Family *Theotima* (1)

Total Body Length 1.0–2.0 mm

Description These small haplogyne and ecribellate spiders tend to be dark, sometimes with a pattern on the opisthosoma. They have six eyes in three diads. The opisthosoma is globular. These spiders lack book lungs and have a single tracheal spiracle located between the epigastric furrow and the spinnerets. They are usually found in leaf litter.

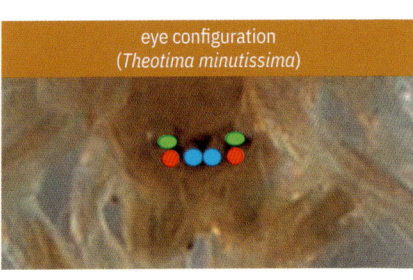

eye configuration
(*Theotima minutissima*)

Females carry eggs in chelicerae; they have been shown to reproduce parthenogenetically. Only one species from this family is found in North America; it is an introduced species.

Similar Families Leptonetidae, Oonopidae, Pholcidae, Telemidae, Trogloraptoridae

How to distinguish Ochyroceratids from other similar families:

Ochyroceratids' tarsal claws are on an onychium (an extension of the tarsus), which helps to distinguish them from leptonetids, oonopids, pholcids, telemids, and trogloraptorids. Ochyroceratids can also be distinguished from leptonetids, oonopids, pholcids, and trogloraptorids by their eye configuration: they have six eyes in three diads. In leptonetids, the posterior medians are usually set back from the other eyes. (Some are eyeless, and a few have contiguous eyes that occupy less than half the width of the prosoma.) Oonopids have their six eyes in a cluster. Pholcids have either eight or six eyes that are in two triads and are usually elevated. Trogloraptorids have six eyes, and the lateral eyes are adjacent to one another, with the posterior median eyes between them. Also note that only one species from this family is known in North America, and it has been documented only in Florida.

Theotima minutissima (Petrunkevitch, 1929)

Common Name This species does not have a common name.

Total Body Length Female 1.0–2.0 mm. Male unknown

Description The prosoma is brown, and a faint pattern is sometimes visible. The six eyes are arranged in three pairs; the anterior median eyes

female
illustration

female

are missing. The opisthosoma varies in color from deep purple to gray. The legs are thin, spineless, and brown. The females carry their eggs in their chelicerae. This species reproduces parthenogenetically, without males, although males may occasionally occur in the population. (More research is needed on this matter.) These small spiders are not easy to photograph, and few images of them are available. Included here are an image of a preserved specimen and Steve Buchanan's illustration from *Common Spiders of North America* (Bradley 2013).

Habitat These spiders tend to build their webs in leaf litter.

Known records from FL
This is an introduced species.

PIMOIDAE ▷ LARGE HAMMOCKWEB SPIDERS

Genera in This Family *Nanoa* (1) and *Pimoa* (14)

Total Body Length 1.4–12.0 mm

Description These spiders are entelegyne and ecribellate. They have eight eyes. They have an oval opisthosoma that tends to be brown to gray, often with light markings. They have long legs, and autospasy occurs at the patella-tibia joint. The fovea is usually distinct and ovate. They build a large sheet web close to the ground, usually in sheltered habitats, such as in hollow logs or under banks or other overhangs.

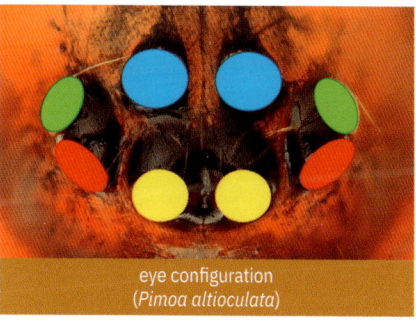
eye configuration
(*Pimoa altioculata*)

ovate fovea

Similar Families Cybaeidae, Linyphiidae, Tetragnathidae

How to distinguish Pimoidae from other similar families:

Pimoids have a fovea that is a distinctly ovate depression; in contrast, the fovea is usually longitudinal in cybaeids. Pimoids can be difficult to distinguish from other similar families unless you are able to look at the genitalia. Pimoid male pedipalps have an integral retrolateral paracymbium, a cymbial sclerite, and a dorsoectal cymbial process. The

Pimoa altioculata

female epigynum protrudes more than its width and has a fold or groove with the copulatory openings at the distal end. Autospasy that occurs at the patella-tibia joint will distinguish pimoids from cybaeids and tetragnathids. Also, pimoids build a sheet web, tetragnathids build an orb web, and the similar-looking cybaeids (*Calymmaria*) build a sheet web combined with a cone or bowl.

Pimoa altioculata (Keyserling, 1886)

Common Name This species does not have a common name.

Total Body Length Female 8.8 mm, Male 6.5 mm

Description The prosoma is brown to tan, with a dark median stripe and edges. The opisthosoma is tan with a dark heartmark and a series of dark

irregular chevrons down its length. The legs are long, thin, and reddish brown with dark banding. These spiders are very similar to species in the family Linyphiidae, and examination of the genitalia is needed to confirm species-level identification.

Habitat This species is found in damp forested areas. Their webs are usually placed at the base of stumps or trees.

Known records from BC (Canada); AK, CA, OR, UT, WA (United States)

female

male

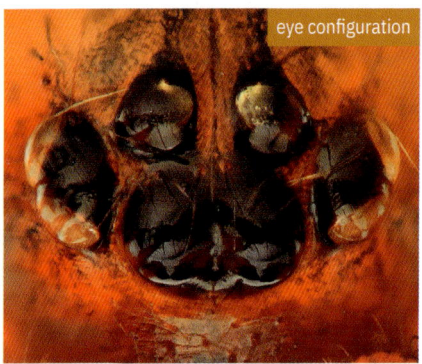

eye configuration

male ventral

TELEMIDAE ❯ LONGLEGGED CAVE SPIDERS

Genera in This Family *Usofila* (4)

Total Body Length 1.0–1.7 mm

Description These very small haplogyne and ecribellate spiders are uniformly pale, usually somewhat orange. They have six eyes in three diads; the anterior median eyes are missing. They lack book lungs and have two pairs of tracheal spiracles. The first pair is located equidistant from the epigastric furrow and spinnerets, and the second pair is anterior to the epigastric furrow (in at least some species). There is usually an anteriodorsal sclerotized area in the shape of a zigzag on the

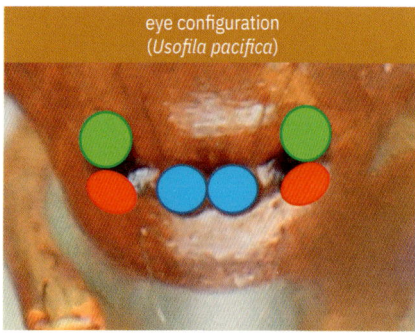

eye configuration
(*Usofila pacifica*)

opisthosoma (although this is often difficult to see). These spiders are found most often in moist habitats, such as in caves, in leaf litter, or under rocks.

Similar Families Ochyroceratidae, Leptonetidae, Oonopidae, Pholcidae, Trogloraptoridae

How to distinguish Telemidae from other similar families:

Telemids can be distinguished from ochyroceratids, as their tarsal claws are not on an onychium. Also, telemids usually have two pairs of tracheal spiracles (one pair anterior of the epigastric furrow, the other pair between the epigastric furrow and the spinnerets), compared to the single tracheal spiracle seen in ochyroceratids. The distinctive eye configuration in leptonetids (usually with the posterior medians set back from the other eyes) can distinguish them from telemids. Oonopids, pholcids, and trogloraptorids have a pair of book lungs, which are lacking in telemids. Trogloraptorids can be distinguished from telemids by the presence of their unique subsegmented raptorial tarsi.

lacks book lungs

Usofila pacifica

Usofila pacifica (Banks, 1894)

Common Name This species does not have a common name.

Total Body Length Female 1.5 mm, Male 1.5 mm

Description The prosoma is orange brown, with some slight shading but no distinct markings. The six eyes are arranged in three pairs. (The anterior median eyes are absent.) The opisthosoma is orange brown, with long pale hairs. The book lungs are absent. The legs are long, thin, and orange brown, similar to the rest of the spider. Oddly, when preserved in ethanol, the opisthosoma turns brilliant green.

Habitat These spiders usually build their webs in leaf litter or under rocks in moist habitats.

Known records from AB, BC (Canada); AK, WA (United States)

female

female

female ventral

eye configuration

The spiders of the Orb Web Weaver Guild construct an orb web (or a modified orb web) that is used for prey capture. The orb web is what most people imagine when visualizing a spider web: a wheel- or circular-shaped web. The web is composed of radial lines (like spokes of a wheel) and a sticky spiral in which prey becomes entrapped. There are some modified orb webs: some look like a couple sections of the orb web have been removed, like a slice of pie, while other spiders pull the orb into a cone shape. All spiders in this guild are Araneomorphae. This guild includes the following families: Anapidae, Araneidae, Symphytognathidae, Tetragnathidae, Theridiosomatidae, and Uloboridae.

KEY TO FAMILY

1a. With cribellum, calamistrum a single row of setae, 4.0–8.0 mm total body length -- **Uloboridae** (page 211)
1b. Without cribellum-- **2**

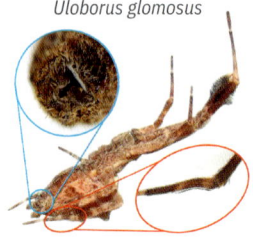
Uloborus glomosus

a

b

Araneus thaddeus

2a. Female lacks pedipalps --- **3**
2b. Female with pedipalps -- **4**

a
Gertschanapis shantzi

b
Acanthepeira cherokee

3a. Four eyes, very small, approximately 0.5 mm total body length --- **Symphytognathidae** (page 190)
3b. Six or eight eyes, pitting on dorsal opisthosoma and lateral prosoma, 1.0–1.5 mm total body length ---------------------------------- **Anapidae** (page 125)

Anapistula

a

b

Gertschanapis shantzi

4a. Prolateral sternum with pits, web pulled into cone shape,
1.3–2.4 mm total body length ----------------------------------- **Theridiosomatidae** (page 208)
4b. Lacks pits on prolateral sternum, web not pulled into cone shape ----------------------------- **5**

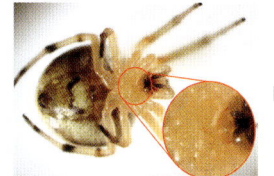

a

b

Theridiosomatidae   *Araneus thaddeus*

Theridiosoma gemmosum

5a. Endites square, web tends to be oriented vertically or nearly vertically,
1.0–30.0 mm total body length --- **Araneidae** (page 127)
5b. Endites long and narrow, web tends to be oriented horizontally or
nearly horizontally, 1.2–25.0 mm total body length ------------------ **Tetragnathidae** (page 191)

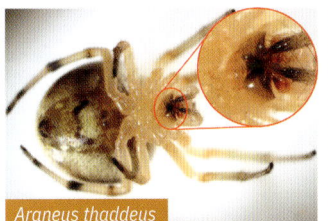

Araneidae

a

Araneus thaddeus

Leucauge venunsta

b

Tetragnatha elongata

ANAPIDAE ▸ GROUND ORBWEAVERS

Genera in This Family *Comaroma* (1) and
Gertschanapis (1)

Total Body Length 1.0–1.5 mm

Description These small entelegyne and
ecribellate spiders tend to be reddish. They
have six to eight eyes. The male opisthosoma
is covered with a large scutum. There are
small pits or sclerotized dots on the dorsal
opisthosoma and lateral sides of the prosoma.
The females lack pedipalps. These spiders
spin a small horizontal web in leaf litter or
low herbaceous vegetation.

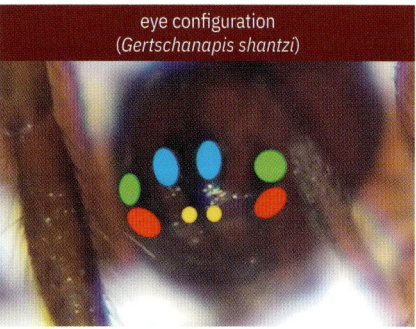

eye configuration
(*Gertschanapis shantzi*)

Similar Families Mysmenidae, Symphytognathidae, Theridiidae, Theridiosomatidae

How to distinguish Anapidae from other similar families:

The presence of a scutum on the male opisthosoma and lack of pedipalps in the females will distinguish anapids from mysmenids, theridiids, and theridiosomatids. Also note that theridiids and mysmenids weave a tangle web, whereas anapids build a small orb web. Symphytognathids have only four eyes, and males do not have a dorsal scutum. Anapids have six or eight eyes.

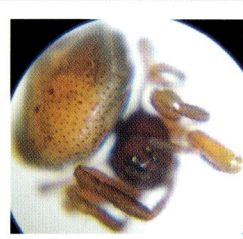

female lacks pedipalps (both images *Gertschanapis shantzi*)

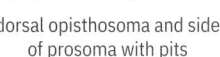

dorsal opisthosoma and side of prosoma with pits

Gertschanapis shantzi (Gertsch, 1960)

Common Name This species does not have a common name.

Total Body Length Female 1.0–1.5 mm, Male 1.0–1.5 mm

Description The prosoma is brown and elevated in the eye region (more so in the males). Of the eight eyes, the anterior medians are the smallest. The female opisthosoma is brown to yellow, sometimes gray. The male opisthosoma has a large scutum covering the dorsal surface and is usually reddish brown; the lateral edges of the opisthosoma are covered with sclerotized dots. The females lack pedipalps. The legs are yellow brown, with dark rings on the proximal ends of the distal segments.

Habitat These spiders often build their webs in leaf litter or similar moist habitats. The small orb webs are oriented horizontally, and the radial lines are often outside the plane of the web.

Known records from CA, OR

female
female
female

ARANEIDAE > ORBWEAVERS

Other Colloquial Names Wheel spiders

Genera in This Family *Acacesia* (1), *Acanthepeira* (4), *Aculepeira* (3), *Allocyclosa* (1), *Araneus* (51), *Araniella* (2), *Argiope* (5), *Cercidia* (1), *Colphepeira* (1), *Cyrtophora* (1), *Cyclosa* (5), *Eriophora* (2), *Eustala* (13), *Gasteracantha* (2), *Gea* (1), *Hypsosinga* (5), *Kaira* (4), *Larinia* (3), *Larinioides* (3), *Mangora* (7), *Mastophora* (15), *Mecynogea* (1), *Metazygia* (4), *Metepeira* (13), *Micrathena* (4), *Neoscona* (7), *Ocrepeira* (4), *Scoloderus* (2), *Singa* (3), *Trichonephila* (3), *Verrucosa* (1), *Wagneriana* (1), and *Zygiella* (5)

Total Body Length 1–30 mm

Description These spiders are entelegyne and ecribellate. They have eight eyes; the four median eyes are situated close to one another in a trapezoid, and the lateral eyes are adjacent to each other some distance from the median eyes. The spiders have a low clypeus, and their endites are squarish. They have a pair of book lungs and a single tracheal spiracle located near the spinnerets. They usually build a vertical orb web, some of which can be quite large.

Similar Families Mimetidae, Mysmenidae, Tetragnathidae, Theridiosomatidae, Uloboridae

How to distinguish Araneidae from other similar families:

Araneids can be distinguished from mimetids, as they lack the diagnostic leg spination that is seen in mimetids. Araneids can be distinguished from mysmenids by the lack of the sclerotized spot on femur I that is present in mysmenids. Also, mysmenids build a tangle web, and araneids build an orb web. Araneids can be distinguished from tetragnathids by their somewhat square-shaped endites; tetragnathids' endites are longer than they are wide. They can be distinguished from theridiosomatids by the lack of sternal pits on the anterior edge; the pits are seen only in theridiosomatids. Additionally, theridiosomatids also have a robust femur I, tend to be very small (1.3–2.4 mm total body length), and build an orb web that is pulled into a cone shape; these traits are not usually seen in araneids. Uloborids are cribellate, whereas araneids are ecribellate, so the presence of a cribellum and calamistrum distinguishes the uloborids. Uloborids have elongated endites, usually 1.5–2 times the width, whereas the endites of araneids are squarish.

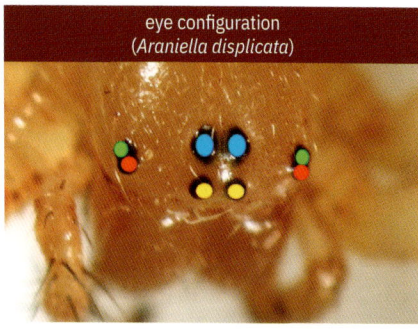

eye configuration
(*Araniella displicata*)

square endites

Araneus thaddeus

orb web

Acacesia hamata (Hentz, 1847)

Common Name Difoliate orbweaver

Total Body Length Female 4.7–9.1 mm, Male 3.6–4.8 mm

Description The prosoma is pale yellow to orange or red. The opisthosoma is a teardrop shape, with the widest area about one-third of the length back. The heartmark is dark, bordered by a thin dark and then a thin pale line. Another set of thin dark and pale lines runs down the opisthosoma, tapering as they near the posterior end. Sometimes the dark portion of these lines appears bright red. The area within the lines is darker than the lateral edges of the spider. The coloration can vary, sometimes appearing silvery gray, golden, and brown, and other times bright orange to deep red.

Habitat These spiders are often found in shrubs. They have been found in cranberry bogs, oak-hickory woods, or tall weeds. This species usually constructs a new web every night, which is then removed at daybreak. These spiders can be found sitting in the hub during the night.

Known records from AL, AR, AZ, CT, DE, FL, GA, IA, IL, IN, KS, KY, LA, MD, MO, MS, NC, NJ, NY, OH, OK, OR, PA, TN, TX, VA, WV

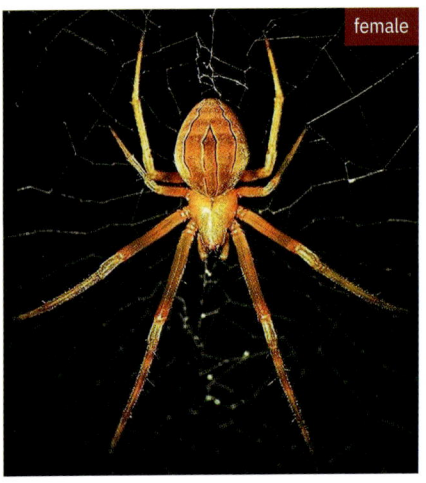
female

male

Acanthepeira cherokee Levi, 1976

Common Name This species does not have a common name.

Other Colloquial Names Cherokee orbweaver

Total Body Length Female 8.3–10.4 mm, Male 6.5–11.0 mm

Description The prosoma is dark brown, with tan hairs from the fovea to the eye region. There is a thin tan border around the edge. The prosoma is broad, with the lateral eyes on tubercles. The prosoma is covered in thick hair. The projections from the anterior face of the prosoma give it an overall squared-off appearance. The anterior of the opisthosoma has a short conical point that overhangs the prosoma. A series of conical humps lines the lateral edges of the opisthosoma, and a final one appears at the posterior end. The legs are banded. The overall color can vary from yellow to tan, shades of brown, red, and even dark brown. The darkest area tends to be the middle of the posterior opisthosoma. Four species of *Acanthepeira* are known from North America. In the Southeast, it can be difficult to make a species-level identification without looking at genitalia.

female

female ventral

female

female

female

Habitat These spiders like to build their webs in bottomland hardwood forests and swamps and are often found near bodies of water.

Known records from AL, AR, FL, GA, IL, KS, LA, MD, MO, MS, NC, OH, TX, VA

Acanthepeira stellata (Walckenaer, 1805)

Common Name Starbellied orbweaver

Total Body Length Female 7.0–15.1 mm, Male 5.1–8.1 mm

Description The prosoma is dark brown, with tan hairs from the fovea to the eye region. There is a thin tan border around the edge. The prosoma is broad, with the lateral eyes on tubercles. The prosoma is covered in thick hair. The projections from the anterior face of the prosoma give it an overall squared-off appearance. The anterior of the opisthosoma has a conical point that overhangs the prosoma. A series of conical points lines the lateral edges of the opisthosoma, and one final one appears at the posterior end. This creates a starlike appearance that gives this species the common name "starbellied orbweaver." The legs are banded. The overall color can vary from yellow to tan, shades of brown, red, and even dark brown. The darkest area tends to be the middle of the posterior opisthosoma. Four species of *Acanthepeira* are known from North America. In the Southeast, it can be difficult to make a species-level identification without looking at genitalia.

female

female

female

juvenile

juvenile

juvenile

female *Allocyclosa bifurca* with egg sacs

Habitat These spiders like to build their webs in grasslands, meadows, and edge habitat.

Known records from MB, NS, ON, PE, QC (Canada); AL, AZ, CO, CT, DE, FL, GA, IL, IN, KS, KY, LA, MA, MD, ME, MI, MN, MO, MS, NC, NH, NJ, NM, NY, OH, OK, PA, RI, SC, TN, TX, VA, VT, WI, WV (United States)

Allocyclosa bifurca
(McCook, 1887)

Common Name This species does not have a common name.

Other Colloquial Names Bifurcate trashline orbweaver

Total Body Length Female 5.1–9.0 mm, Male 1.8 mm

Description The prosoma is usually pale yellow to off-white. There are pale green to tan submarginal bands. The opisthosoma is pale green and white with a biforked posterior end and anterior humps. The legs are banded in white and green, often with dark brown bands at the distal tibiae.

Habitat These spiders are often found under the eaves of buildings, in areas with palmettos and palms, and in other forested habitats. The egg sacs resemble the adult spider in color and shape and are suspended in the upper half of the web in a central line. The bottom half of the web usually has a debris line of wrapped prey.

Known records from AL, FL, GA, LA, MS, TX

Araneus bicentenarius (McCook, 1888)

Common Name Lichenmarked orbweaver

Other Colloquial Names Giant lichen orbweaver

Total Body Length Female 21.0–28.0 mm, Male 7.0–11.5 mm

Description The prosoma can vary in color from brown to green. The opisthosoma is overall triangular, with prominent humps on the anterior (sometimes called shoulders) and sometimes with a white spot between them. The legs are banded and covered with short spines; the femurs tend to be red to orange. The overall color tends to be many shades of green and brown, creating the appearance that the spider is covered in lichen. This provides the spider with perfect camouflage on lichen-covered tree trunks. The name "bicentenarius" was chosen because the first record was described in Philadelphia during the bicentennial of the city.

female

Habitat These spiders tend to build their webs in forested areas.

Known records from NS, ON, QC (Canada); AL, FL, GA, IL, IN, LA, MA, ME, MI, MN, NC, ND, NH, NJ, NM, NY, OH, PA, SC, TN, TX, VA, WI, WV (United States)

female

Araneus bonsallae (McCook, 1894)

Common Name The genus has the common name of "angulate and round-shouldered orbweavers".

Total Body Length Female 4.2–5.8 mm, Male 3.3–3.8 mm

Description The prosoma is vivid green. There is often a slightly darker V-shaped marking from the fovea to the posterior lateral eyes. The eyes are ringed in yellow. The opisthosoma is vivid green. Paired red spots, outlined in yellow, are on either side of the dorsal surface. A transverse white to yellow band is near the anterior; this often has orange or brown speckling within it. Posterior to the white or yellow marking is a transverse red band. The opisthosoma markings can vary somewhat and may include

male

juvenile

male

male

black markings. The first two pairs of femurs are green proximally and red distally; this is most apparent in the males. All legs have red distally on the tibia and orange to yellow metatarsi and tarsi. These spiders are tricky to identify and easy to confuse with the other small greenish *Araneus* species.

Habitat Not much is published about the habitat preferences of this species, but it does seem to be associated with forested habitats.

Known records from AR, FL, GA, IL, KS, LA, MD, ME, MO, MS, NC, NE, NY, TX, VA

Araneus cavaticus (Keyserling, 1881)

Common Name Barn orbweaver

Other Colloquial Names Barn spider, Charlotte (the character from E. B. White's book *Charlotte's Web*)

Total Body Length Female 13.0–22.0 mm, Male 10.0–19.0 mm

Description The prosoma is tan to light brown. The opisthosoma is overall triangular, with a pair of humps on the anterior (sometimes called shoulders). The ventral opisthosoma has a black central band that is bordered with light markings shaped like square brackets. The lateral edges are mottled. The legs are banded and covered in short spines. The overall coloration can vary from very pale yellow to rich browns. Their webs can be quite large. Author E. B. White based the character Charlotte in the book *Charlotte's Web* on this species. In the book, she introduces herself as Charlotte A. Cavatica, a reference to the scientific name.

female

female

female with prey

female

Habitat These spiders are often found near human structures, including barns.

Known records from NB, NS, ON, QC (Canada); AL, AZ, CA, CT, KY, MA, ME, NH, NY, OH, PA, TN, TX, VA, VT, WV (United States)

Araneus cingulatus
(Walckenaer, 1841)

Common Name
The genus has the common name of "angulate and round-shouldered orbweavers".

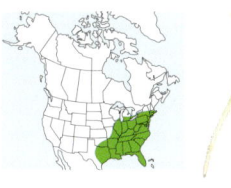

subadult male

Other Colloquial Names
Red spotted orbweaver

Total Body Length
Female 4.6–6.0 mm,
Male 2.7–3.5 mm

male

Description The prosoma is uniformly pale green. The opisthosoma has a green background, and the markings can be quite variable. All seem to have a paired row of red spots that run from the widest part of the opisthosoma (about one-third back) to the posterior end; these may be ringed in yellow. There are usually no black markings on the opisthosoma. The legs are green with pale distal segments, no red banding. These spiders are tricky to identify and easy to confuse with the other small greenish *Araneus* species.

133

Habitat These spiders are often found in wooded habitats and orchards. Evidence suggests they prefer building their webs in the upper canopy.

Known records from AR, FL, GA, IL, KS, KY, LA, MA, MD, MO, NC, NJ, NY, OH, OK, PA, TN, TX, VA

Araneus detrimentosus (O. Pickard-Cambridge, 1889)

Common Name The genus has the common name of "angulate and round-shouldered orbweavers".

Total Body Length Female 4.0–6.0 mm, Male 2.5–4.2 mm

Description The prosoma is brown and carpeted with long white hairs. The opisthosoma is brown, with a large lime-green marking that covers most of the dorsal surface; this marking has a cream-white border. The legs are tan to light brown and faintly banded. The ventral surface is pale tan to yellow. There are two bright yellow spots just before the spinnerets.

juvenile

Habitat These spiders often build their small webs in fields and shrubs.

Known records from AL, AZ, CA, FL, MS, TX

juvenile

juvenile ventral

Araneus diadematus Clerck, 1757

Common Name Cross orbweaver

Other Colloquial Names European garden spider, diadem spider, orangie, cross spider, crowned orb weaver

Total Body Length Female 6.5–20.0 mm, Male 5.7–13.0 mm

Description The prosoma is usually fairly light in color, with a dark midline and dark borders. The opisthosoma is ovoid, with a series of white dots that make a cross pattern on the anterior portion; a line of dots usually continues down the midline of the opisthosoma. The anterior humps (sometimes called shoulders) on the opisthosoma are less obvious in this species than they are in many *Araneus* species. There is often a dark zigzag line on either side of the white dots that tapers down the length of the opisthosoma; the lateral edges tend to be darker. The ventral opisthosoma has a black central band that is bordered with light markings shaped like square brackets. The lateral edges are mottled. The legs are banded. The overall color varies highly from pale cream, yellows, and browns to deep reds. The females often sit in the hub of the web during the daytime.

female

female

female

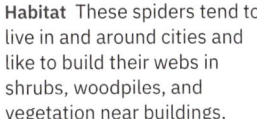

female

female

female ventral

Habitat These spiders tend to live in and around cities and like to build their webs in shrubs, woodpiles, and vegetation near buildings.

Known records from BC, NB, NL, NS, ON, PE, QC (Canada); IL, MA, ME, MI, NY, OH, OR, PA, RI, VT, WA (United States) This is an introduced species and, based on the distribution, has likely been introduced at least twice, as there are populations on the West Coast and East Coast and limited records in the middle.

Araneus gemmoides
Chamberlin & Ivie, 1935

Common Name The genus has the common name of "angulate and round-shouldered orbweavers".

Other Colloquial Names Cat-faced spider, plains orbweaver, jewel spider

Total Body Length Female 13.0–25.0 mm, Male 5.4–7.9 mm

Description The prosoma is fairly uniform in color, although sometimes a faint dark midline can be seen. It is coated with long hairs. The opisthosoma has a pair of humps on the anterior end (sometimes called shoulders); between them is usually a short white line crossed by two short Vs. These white markings are variable. The overall coloration varies highly from creams and yellows, oranges and reds, to dark, almost black, and browns. The ventral opisthosoma has a black central band that is bordered with light markings shaped like square brackets. The lateral edges are mottled. They can look very similar to some of the other *Araneus* species.

female

Habitat These spiders often build their webs near human structures but can also be found in forested and shrubby areas.

Known records from AB, BC, MB, SK (Canada); AL, AZ, CA, CO, IA, ID, IL, MI, MN, MO, MT, ND, NE, NM, NV, OR, SD, UT, WA, WI, WY (United States)

Araneus guttulatus (Walckenaer, 1841)

Common Name The genus has the common name of "angulate and round-shouldered orbweavers".

Other Colloquial Names Red-backed orbweaver

Total Body Length Female 3.8–6.0 mm, Male 3.9–4.8 mm

Description The prosoma is pale yellow to green and uniform in color. The opisthosoma usually has a dark reddish dorsal surface that contains a pattern of white markings. These spiders sometimes have a black triangular marking posteriorly and/or a black anterior band. The legs are uniformly pale green to greenish yellow and have no red banding. These spiders are tricky to identify and easy to confuse with the other small greenish *Araneus* species.

Habitat These spiders tend to be found in wetlands, bogs, swamps, and other wet habitats.

Known records from ON, NB, NS, QC (Canada); AL, AR, GA, IL, ME, MI, MN, NC, NH, NY, SC, TX, VA, WI (United States)

Araneus juniperi (Emerton, 1884)

Common Name The genus has the common name of "angulate and round-shouldered orbweavers".

Other Colloquial Names Juniper orbweaver

Total Body Length Female 2.5–5.2 mm, Male 3.2–4.6 mm

Description The prosoma is vivid green to greenish yellow. The posterior median eyes are ringed in yellow or off-white. There may be a short dark dash at the fovea, but the prosoma is otherwise unmarked. The opisthosoma is usually green or greenish yellow, with a white heartmark; sometimes the heartmark is part of a longitudinal white stripe that runs the entire length of the opisthosoma. A pair of white longitudinal stripes go from the anterior end to the spinnerets on each side. There may or may not be paired red or black dots in the posterior half of those stripes. The legs are green to yellow green.

male

Habitat The species is aptly named, as it seems to have an affinity for junipers, but it is also often found on cedars and other coniferous plants.

Known records from NS, QC (Canada); AL, AR, CT, FL, GA, IL, IN, MA, MD, ME, NC, NH, NY, OH, TX (United States)

Araneus marmoreus
Clerck, 1757

female

female

Common Name Marbled orbweaver

Other Colloquial Names Pumpkin spider

Total Body Female 9.0–18.0 mm, Male 5.9 mm

Description The prosoma is typically orange, usually with a visible darker midline. The opisthosoma is overall pale yellow to orange, with markings that can range in color from orange to black. The legs are banded in black or brown and white, but the femurs are orange. There are several color variations of this species: they can be almost completely cream, they can have a cream opisthosoma with a dark keyhole mark on the posterior opisthosoma, and they can also be shades of gray. The ventral opisthosoma has a black central band that is bordered with light markings shaped like square brackets. The lateral edges are mottled.

Habitat The webs can be found in forested and open habitats, and the spider will often have a retreat in a curled leaf at the edge of the web.

female

female in retreat

female ventral

female (keyhole variant)

Known records from AB, BC, MB, NB, NL, NS, NT, ON, PE, QC, SK, YT (Canada); AK, AL, AR, AZ, CO, CT, GA, IA, IL, IN, KS, KY, MA, ME, MI, MN, MO, MS, MT, NC, ND, NE, NH, NJ, NY, OH, OR, PA, SC, TN, TX, UT, VA, VT, WA, WI, WV, WY (United States)

male

male

Araneus miniatus (Walckenaer, 1841)

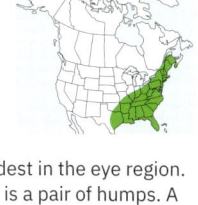

Common Name The genus has the common name of "angulate and round-shouldered orbweavers".

Other Colloquial Names Black-spotted orbweaver

Total Body Length Female 3.0–4.7 mm, Male 2.5–3.7 mm

Description The prosoma is tan to green, with a dark central area that is widest in the eye region. The opisthosoma is brown to red and is much wider anteriorly, where there is a pair of humps. A white line usually runs between these humps (or two white patches appear there). Sometimes, this white mark is extended and looks almost like the "bat symbol." Paired black spots run down the opisthosoma from the humps to the posterior end. The legs are banded and spotted. This species is often mistaken for *Araneus partitus*. They are similar in size and overall shape. The markings on the two species are somewhat different, but they are similar in general appearance.

Habitat These spiders are usually found in wooded habitats. They are known to capture mosquitoes frequently.

female

Known records from NS (Canada); AL, CT, DC, FL, GA, LA, MA, MS, NC, NJ, NY, OH, OK, TN, TX (United States)

female with egg sac

Araneus niveus (Hentz, 1847)

Common Name The genus has the common name of "angulate and round-shouldered orbweavers".

Total Body Length Female 3.2–5.0 mm, Male 2.9–4.9 mm

Description The prosoma is pale yellow to green. Dark yellow to tan rings are around the eyes. Markings on the opisthosoma vary highly, which adds to the difficulty in identification. Usually, a central red area appears on a pale yellow background on the dorsal surface. Within the red area, there may be some white spots or a series of dark spots. Black markings are often on the opisthosoma, often in the median area. The legs are pale green; the males often have a set of red bands at the distal end of the tibiae. This species can look very much like some of the other small greenish *Araneus* species.

Habitat These spiders are often found in wooded habitats, including orchards. They may prefer to build their webs in the top of the tree canopy.

Known records from AL, AR, CT, FL, GA, IL, KY, LA, MD, MO, NC, NY, OH, TN, VA

female with egg sacs

male

male

juvenile

Araneus nordmanni
(Thorell, 1870)

Common Name The genus has the common name of "angulate and round-shouldered orbweavers".

Other Colloquial Names Nordmann's orbweaver

Total Body Length Female 7–19 mm, Male 6–10 mm

Description The prosoma is tan to brown and usually appears quite hairy. The opisthosoma is highly variable in coloration: some have a dramatic black-and-white appearance, while others appear in shades of tan, brown, red, or green. The heartmark tends to be white and sits between the two anterior humps (sometimes called shoulders). A pair of zigzag lines that run down the dorsal surface taper toward the posterior end. The ventral opisthosoma has a dark black area in the middle surrounded by yellow. There are two light dots in the black area. The legs are banded. This species could easily be confused with the other larger *Araneus* species.

Habitat These spiders can be found in a variety of habitats. Some are found in shrubby areas, others near mature trees.

female ventral

female

female

female

Known records from AB, BC, LB, MB, NB, NL, NS, NT, ON, PE, QC, SK, YT (Canada); AK, AZ, CA, CO, CT, FL, ID, MA, ME, MI, MN, MT, NC, NH, NM, NY, OR, PA, TN, TX, UT, VA, VT, WA, WI, WV, WY (United States)

Araneus partitus
(Walckenaer, 1841)

Common Name The genus has the common name of "angulate and round-shouldered orbweavers".

Total Body Length
Female 3.3–4.3 mm, Male 2.7–3.3 mm

Description The prosoma is tan yellow to pale green, with a goblet-shaped brown mark that points to the posterior lateral eyes. The edges are dark. The opisthosoma is usually brown to red, with an anterior pair of humps connected by a yellow to tan band. A series of paired dark red to black marks taper as they approach the posterior end. The legs are tan and have visible

female

female

141

female

female

spines. *A. partitus* can easily be confused with *Araneus miniatus*; the markings on the two species are somewhat different, but they are similar in general appearance.

Habitat These spiders tend to build their webs in forested habitats.

Known records from AR, CT, FL, GA, IL, MA, MD, MS, NC, NJ, NY, OH, OK, SC, VA, WV

Araneus pegnia (Walckenaer, 1841)

Common Name The genus has the common name of "angulate and round-shouldered orbweavers".

Other Colloquial Names Butterfly orbweaver, butterfly-marked orbweaver

Total Body Length Female 3.5–8.2 mm, Male 2.7–3.3 mm

Description The prosoma is light tan to pale orange. There is usually a visible dark midline, although it often does not run the entire length of the prosoma. The opisthosoma is tan to brown. At the anterior end, there is a marking that many observers compare to outspread butterfly wings. The marking is usually tan to yellow but can also be deep pink in some cases. The legs are banded.

Habitat These spiders often build their webs in low shrubby vegetation, but webs can also be found under the eaves of buildings or on fences. The spider's retreat looks like a small *Araneus thaddeus* (lattice orbweaver) retreat but is much smaller and less structured. The web will sometimes have a tangle web attached, similar to that built by *Metepeira labyrinthea* (labyrinth orbweaver). The webs sometimes lack a sector, similar to *Zygiella x-notata*. There is usually a line that runs from the retreat to the hub of the web.

female

female

female ventral

female in retreat

Known records from AL, AZ, CA, CT, FL, GA, IN, LA, MA, MD, MS, NC, NH, NJ, NM, NY, OH, OK, TX, VA, WV

juvenile

male

male

Araneus pratensis
(Emerton, 1884)

Common Name Openfield orbweaver

Total Body Length
Female 3.8–5.0 mm, Male 3.0–3.5 mm

female

Description The prosoma is orange, usually with a slightly paler area near the fovea and forward to the eye region. The opisthosoma is cream to white, with a dark midline and a pair of orange to red or even brown stripes. The edges are white to cream, giving the opisthosoma an overall striped appearance. The ventral opisthosoma has dark midline stripes bordered by yellow lines. The spinnerets are dark. The sternum is also dark. The legs are yellow to orange.

Habitat These spiders often build their webs in fields or grasslands. The diameter of the web is usually about 7–8 in.

female

female ventral

Known records from ON (Canada); AR, CT, FL, IA, IL, MA, MD, MI, MN, MO, NC, NH, NY, OH, PA, TX, VA, VT, WV (United States)

Araneus saevus (L. Koch, 1872)

Common Name The genus has the common name of "angulate and round-shouldered orbweavers".

Other Colloquial Names Fierce orbweaver

Total Body Length Female 11.0–17.0 mm, Male 9.0–11.0 mm

Description The prosoma is dark brown and covered with lighter-colored hairs. The opisthosoma is usually quite dark, sometimes almost black. A white mark is situated between the anterior humps (sometimes called shoulders). Sometimes, a white to orange mark bridges the gap between the two humps. The ventral opisthosoma has a black central band that has two light spots toward the posterior end. The lateral edges also tend to be dark. The legs are banded.

Habitat These spiders are usually found in forested habitats and tend to build their webs near large trees.

Known records from AB, BC, LB, MB, NB, NL, NS, NT, ON, PE, QC, SK, YT (Canada); AK, CO, CT, ID, MA, ME, MI, MN, MT, NH, NY, OH, OR, PA, SD, UT, VT, WA, WI, WY (United States)

female

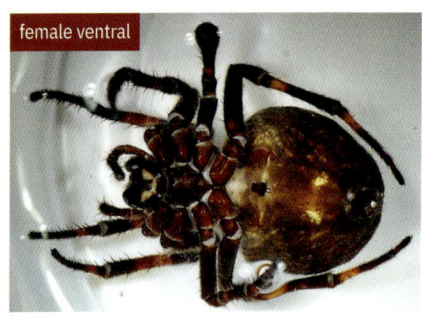

female ventral

Araneus thaddeus (Hentz, 1847)

Common Name Lattice orbweaver

Total Body Length Female 5.9–8.0 mm, Male 3.7–5.7 mm

Description The prosoma is orange, with little variation. The opisthosoma of the female is usually a pale cream-white color, although females seem to darken to deep orange toward the end of the summer. There is a pair of black bands on the lateral aspects of the anterior of the opisthosoma. The pattern on the male can resemble the butterfly marking of *Araneus pegnia*. The ventral opisthosoma has a yellow trapezoid surrounded by black that is bordered in yellow. The femur and patella are deep orange, and the distal segments of the legs are banded in white and black.

Habitat These spiders tend to build their webs in shrubby vegetation and can also be found in landscaping plants. The spider is best known for the lattice tubular retreat that it builds.

Known records from MB, ON, QC (Canada); AL, AR, AZ, CO, CT, GA, IL, KY, MA, MD, MN, MO, NC, NE, NH, NJ, NM, NY, OH, PA, TN, TX, UT, VA, VT, WI, WV (United States)

female

female

female

female

female in retreat

female ventral

juvenile

male

Araneus trifolium (Hentz, 1847)

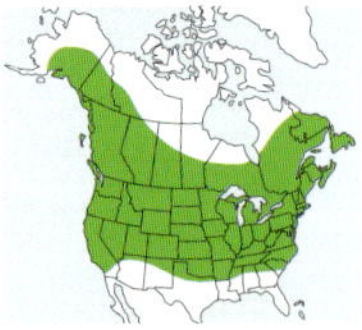

Common Name Shamrock orbweaver

Other Colloquial Names Pumpkin spider, shamrock spider

Total Body Length Female 9.0–20.0 mm, Male 4.5–8.0 mm

Description The female prosoma is tan to brown, with a dark median stripe that is widest in the eye region. The female opisthosoma can vary greatly in color. In some specimens, it is almost completely cream white to deep yellow; in others, it is orange or deep purple red, with a cluster of white markings on the dorsal surface, occasionally in shades of green. This species can change color, and it is thought the spiders may do so to match their habitat. The ventral opisthosoma has a dark black median band with a pair of yellow to orange markings in the shape of parentheses on either side. The lateral edges are mottled with the color of the dorsal opisthosoma.

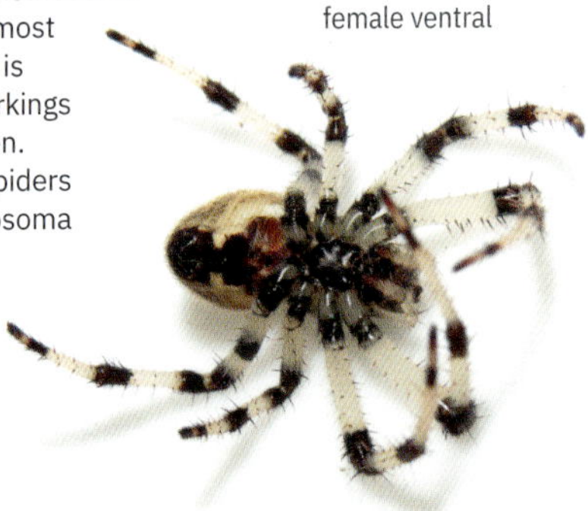

female ventral

female

female

female

female ventral

female in retreat

male

The spinnerets are dark. The legs are banded in highly contrasting colors: black brown or deep red and white to yellow. The male prosoma is off-white to orange. There may be a thin brown median stripe; often, this does not reach the eye region. The male opisthosoma is cream to yellow. The male legs tend to be banded in yellow and orange.

Habitat These spiders tend to build their webs in low shrubby and herbaceous vegetation. Often, their web has a curled leaf retreat at the side. An egg sac of this species was documented to contain more than 2,600 eggs, which may be the most of any spider egg sac.

Known records from AB, BC, LB, MB, NB, NL, NS, NT, ON, PE, QC, SK, YT (Canada); AK, AL, CA, CO, CT, IA, ID, IL, IN, MA, MD, ME, MI, MN, MT, NC, ND, NH, NJ, NM, NY, OH, OR, PA, SD, TN, UT, VA, VT, WA, WI, WV, WY (United States)

Araniella displicata (Hentz, 1847)

Common Name Sixspotted orbweaver

Total Body Length Female 4.8–7.2 mm, Male 4.0–5.0 mm

Description The prosoma is yellow to orange and uniformly colored. The opisthosoma is usually pale yellow or cream on the dorsal surface (although some are orange to red), with darker lateral edges, similar to *Araneus thaddeus*. Starting just over halfway back on the opisthosoma are three to four pairs of dark spots that are bordered in yellow. These taper, and the last pair is almost adjacent on the posterior end. The ventral opisthosoma has a broad dark brown to reddish stripe bordered in yellow. The spinnerets are dark. The males tend to be darker than the females. The males have dark orange-banded legs. The legs of females are usually the same color as the prosoma and are not banded, or they have dusky bands at the proximal ends of each segment.

female

female

Habitat These small spiders often build their web within the confines of a single leaf or at the tip of coniferous branches.

Known records from AB, BC, LB, MB, NB, NL, NS, NT, ON, PE, QC, SK, YT (Canada); AK, AZ, CA, CO, CT, IA, ID, IL, IN, KY, MA, MD, ME, MI, MN, MT, NC, ND, NH, NJ, NM, NV, NY, OH, OR, PA, RI, TX, UT, VA, VT, WA, WI, WV, WY (United States)

female ventral

female

female

male

in web

with prey

Argiope argentata (Fabricius, 1775)

Common Name Silver garden spider

Other Colloquial Names Silver argiope, silver garden orbweaver

Total Body Length Female 12.0–16.0 mm, Male 5.0–8.0 mm

Description The prosoma is white or silver and covered with a thick blanket of hairs. The anterior half of the opisthosoma of the female is white. The posterior is yellow, orange, or brown, sometimes with white spots. There are three pairs of lobes on the lateral edge, and the posterior end comes to a point. The ventral opisthosoma is dark, with a yellow triangle pointing at the epigastric furrow. The legs have pale femurs and are banded distally. The males are much smaller than the females. They lack the lobes on the opisthosoma and have a silvery-colored opisthosoma with a dark heartmark. A pale median line that is bordered by two bands of slightly darker coloration runs the length of the opisthosoma.

male and female

spiderlings

male and female

juvenile female

juvenile male

Habitat These spiders are often found in grassy and shrubby habitats, and the web can be quite large. Often, a stabilimentum, or web decoration, makes a cross in the middle of the web where the spider rests, its legs positioned to almost match the stabilimentum. The purpose of this structure has been debated: it may help camouflage the spider in the web; provide web stabilization; make the web obvious to birds, so they don't fly into it; and/or attract insects to the web.

Known records from AZ, CA, FL, TX

Argiope aurantia Lucas, 1833

Common Name Yellow garden spider

Other Colloquial Names Black and yellow garden spider, black and yellow argiope, golden garden spider, golden orbweaver, writing spider, zigzag spider, zipper spider, corn spider, banana spider, Steeler spider, McKinley spider

Total Body Length Female 19.0–28.0 mm, Male 5.0–8.0 mm

Description The prosoma is white to silver and covered in a thick blanket of hairs. The opisthosoma of the female is usually black (occasionally deep golden), with two to four pairs of white to yellow spots down the median line. The edges have yellow to white markings in various shapes, giving a somewhat striped appearance. There is a pair of humps near the anterior end of the opisthosoma. The ventral opisthosoma is black, with a series of small paired yellow to white dots in the central area. Two white to yellow zigzag stripes are on either side of these dots. A bright yellow stripe runs down the middle of the sternum. The spinnerets are dark and surrounded by yellow to white dots. The proximal femurs are pale, sometimes almost pale orange. The distal segments of the legs are black, and they are occasionally banded.

female ventral

The males are much smaller than the females. They can sometimes be found in a web built near the female's web. They have a similar pattern on the opisthosoma, but it is less brilliantly colored. The legs are dark proximally and tend to be pale distally.

Habitat These spiders are often found in humid habitats, and the web can be quite large. Often a stabilimentum, or web

female

female

male

male ventral

male

decoration, makes a cross in the middle of the web where the spider rests, its legs almost positioned to match the stabilimentum. The purpose of this structure has been debated: it may help camouflage the spider in the web; provide web stabilization; make the web obvious to birds, so they don't fly into it; and/or attract insects to the web. The egg sacs of these spiders are quite large, almost the size of a ping-pong ball, and look like a clay water pot. They have a papery outer covering that tapers to a lip at the top. Some females make multiple egg sacs in the fall. They usually suspend their egg sacs in a protected area. The young will overwinter in the egg sac and emerge in the spring.

juvenile in web

juvenile in web

egg sac

Known records from NB, NS, ON, PE, QC (Canada); AL, AR, AZ, CA, CT, FL, GA, IA, IL, IN, KS, KY, LA, MA, MD, ME, MI, MO, MS, NC, ND, NE, NH, NJ, NM, NY, OH, OK, OR, PA, RI, TN, TX, VA, WI, WV (United States)

Argiope trifasciata (Forsskål, 1775)

Common Name Banded garden spider

Other Colloquial Names Banded argiope, banded orbweaver

Total Body Length Female 15.0–25.0 mm, Male 4.0–5.5 mm

Description The prosoma is silver to white and covered in a thick blanket of hairs. The female opisthosoma is striped, varying in color from white to yellows, oranges, and black. The ventral opisthosoma is black, with a series of small paired yellow to white dots in the central area. Two long white to yellow zigzag stripes are on either side of these dots. The proximal femurs are orange to yellow, and the distal legs are banded in yellow to orange and black. The male opisthosoma is yellow to cream; a faint pattern may be visible. The legs are solidly colored and darken toward the distal end.

female

female

female

female ventral

Habitat These spiders are often found in fields, grassy areas, and gardens, and the web can be quite large. Often a stabilimentum, or web decoration, makes a cross in the middle of the web where the spider rests, its legs almost positioned to match the stabilimentum. The purpose of this structure has been debated: it may help camouflage the spider in the web; provide web stabilization; make the web obvious to birds, so they don't fly into it; and/or attract insects to the web.

male

male

female & male

female & male

female & male

male

juvenile

Known records from AB, BC, MB, NB, NS, ON, PE, QC, SK (Canada); AL, AR, AZ, CA, CO, CT, FL, IA, ID, IL, IN, KS, MA, MD, ME, MI, MN, MO, MS, MT, NC, ND, NE, NH, NJ, NM, NV, NY, OH, OK, OR, PA, RI, TN, TX, UT, VA, VT, WA, WI, WV, WY (United States)

Cyclosa conica (Pallas, 1772)

Common Name The genus *Cyclosa* is known as "trashline orbweavers."

Other Colloquial Names Conical trashline orbweaver

Total Body Length Female 3.6–7.9 mm, Male 3.5–4.9 mm

Description The prosoma is yellow to dark brown. The opisthosoma is teardrop shaped and tapers to an extended hump on the posterior end; this hump is less obvious in immature spiders. The opisthosoma can vary in color and markings, and it lacks the tubercles that are seen in *Cyclosa turbinata*. The ventral opisthosoma has a dark area posterior of the epigastric furrow, and two yellow spots sit adjacent to the epigastric furrow. The males are similarly colored and marked but tend to lack the posterior hump that is seen on the female.

Habitat These spiders tend to be found in open woodlands and suspend their webs between trees. The web is usually located 1.5–2.0 m above the ground and usually seen more often than the spider. The vertically oriented orb web is adorned with the remains of killed prey and other debris arranged in a vertical line down the middle of the web. The spider is often camouflaged within this trash line. The egg sacs are attached to twigs and leaves near the web. The spiders will often shake the web rapidly if disturbed.

female

female

female

female

male

subadult male

Known records from AB, BC, LB, MB, NB, NL, NS, NT, ON, PE, QC, SK, YT (Canada); AK, AZ, CA, CO, CT, IA, ID, IL, MA, ME, MI, MN, MT, ND, NH, NJ, NM, NY, OH, OR, PA, RI, SD, TX, UT, VT, WA, WI, WV, WY (United States)

Cyclosa turbinata (Walckenaer, 1841)

Common Name Humped trashline orbweaver

Total Body Length Female 3.3–5.2 mm, Male 2.1–3.2 mm

Description The female prosoma is dark brown to orange. The female opisthosoma is patterned with white, brown, and orange. The opisthosoma is teardrop shaped, and a pair of tubercles are on the opisthosoma at the widest point. The males of this species are easily confused with the theridiids in the genus *Asagena*, which look similar. They have a very dark prosoma and a red opisthosoma with a pair of white spots toward the anterior end. The web is usually seen more often than the spider.

female
ventral

Habitat These spiders tend to build their webs in fields, gardens, and open woodlands. The vertically oriented orb web

female

female

155

is adorned with the remains of killed prey and egg sacs arranged in a vertical line down the middle of the web. The spider is often camouflaged within this trash line. The spiders will often shake the web rapidly if disturbed.

Known records from ON (Canada); AL, AR, AZ, CA, CT, FL, GA, IA, IL, IN, KS, LA, MA, MD, MI, MO, MS, MT, NC, NE, NJ, NM, NY, OH, OK, OR, PA, SC, SD, TN, TX, VA, WA, WI, WV, WY (United States)

female in web

female in web

male

male ventral

Cyrtophora citricola (Forsskål, 1775)

Common Name The genus *Cyrtophora* has the common name "tentweb weavers."

Other Colloquial Names Tropical tent web spider

Total Body Length Female 8.8–15.2 mm, Male 2.5–3.0 mm

Description The prosoma is uniformly orange to brown and somewhat hairy in appearance. The opisthosoma is irregularly shaped, sometimes with conspicuous lobes. The color can vary from pale yellow tan to almost black with white markings. This species is able to shift its overall color to suit the habitat. The posterior end of the opisthosoma is bifurcate (two lobed).

Habitat These spiders tend to build their webs on guardrails and bridges, and many individual webs can often be found close to one another. The web consists of a platform that has a fine

female

female ventral

female in web

mesh-like appearance, surrounded by knockdown lines that create a tentlike appearance. The spider usually hangs under the platform and often looks like a dead leaf in the web.

Known records from FL
This is an introduced species.

Eriophora ravilla (C.L. Koch, 1844)

Common Name The genus *Eriophora* has the common name "tropical orbweavers."

Total Body Length Female 12.0–24.0 mm, Male 9.0–13.0 mm

Description The prosoma is tan to brown and covered with hairs. The opisthosoma is highly variable. Some are almost uniformly yellow, or tan to brown with yellow shoulders. Others have dark lateral edges and a bright green area in the middle. In others, a series of dots on the posterior end can be red to black and outlined in pale colors. In others, the background is brown to tan with a vague cross-type pattern of pale cream spots. The ventral opisthosoma has a yellow line near the epigastric furrow; a half circle area of black is posterior to this. The long scape of the female epigynum is usually quite visible in mature specimens and points to the posterior end. The legs tend to be dark red proximally and paler at the distal ends. Some mature females can look very similar to *Neoscona crucifera*, but the ventral markings are distinctly different.

Habitat These spiders build their large webs in shrubs, trees, and other low vegetation.

Known records from FL, LA, MD, TX

female

female

juvenile

Eustala anastera (Walckenaer, 1841)

Common Name Humpbacked orbweaver

Total Body Length Female 5.4–10.0 mm, Male 3.9–9.5 mm

Description The prosoma is variable in color, usually matching the opisthosoma. The opisthosoma is triangular and tallest at the posterior end; this humped appearance leads to the common name. The coloration and pattern can vary from shades of tan and brown to vivid greens. Some have a pair of spots on the lateral sides of the humped area of the

female

female

opisthosoma; some have a dark line down the median. Many have a foliate pattern. The ventral opisthosoma has a light central mark that runs from the epigastric furrow to the spinnerets; the mark is bordered in black. The scape of the female epigynum is directed anteriorly. The legs also vary in color and are covered with short spines in the female; the spines are more pronounced in the mature males. More than a dozen species of *Eustala* are in North America, and identification to species level is difficult without a specimen.

Habitat These spiders often build their webs among dead branches in forested habitat. Typically, the webs are removed at dawn and rebuilt each night.

Known records from AB, MB, NB, NS, ON, QC, SK (Canada); AL, AR, AZ, CA, CT, FL, GA, ID, IL, IN, KS, KY, LA, MA, MD, ME, MI, MN, MO, MS, MT, NC, NH, NJ, NM, NY, OH, OK, OR, PA, RI, SC, SD, TN, TX, UT, VA, WA, WI, WV (United States)

female

female ventral

female

female

female

male subadult

male subadult

male

male

Eustala cepina
(Walckenaer, 1841)

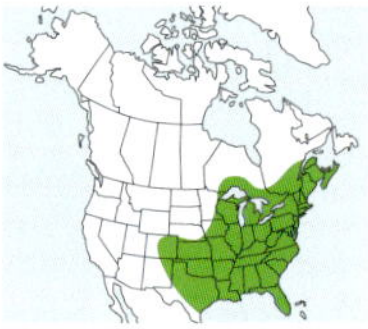

female

Common Name This species does not have a common name.

Other Colloquial Names
Riparian duncecap orbweaver

Total Body Length Female 3.4–7.9 mm, Male 2.5–4.3 mm

Description The prosoma varies in color, usually matching the opisthosoma. The opisthosoma is oval to somewhat triangular and usually lacks the hump seen in *Eustala anastera*. The coloration and pattern can vary from shades of tan and brown to olive greens. Most have a foliate pattern. The ventral opisthosoma has a light central mark that runs from the epigastric furrow to the spinnerets; the mark is bordered in black. The scape of the female epigynum is directed anteriorly. The legs also vary in color and are covered with short spines in the female; the spines are more pronounced in the mature males (but the males do not have any ventral spines on femur II). The legs may be faintly banded. More than a dozen species of

female

Eustala are in North America, and identification to species level is difficult without a specimen.

Habitat: These spiders can be found in grassy habitats or in understory vegetation.

Known records from NS, ON, QC (Canada); AL, AR, CO, CT, FL, GA, IL, IN, KS, KY, LA, MA, MD, ME, MI, MN, MO, MS, NC, NJ, NY, OH, OK, PA, SC, TN, TX, VA, WI, WV (United States)

female
ventral

male

male

male

Gasteracantha cancriformis
(Linnaeus, 1758)

Common Name Spinybacked orbweaver

Other Colloquial Names Crab spider, spiny orbweaver spider, crab-like orbweaver spider, crab-like spiny orbweaver spider, jewel spider, spiny-bellied orbweaver, jewel box spider, smiley face spider, crablike spiny orbweaver

Total Body Length Female 5.8–8.6 mm, Male 1.9–2.7 mm

female

female

female

female

female ventral

male

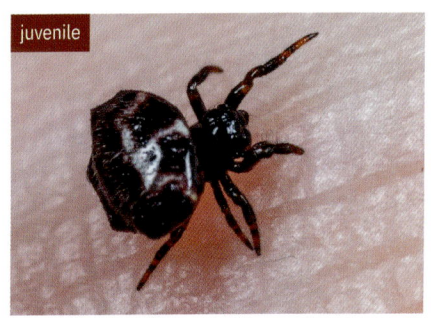
juvenile

Description The prosoma is dark, usually black. The opisthosoma of the female has prominent spines: two on each side and two directed posteriorly at the end. The color can vary from white yellow to deep red. The spines are either black or brilliant red. Black spots on the dorsal surface look similar to a hockey mask. The legs are dark, sometimes with some visible banding. The males are much smaller than the females. They lack the spines and are generally dark, with a faint pattern of lighter markings.

Habitat These spiders often build their webs in tree branches or on human structures. They are often observed sitting in the hub of the web during daytime. The egg sacs are usually attached to the vegetation or other structures and are covered in neon-green to bright yellow silk, sometimes with a dark longitudinal line.

Known records from AL, CA, FL, GA, LA, MS, NC, OK, SC, TX, VA

Gea heptagon (Hentz, 1850)

Common Name This species does not have a common name.

Other Colloquial Names Heptagonal orbweaver

Total Body Length Female 4.5–5.8 mm, Male 2.6–4.3 mm

Description The prosoma is brown to orange. The opisthosoma of the female has paired humps: one pair on the anterior end pointing forward, and two pairs on the lateral edges. The color varies from browns and tans to yellows and oranges, usually with a dark triangular area toward the posterior and lighter-colored humps (especially the anterior pair). The legs are banded. The males tend to be overall orange to yellow, with a similar dark triangle. Both males and females often sit holding their legs in pairs, giving the appearance of having only four legs.

female

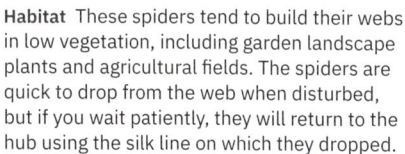
male

Habitat These spiders tend to build their webs in low vegetation, including garden landscape plants and agricultural fields. The spiders are quick to drop from the web when disturbed, but if you wait patiently, they will return to the hub using the silk line on which they dropped.

male

Known records from ON (Canada); AL, AR, AZ, CA, FL, GA, IL, IN, KS, LA, MD, MI, MO, MS, NC, NE, NJ, OH, OK, SC, TN, TX, VA, WI (United States)
This is an introduced species.

Hypsosinga pygmaea
(Sundevall, 1831)

Common Name This species does not have a common name.

Other Colloquial Names Small orbweaver

Total Body Length Female 3.1–4.0 mm, Male 2.9–2.8 mm

Description The prosoma is orange, with a black marking around the eyes. Sometimes a dark triangular marking is also near the fovea. The opisthosoma is variable in its coloration. It can be completely black, it may have dark stripes on a cream-colored background, or it may have dark spots on a cream background. The legs are orange to dark yellow and lack markings.

female

female

Habitat These spiders tend to build their webs in herbaceous vegetation in grasslands, meadows, and fields. They have been found in roadside weeds.

Known records from AB, BC, LB, MB, NB, NL, NS, NT, ON, PE, QC, SK, YT (Canada); AK, CO, CT, FL, GA, IL, IN, MA, ME, MI, MN, MT, ND, NH, NJ, NY, OH, PA, RI, SD, TN, UT, VT, WA, WY (United States)

Hypsosinga rubens (Hentz, 1847)

Common Name This species does not have a common name.

Other Colloquial Names Rubens orbweaver

Total Body Length Female 7.4–8.7 mm, Male 6.0–6.5 mm

Description The prosoma is red to reddish orange and darker in the eye region. The opisthosoma is usually bright red-orange, often with paired dark marks on the lateral edges toward the posterior end. The opisthosoma can be completely black, especially for the males. The proximal legs are orange red; often the distal segments are black. This spider can look similar to the theridiid *Thymoites unimaculatus*. The web structure and eye placement are helpful traits to distinguish the two.

Habitat These spiders often build their webs in low vegetation, such as shrubs and forest understories.

Known records from AB, BC, MB, NB, NL, NS, NT, ON, QC, SK (Canada); AK, AL, AR, CT, DE, FL, GA, IL, IN, KS, KY, LA, MA, MD, ME, MI, MO, MS, NC, NH, NJ, NY, OH, OK, PA, RI, SD, TN, TX, VA, WI (United States)

female

female

female with prey

male

Larinia directa
(Hentz, 1847)

Common Name This species does not have a common name.

Total Body Length
Female 4.8–11.7 mm,
Male 4.5–6.5 mm

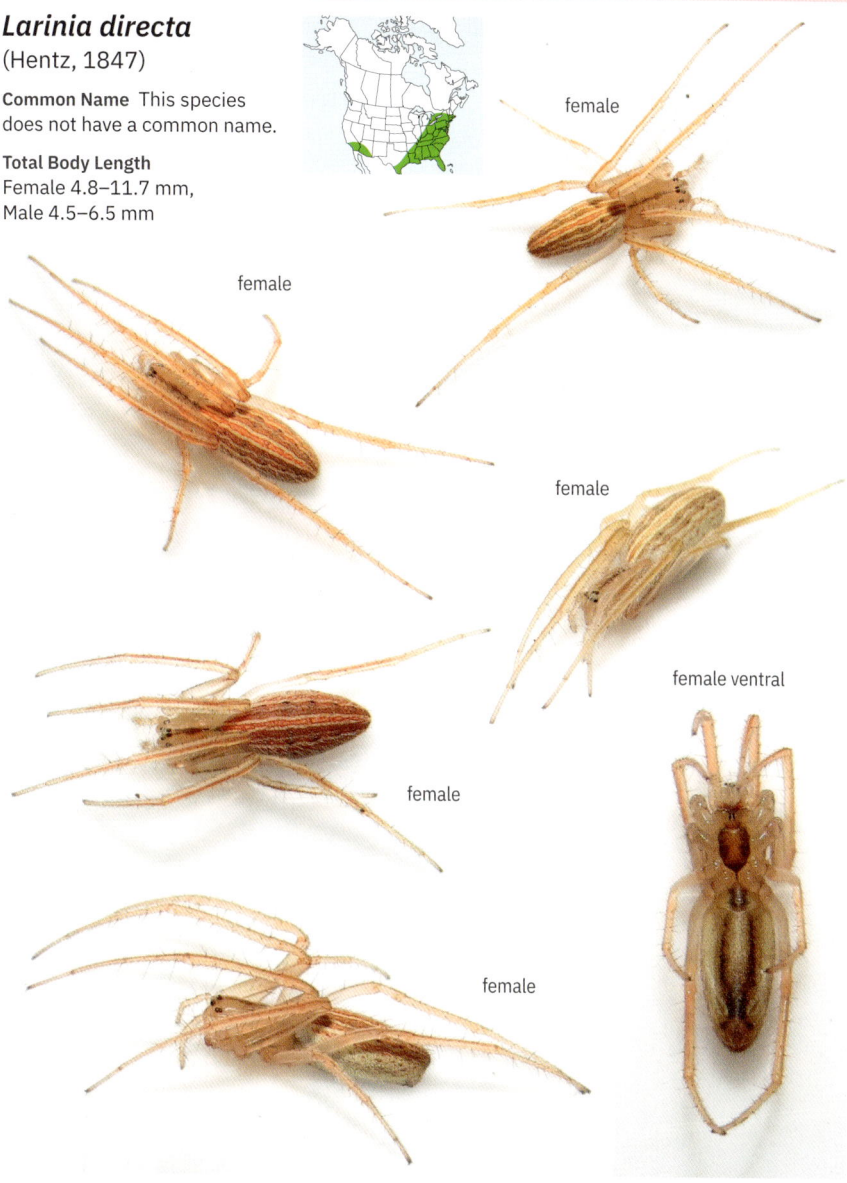

female

female

female

female

female ventral

female

female

Description The prosoma is tan to light orange. There is a faint median line that is widest anterior of the fovea. The opisthosoma is tan, with lengthwise stripes of yellows and reds. The ventral opisthosoma has a central light band bordered with black. Species-level identification can be tricky, as the other species of *Larinia* look very similar.

Habitat These spiders tend to rest on blades of grass, where they are very well camouflaged.

Known records from ON (Canada); AL, AR, AZ, CA, FL, GA, LA, MO, MS, NC, NJ, OH, OK, SC, TN, TX, VA (United States)

Larinioides cornutus (Clerck, 1757)

Common Name Furrow orbweaver

Other Colloquial Names Furrow spider, furrow orb spider, foliate spider

Total Body Length Female 6.5–14.0 mm, Male 4.7–8.5 mm

Description The prosoma is dark brown and lightly covered with light-colored hair, with a pale edge. The opisthosoma is tan to brown, with a foliate pattern bordered in dark brown that fades toward the midline. The heartmark is dark. The lateral edges of the opisthosoma are pale. They can have a glossy appearance. The ventral opisthosoma has a dark semicircular area just posterior of the epigastric furrow, which is bordered on each side by two bright yellow comma-shaped marks. The legs are banded, but this species does not have the mid-metatarsi IV band that is seen in other species of this genus.

female

female

female

female

female ventral

female

Habitat These spiders are often found with their webs attached to human structures, especially near bodies of water.

Known records from AB, BC, LB, MB, NB, NL, NS, NT, ON, PE, QC, SK, YT (Canada); AK, AL, AR, AZ, CA, CO, CT, FL, GA, IA, IL, IN, KS, KY, LA, MA, MD, ME, MI, MN, MO, MS, NC, NE, NH, NJ, NM, NY, OH, OK, OR, PA, RI, SC, TN, TX, UT, VA, VT, WI, WV, WY (United States)

male

male

Larinioides sclopetarius
(Clerck, 1757)

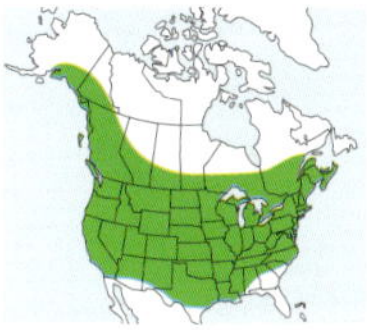

Common Name Bridge orbweaver

Other Colloquial Names Gray cross spider

Total Body Length Female 8.0–14.0 mm, Male 6.0–7.0 mm

Description The prosoma is dark brown and covered in light-colored hair. The pale edges also highlight the eye region. The opisthosoma has a dark heartmark and a foliate marking that is outlined in white to pale yellow. The lateral edges are mottled with dark brown. The ventral opisthosoma has a dark semicircular area just posterior of the epigastric furrow, which is bordered on each side by two bright yellow comma-shaped marks. The legs are banded and have a mid-metatarsi IV band.

Habitat The webs are often found on human structures, especially bridges.

Known records from BC, LB, MB, NB, NL, NS, ON, QC (Canada); AK, CT, ID, IL, IN, KY, MA, ME, MI, NH, NY, OH, OK, PA, RI, TX, UT, VA, VT, WA, WI, WV (United States) This is an introduced species.

female

female

female

Mangora gibberosa (Hentz, 1847)

Common Name Lined orbweaver

Total Body Length Female 3.4–4.8 mm, Male 2.6–3.2 mm

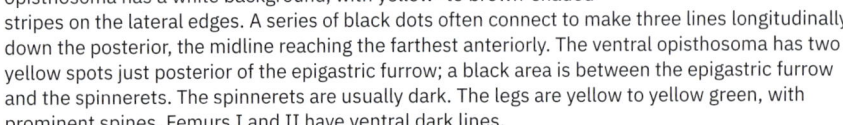

Description The prosoma is pale yellow to yellow green, with a dark median stripe and often darker shading in the submarginal area. The opisthosoma has a white background, with yellow- to brown-shaded stripes on the lateral edges. A series of black dots often connect to make three lines longitudinally down the posterior, the midline reaching the farthest anteriorly. The ventral opisthosoma has two yellow spots just posterior of the epigastric furrow; a black area is between the epigastric furrow and the spinnerets. The spinnerets are usually dark. The legs are yellow to yellow green, with prominent spines. Femurs I and II have ventral dark lines.

Habitat These spiders often build their webs in fields, along roadsides, and in other edge-type habitats. When disturbed, these spiders will drop from their web and darken in color. The webs sometimes have a messy circular stabilimentum at the hub.

Known records from ON, QC (Canada); AL, AR, CO, CT, FL, GA, IA, IL, IN, KS, KY, LA, MA, MD, ME, MI, MN, MO, MS, ND, NE, NH, NY, OH, OK, PA, RI, SC, TN, TX, VA, VT, WI, WV (United States)

female

female

female

egg sac

Mangora maculata (Keyserling, 1865)

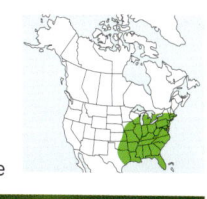

Common Name Greenlegged orbweaver

Total Body Length Female 3.6–5.5 mm, Male 2.7–4.0 mm

Description The prosoma is pale green, with a slightly darker green mark near the fovea and in the submarginal bands. The opisthosoma has a white background. A black spot is at the anterior end. At the posterior end are two to three rows of black spots. The legs are pale green, with yellow banding and prominent spines. There are no ventral lines on femurs I and II.

female

Habitat These spiders often build their horizontal orb webs in understory vegetation of deciduous forests. When disturbed, these spiders will drop from their web and darken in color. The webs sometimes have a messy circular stabilimentum at the hub.

Known records from ON (Canada); AL, AR, CT, FL, GA, IA, IL, IN, KS, KY, LA, MA, MD, MI, MO, MS, NC, NE, NJ, NY, OH, OK, PA, RI, SC, TN, TX, VA, WV (United States)

female

female

female ventral

juvenile

male ventral

male

Mangora placida (Hentz, 1847)

Common Name Tuftlegged orbweaver

Total Body Length Female 2.3–4.5 mm, Male 2.0–2.8 mm

Description The prosoma is tan to light brown, with a dark brown median stripe that is widest in the eye region. There are dark edges to the prosoma. The opisthosoma has a white background and a golden-brown midline stripe that starts narrow and widens about halfway down the opisthosoma. There are four pairs of black dots at the edge of the brown stripe after it widens. A pair of white dots are within the brown stripe near the second pair of black spots. The legs are tan to brown and faintly banded.

Habitat These spiders are often found in the understory of forested habitats and grassy fields.

female

female

female ventral

female

Known records from MB, NB, NS, ON, PE, QC (Canada); AL, AR, CO, CT, FL, GA, IA, IL, IN, KS, LA, MA, MD, ME, MI, MN, MO, MS, NC, ND, NE, NH, NJ, NY, OH, OK, PA, RI, SC, TX, VA, VT, WI, WV (United States)

Mastophora hutchinsoni Gertsch, 1955

Common Name Cornfield bolas spider

Total Body Length Female 6.2–10.4 mm, Male 1.7 mm

Description The prosoma is dark brown, with a crown of pale antler-like protrusions. The opisthosoma is white and mottled with browns and black. A pair of humps near the anterior end are usually brown. The legs

female

egg sac

female
ventral

female

female

are covered with hairs and banded in yellow and brown. When at rest, the spider resembles bird scat. The males are dull red, with a triangular opisthosoma.

Habitat These spiders can be found in gardens, orchards, and other areas with trees. They are often found on hackberry and wild cherry trees. These spiders specialize in capturing moths. Rather than building the traditional orb web, they build a scaffold of a few lines. The female then constructs a bolas, a sticky silk ball, that is suspended from one of her front legs. These spiders mimic the sex pheromones of their target prey, and when a moth approaches, the bolas is swung around to secure the prey. Once trapped, the moth is pulled close enough for the spider to grasp and bite it. The males do not use this bolas technique but do utilize pheromone mimicry to attract psychodid flies. The egg sacs are shaped like a vase and are attached to tree branches.

Known records from QC (Canada); CT, DC, IL, IN, KY, MA, MD, MI, MN, NH, NJ, NY, OH, RI, SC, TN, VA (United States)

Mastophora phrynosoma Gertsch, 1955

Common Name Toadlike bolas spider

Total Body Length Female 8.3–12.3 mm, Male 1.5–1.7 mm

Description The prosoma is orange to orange brown, with a crown of bifurcate antler-like protrusions. The opisthosoma of the female is mottled with brown above, with a band of white on the lateral edges. The opisthosoma is widest at the anterior end, where a pair of humps can be clearly seen. The legs have a fuzzy appearance. When at rest, the spider resembles bird scat. The males are tiny and deep red. Their shape is similar to that of the females, just on a micro scale.

Habitat These spiders are often found on shrubs or low tree branches. They specialize in capturing moths. Rather than building the traditional orb web, they build a scaffold of a few lines. The female then constructs a bolas, a sticky silk ball, that is suspended from one of her front legs.

female

female

female

female
ventral

female with egg sac

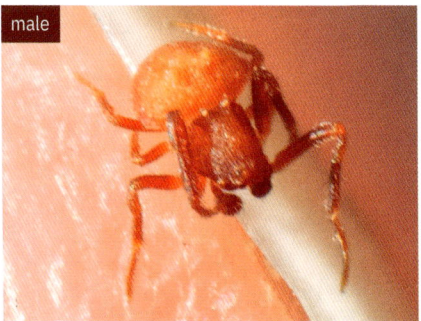

male

These spiders mimic the sex pheromones of their target prey, and when a moth approaches, the bolas is swung around to secure the prey. Once trapped, the moth is pulled close enough for the spider to grasp and bite it. The males do not use this bolas technique but do utilize pheromone mimicry to attract psychodid flies. The egg sacs are uniquely shaped, almost like a flower on a stem, and are usually tucked into a folded leaf.

Known records from AL, CT, FL, IL, IN, KY, MD, MO, NY, OH, TX, VA

Mecynogea lemniscata (Walckenaer, 1841)

Common Name Basilica orbweaver

Total Body Length Female 6.0–9.0 mm, Male 5.0–6.9 mm

Description The prosoma is yellow to orange, with a median black stripe and shaded edges. The opisthosoma is oblong, with anterior humps. The edges are brilliant green; the central area is white with a pattern of goldens, reds, and browns. The ventral opisthosoma is black, with bright yellow borders. The legs are green, and the femurs have a dorsal longitudinal stripe. The spiders look similar to *Leucauge* and *Mangora* species, but the web structure and opisthosoma markings are unique.

Habitat These spiders tend to build their webs in low woody vegetation, usually in or near forested habitats. They can also be found in urban habitats. The web resembles the dome of

female

female

female
ventral

173

male

web

male

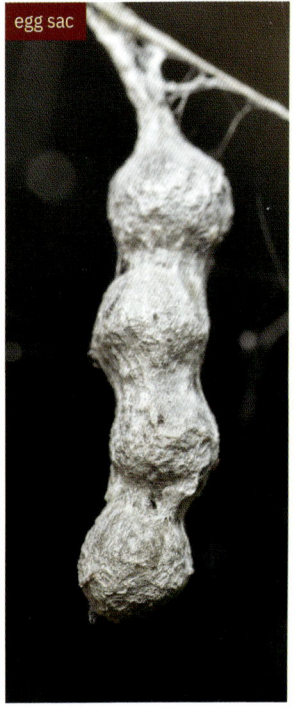

egg sac

Neriene radiata, but on closer inspection, the fine mesh structure would rule out a sheetweb weaver. The common name comes from the fact that the web resembles the apse of cathedrals. The spider is found inverted at the top of the dome. The egg sacs of these spiders are created in a string within the top of the dome and resemble a string of pearls.

Known records from AL, FL, GA, LA, MD, MO, NC, OK, SC, TN, TX, UT, VA, WV

Metazygia wittfeldae (McCook, 1894)

Common Name This species does not have a common name.

Total Body Length Female 6.0–10.2 mm, Male 5.0–7.0 mm

Description The prosoma is pale orange, with a median stripe that starts at the fovea and widens to the eye region, making the entire anterior portion of the prosoma look dark. The opisthosoma is cream to light gray, with a foliate marking in brown or dark gray. The heartmark is brown to dark gray.

female

female with prey

male

The legs are orange to brown and unbanded. The stripe on the prosoma is wider than what is usually seen in other species of *Metazygia*.

Habitat These spiders are often found under the eaves of houses, under bridges, and in citrus tree groves.

Known records from AL, FL, GA, LA, MS, NC, TX, VA

Metepeira labyrinthea
(Hentz, 1847)

Common Name Labyrinth orbweaver

Total Body Length Female 4.0–8.6 mm, Male 3.0–6.8 mm

Description The prosoma is dark brown and covered with thick white hairs. The opisthosoma is tan to red. Down the dorsal surface, a foliate pattern that is outlined in white to cream creates one or two bright white, lobed T shapes or fleur-de-lis shapes. The ventral opisthosoma has a bright yellow midline surrounded by black. Yellow spots are near the spinnerets. The sternum also has a yellow midline. The legs are tan to brown and banded.

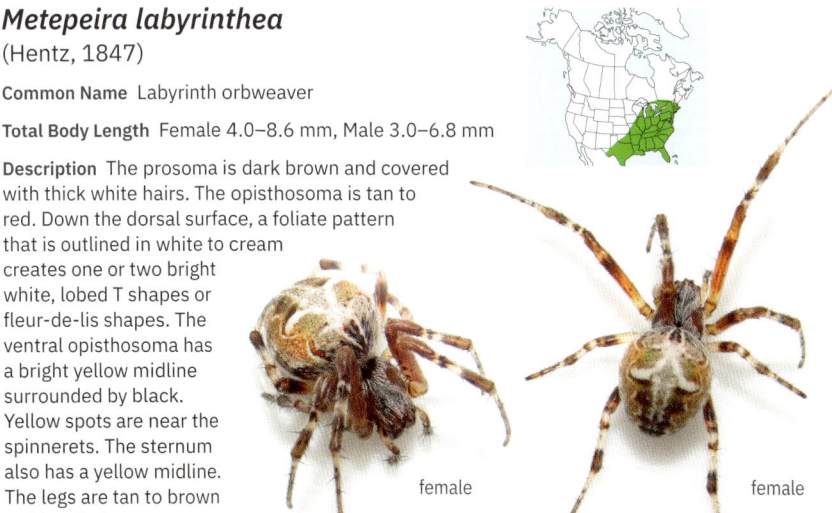

female

female

female with egg sacs

female & male

web

Habitat These spiders tend to be found in forested habitats, where they build their webs in shrubs or low tree branches. One of the most distinctive features of this species is the web. The orb web is located below a tangle web that resembles the constructions of space web weavers. The spider hides in a debris-covered retreat in the tangle web and has a line that leads to the hub of the orb web. Its disc-shaped egg sacs are often stacked on top of the retreat.

male

Known records from ON (Canada); AL, AR, CO, CT, FL, GA, IL, IN, KS, KY, LA, MA, MI, MO, MS, NC, NJ, NY, OH, OK, PA, SC, TN, TX, VA, WI, WV (United States)

Micrathena gracilis (Walckenaer, 1805)

Common Name Spined micrathena

Other Colloquial Names Castleback orbweaver

Total Body Length Female 7.0–11.0 mm, Male 4.2–5.1 mm

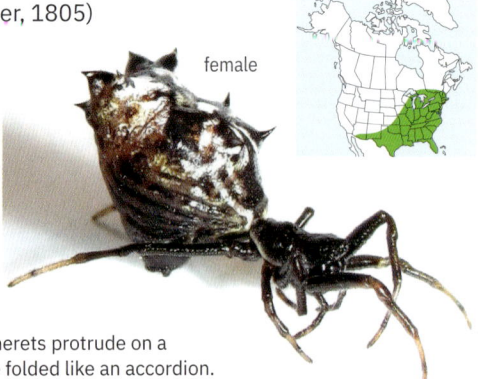
female

Description The prosoma is brown to black, with pale edges. The opisthosoma of the female is white to tan (rarely yellow), with ten stout spines that are brown to black (rarely red). The ventral opisthosoma is usually dark, and the spinnerets protrude on a tubercle. The sides often look like they are folded like an accordion.

female
ventral

female

female in web

female with prey

female & male

female & male mating

male

juvenile

177

The sternum has a central yellow to orange marking. The legs are brown to black and uniformly colored. The males are much smaller than the females and lack the spines. The opisthosoma of the males is widest about halfway down and pale. The legs of the male sometimes have banding.

Habitat These spiders like to build their webs between trees in forested habitats; the webs often seem to be at about face level across hiking trails. There is usually an open area at the hub, and the spiral is finely spaced.

Known records from AL, AR, AZ, CT, DE, FL, GA, IA, IL, IN, KS, KY, LA, MA, MD, MI, MO, MS, NC, NE, NJ, NY, OH, OK, PA, RI, SC, TN, TX, VA, WI, WV

egg sac

Micrathena mitrata (Hentz, 1850)

Common Name White micrathena

Total Body Length Female 4.7–6.0 mm, Male 3.0–3.7 mm

Description The prosoma is brown to black, with pale edges. The opisthosoma of the female is white, with four points at the posterior end. Usually, an assortment of dark spots and markings is on the dorsal surface. I have heard people comment that the points at the posterior end and the dark markings make the opisthosoma look like a cat's face. The male opisthosoma lacks the points, or they are much less prominent; it can be black and white to red and brown.

Habitat The web is often constructed in the understory of woodlands.

Known records from AL, AR, CT, FL, GA, IA, IL, IN, KS, KY, MD, ME, MI, MO, MS, NC, NH, NJ, NY, OH, OK, PA, RI, SC, TN, TX, VA, WI, WV

female

female

female ventral

male

male

male

Micrathena sagittata (Walckenaer, 1841)

Common Name Arrowshaped micrathena

Total Body Length Female 5.4–8.6 mm, Male 4.2–5.9 mm

Description The prosoma is red to brown, with yellow edges. The opisthosoma in the female is bright yellow, with two large points at the posterior end that darken at the tips. Four smaller points are on the lateral edges of the opisthosoma. The opisthosoma is widest at the posterior end and tapers in a triangular shape, which gives it the common name "arrowshaped." The ventral opisthosoma is dark to black, with yellow spots. The legs and the prosoma are similar in color. The males are overall dark and sometimes have

female ventral

female

female

female

juvenile

juvenile

male

male

paired yellow dots at the posterior end. The opisthosoma is widest at the posterior end. The first pair of legs tends to be darker than the others.

Habitat The web is often built in humid forested areas.

Known records from ON (Canada); AL, AR, CT, FL, GA, IA, IL, IN, KS, KY, LA, MA, MD, MI, MN, MO, MS, NC, NE, NH, NJ, NY, OH, OK, PA, RI, SC, TN, TX, VA, WI, WV (United States)

Neoscona arabesca
(Walckenaer, 1841)

Common Name
Arabesque orbweaver

Total Body Length
Female 5.2–7.7 mm,
Male 4.2–5.9 mm

Description The prosoma is yellow to brown. When it is paler, you can make out a median stripe and darker edges. It is often covered with hair. The opisthosoma is highly variable in color: some are pale yellows and tans, while others are rich browns. Paired dark brown to black dashes are on the posterior half of the opisthosoma. The ventral opisthosoma is black, with horizontal yellow dashes at the epigastric furrow and paired longitudinal yellow dashes toward the spinnerets.

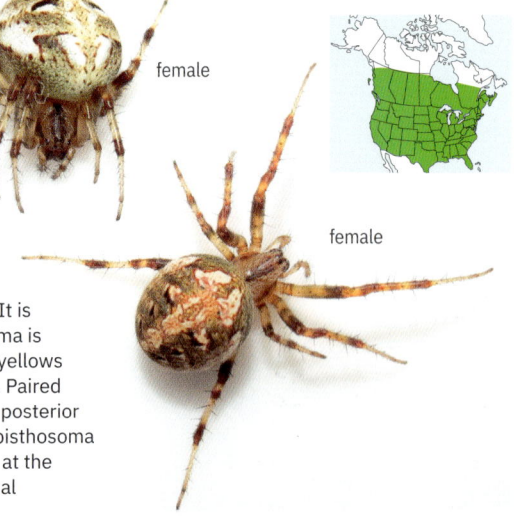

female

female

female

female

Habitat The webs are built in low vegetation in a variety of habitats.

Known records from AB, BC, MB, NB, NL, NS, ON, PE, QC, SK (Canada); AL, AR, AZ, CA, CO, CT, DC, FL, GA, IA, ID, IL, IN, KS, LA, MA, MD, ME, MI, MN, MO, MS, MT, NC, ND, NH, NJ, NM, NY, OH, OK, OR, PA, RI, SC, SD, TN, TX, UT, VA, VT, WA, WI, WV, WY (United States)

female with egg sac

female ventral

male

male

Neoscona crucifera (Lucas, 1838)

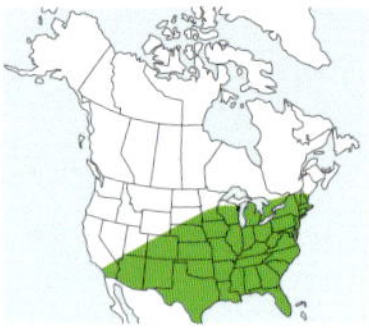

Common Name The genus *Neoscona* has the common name "spotted orbweavers."

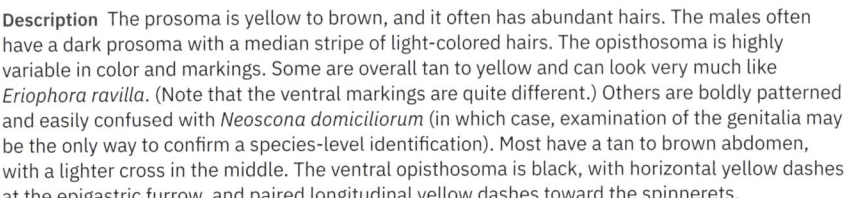

Other Colloquial Names Cross orbweaver, Hentz orbweaver

Total Body Length Female 8.5–19.7 mm, Male 4.5–15.0 mm

Description The prosoma is yellow to brown, and it often has abundant hairs. The males often have a dark prosoma with a median stripe of light-colored hairs. The opisthosoma is highly variable in color and markings. Some are overall tan to yellow and can look very much like *Eriophora ravilla*. (Note that the ventral markings are quite different.) Others are boldly patterned and easily confused with *Neoscona domiciliorum* (in which case, examination of the genitalia may be the only way to confirm a species-level identification). Most have a tan to brown abdomen, with a lighter cross in the middle. The ventral opisthosoma is black, with horizontal yellow dashes at the epigastric furrow, and paired longitudinal yellow dashes toward the spinnerets.

Habitat These spiders often suspend their webs in low vegetation or on human structures. The webs can be quite large. The spiders often build a new web each evening, removing that web and consuming the silk in the morning.

Known records from ON (Canada); AL, AR, AZ, CA, CO, CT, FL, GA, IA, IL, IN, KS, KY, LA, MA, MD, MI, MN, MO, MS, NC, NE, NJ, NM, NY, OH, OK, PA, SC, SD, TN, TX, VA, WI, WV, WY (United States)

female

female

female

female

female in retreat

male

male

female ventral

Neoscona domiciliorum (Hentz, 1847)

Common Name The genus *Neoscona* has the common name "spotted orbweavers."

Other Colloquial Names Red-femured spotted orbweaver

Total Body Length Female 7.2–16.2 mm, Male 8.0–9.0 mm

Description The prosoma is orange to brown and often covered with white hairs. The opisthosoma is boldly marked with a pale cross shape surrounded by dark brown, and it is tan to cream on the lateral edges. The ventral opisthosoma is black, with horizontal yellow dashes at the epigastric furrow and paired longitudinal yellow dashes toward the spinnerets. The legs have red to orange

female

female

femurs, and the distal segments are banded. These spiders can be difficult to distinguish from other *Neoscona* species.

Habitat These spiders tend to build their webs in shrubby habitats and near human structures.

Known records from AL, CT, FL, GA, IL, IN, KS, KY, LA, MA, MD, MO, MS, NC, NJ, NY, OH, OK, PA, RI, TN, TX, VA, WV

Neoscona oaxacensis (Keyserling, 1864)

Common Name Western spotted orbweaver

Total Body Length Female 8.9–18.0 mm, Male 6.3–12.7 mm

Description The prosoma is tan to orange, with a dark median line and slightly darker edges. The opisthosoma is usually dark (on some specimens, it is almost black), with a central cream or yellow-colored band with more

female

female ventral

juvenile

juvenile

juvenile

than five lobes down the middle. The markings are very similar to *Aculepeira packardi*, but the ventral markings are quite different, and *A. packardi* usually has only five lobes in the central dorsal band. The ventral opisthosoma is black, with horizontal yellow dashes at the epigastric furrow and paired longitudinal yellow dashes toward the spinnerets. The legs are banded.

Habitat These spiders build their webs in fields and shrubs and near human structures.

Known records from AK, AZ, CA, CO, IN, KS, NM, NV, OK, RI, TX, UT, WA

Ocrepeira ectypa (Walckenaer, 1841)

Common Name Asterisk spider

Total Body Length Female 5.2–9.4 mm, Male 5.4–7.2 mm

Description The prosoma is tan to brown and covered with hairs. The posterior median eyes are elevated on tubercles; this seems to be more pronounced in immature specimens. The opisthosoma is colored with tans and browns and has a highly elevated anterior pair of humps that sit high above the prosoma. The legs are banded and hairy. When resting, this spider looks very much like a branch bud and can be very difficult to find.

female

female

female

female

female

female ventral

web

Habitat These spiders are often found in wooded habitats but can occasionally be found on human structures. The web differs from a standard orb web, as it lacks the sticky spiral; it is just the radial lines and a hub.

Known records from AR, CT, FL, GA, MA, MD, MI, MO, NC, NJ, OH, RI, TX, VA

Scoloderus nigriceps (O. Pickard-Cambridge, 1895)

Common Name This species does not have a common name.

Other Colloquial Names Ladder spider

Total Body Length Female 2.9–4.3 mm, Male 1.9–2.6 mm

Description The prosoma is red to brown and highly elevated in the middle. The opisthosoma is colored with tans to browns and has a highly elevated anterior pair of humps that sit high above the prosoma. Overall, these spiders look very similar to *Ocrepeira* species, but the shape of the prosoma is quite different.

Habitat These spiders tend to build their webs in forested habitats and can be found year-round. This species is easiest to detect by its unique orb web. The round orb portion is at the bottom. Above it is a lattice or ladder structure that is up to 1 m in height. When a moth strikes the upper portion of the web, it tumbles down, losing scales to the sticky webbing as it falls. It is thought that by the time the moth reaches the circular orb web, it will have lost enough scales to become trapped.

Known records from FL, GA, TX

web

female

female

female

female ventral

Trichonephila clavata (L. Koch, 1878)

Common Name This species does not have a common name.

Other Colloquial Names Joro spider

Total Body Length Female 17.0–30.0 mm, Male 4.0–8.0 mm

Description The prosoma is dark but covered with a thick coating of silver to white hairs. The opisthosoma is blue gray, with broad horizontal yellow bands. The ventral opisthosoma is black and yellow with a large red spot encompassing the spinnerets. The legs are dark, with occasional yellow bands.

Habitat This species seems to be associated with human structures. The massive orb webs are spun using a rich golden-colored silk.

Known records from GA
This is an introduced species and a recent introduction, as the first confirmed record was in 2014. This species seems to be spreading from the introduced area, but there is no scientific evidence at this time that the spider is invasive. The species may be filling an unused niche in the environment.

female

Trichonephila clavipes (Linnaeus, 1767)

Common Name Golden silk orbweaver

Other Colloquial Names Banana spider, golden orbweaver, golden silk spider, calico spider

Total Body Length Female 19.0–34.0 mm, Male 5.0–8.0 mm

Description The prosoma is dark but covered with a thick coating of silver to white hairs. The opisthosoma is elongated and yellow to orange, and paired white spots run down the median line. The legs are pale proximally and darker distally. Legs I, II, and IV have tufts of stiff hairs at the

male & female

male & female

juvenile female

female ventral

distal ends of the femurs and tibiae. The males are tiny in comparison to the females and lack the tufts of hairs on the legs. Often, multiple males will inhabit the web with the female.

Habitat These spiders are often found in moist forested habitats. The webs can be massive and are spun using a silk that is a rich golden color.

Known records from AL, AZ, FL, GA, LA, MS, NC, OH, SC, TX

Verrucosa arenata (Walckenaer, 1841)

Common Name Triangulate orbweaver

Other Colloquial Names Arrowhead spider, triangle orbweaver, arrowhead orbweaver

Total Body Length Female 5.0–9.5 mm, Male 4.0–6.1 mm

Description The prosoma is red to very dark brown. The female opisthosoma is triangular. The lateral edges are similar in color to the prosoma, but the area in the middle is a bold triangular marking in white, yellow, or pink. The posterior edge of the opisthosoma has six to eight tubercles. The ventral opisthosoma is very dark, and there are two pale dashes on either side, but they are much more inconspicuous than the light markings on other species. The female epigynum scape is long (almost as long as the

female

female

female

female ventral

male
subadult

male

opisthosoma) and points posteriorly. The males have a similarly shaped opisthosoma, but the marking is less dramatic and the tubercles are more pronounced. The most dramatic difference between them is seen on leg II, where the male has a large clasping spine on the tibia.

male

Habitat These spiders are often found in wooded habitats, but they can also be found in gardens. They like to build their webs in open areas. Unlike most spiders that sit in their webs with their head facing downward, this species sits in the web with its head facing up.

Known records from ON (Canada); AL, AR, CO, FL, GA, IA, IL, IN, KS, KY, LA, MD, MO, MS, NC, NJ, NY, OH, OK, PA, SC, TN, TX, VA, WV (United States)

Zygiella x-notata (Clerck, 1757)

Common Name Open sector orbweaver

Other Colloquial Names Silver-sided sector spider, missing sector orb weaver

Total Body Length Female 7.4–8.7 mm, Male 6.0–6.5 mm

Description The prosoma is orange to brown, with a dark median stripe that widens to cover the eye region and a thin dark line around the very edge. The sternum is dark at the edges and often very pale in the middle. The opisthosoma is pale and has a brown to tan foliate marking that is darkest at its lateral edges. These markings look similar to those on *Larinioides cornutus*, but the ventral opisthosoma markings are very different. The ventral opisthosoma has a broad black median stripe that is bordered by tan to yellow bands. The legs are banded.

Habitat The orb web of this species often lacks the sticky spiral silk in one or two sections, so it looks like a section of the web has been removed. They often build their webs on human structures.

Known records from BC (Canada); CA, MA, ME, NY, OR, RI, VA, WA (United States)
This is an introduced species.

female

female ventral

SYMPHYTOGNATHIDAE ▶ DWARF ORBWEAVERS

Genera in This Family *Anapistula* (1)

Total Body Length Approximately 0.5 mm

Description These tiny (some of the smallest known spiders) entelegyne ecribellate spiders have only four eyes in two pairs. They have a rounded opisthosoma and tend to be yellowish. The females lack pedipalps. They lack book lungs. They tend to build a small flat orb web in humid habitats usually less than 2 m from the ground.

Similar Families Anapidae, Mysmenidae, Theridiosomatidae

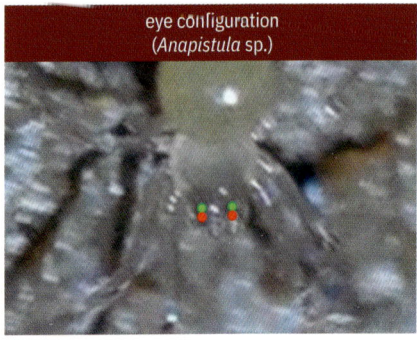
eye configuration
(*Anapistula* sp.)

How to distinguish Symphytognathidae from other similar families:

Symphytognathids have only four eyes arranged in two pairs, distinguishing them from other similar families, which have six to eight eyes. Anapid females lack pedipalps but have six to eight eyes, and the males have a scutum on the opisthosoma that is not found in symphytognathids. Mysmenids build a tangle web, in contrast to the orb web of symphytognathids. Mysmenids also have eight eyes and, usually, a high prosoma with the eyes on a turret, which is distinctly different from the prosoma of symphytognathids. Whereas theridiosomatids pull their orb web into a cone, symphytognathid webs are flat. Also, theridiosomatids have eight eyes and sternal pits, which are absent in symphytognathids.

Anapistula secreta Gertsch, 1941

Common Name This species does not have a common name.

Total Body Length Approximately 0.5 mm

Description The prosoma is pale yellow, and the four eyes are arranged in two pairs on either side. The opisthosoma is pale yellow and much higher than it is long. This species lacks book lungs. The females lack pedipalps. The legs are yellow. Males are extremely rare, and it is hypothesized that this species may be parthenogenetic. The specimen shown here is likely an undescribed species and was found in Texas but looks similar to *A. secreta*.

Habitat These spiders tend to build their webs in extremely humid microhabitats, such as leaf litter, caves, humus, sphagnum moss, and rotting logs. The small orb webs measure only a couple centimeters in diameter and include out-of-plane radial threads.

Known records from FL

female   female

TETRAGNATHIDAE ➤ LONGJAWED ORBWEAVERS

Genera in This Family *Azilia* (1), *Dolichognatha* (1), *Glenognatha* (5), *Leucauge* (3), *Meta* (2), *Metellina* (3), *Metleucauge* (1), *Pachygnatha* (8), and *Tetragnatha* (18)

Total Body Length 1.2–25.0 mm

Description These spiders are entelegyne (or secondarily haplogyne) and ecribellate. They usually have long narrow endites. They have eight eyes; the median eyes form a trapezoid,

eye configuration
(*Tetragnatha laboriosa*)

Tetragnatha elongata

long endites

horizontal or near horizontal web

and the lateral eyes are some distance away. They have a pair of book lungs and a single tracheal spiracle that is usually located near the spinnerets. Their first pair of legs is the longest, and their third pair of legs is the shortest. They usually build an orb web that is oriented horizontally, at times near or over water.

Similar Families Araneidae, Deinopidae, Linyphiidae, Pimoidae, Theridiidae

How to distinguish Tetragnathidae from other similar families:

Both araneids and tetragnathids build an orb web, but they can be distinguished by their endites: araneids have squarish endites, while the endites of tetragnathids are longer than they are wide. Deinopids have extremely large posterior median eyes and are cribellate, whereas tetragnathids have eyes of fairly similar size and are ecribellate. In linyphiids and pimoids, autospasy occurs at the patella-tibia joint, and they build a sheet web, which distinguishes them from tetragnathids. Theridiids build a tangle web, whereas most tetragnathids build an orb web or have adopted a ground-hunting lifestyle.

Dolichognatha pentagona (Hentz, 1850)

Common Name This species does not have a common name.

Total Body Length Female 2.6–4.0 mm, Male 2.6–4.3 mm

Description The prosoma is tan to light brown, with a dark edge and dark markings in the eye region. The opisthosoma is mottled with brown, black, and white and has four humps on the posterior end. The legs are banded in tan and brown.

female

female

Habitat The spiders can be found in forested habitats, where they tend to build their webs between tree roots. The 6–12 cm diameter webs are close to the ground and horizontal (or nearly so), and they often have a line of debris associated with them.

Known records from AL, FL, GA, IL, MS, NC, OH, TN

female

female ventral

Glenognatha foxi (McCook, 1894)

Common Name This species does not have a common name.

Other Colloquial Names Dwarf ground long-jawed spider

Total Body Length Female 1.6–2.6 mm, Male 1.4–2.2 mm

Description The prosoma is orange to red, with a black mask around the eye region. The opisthosoma is orange to red, with faint dark spots or white markings on the lateral edges. The legs are red to orange. The male pedipalps are black and quite large in relationship to the size of the spider.

Habitat These spiders' orb webs are often built within inches of the ground and can be found in lawns. The spiders also like to inhabit other damp habitats.

Known records from ON (Canada); AL, AR, AZ, CA, CT, DC, FL, GA, IL, IN, KS, KY, LA, MA, MD, MO, MS, NC, NM, NY, OH, OK, PA, SC, TN, TX, VA, WI, WV (United States)

female

female ventral

female

female in web

male

male

Leucauge argyra (Walckenaer, 1841)

Common Name This species does not have a common name.

Total Body Length Female 4.5–10.0 mm, Male 4.1–6.6 mm

Description The prosoma is tan to orange, with a dark median tuning-fork mark. The edges are dark. The opisthosoma is silver white, with a series of black lines and yellow spots. The dorsal side has three black lines: the center line runs the entire length of the opisthosoma. The two lateral lines run parallel and, about halfway down the opisthosoma, turn toward the center of the opisthosoma and then continue down the length of the opisthosoma. Faint yellow patches may be within the lines. The ventral opisthosoma has a pale thin central line surrounded in black. Two white to green lines border the black area. Yellow spots are near the spinnerets. The legs are green. A cluster of trichobothria (fine hairs) is on femur IV. This species looks very similar to the other *Leucauge* species in the area. The jagged pattern of the dorsal black lines and the ventral markings help distinguish the species.

female

female

Habitat These spiders build their horizontal (or nearly horizontal) webs in mangroves, forested habitats, and, sometimes, in sugarcane fields.

Known records from FL, TX

Leucauge argyrobapta (White, 1841)

Common Name This species does not have a common name.

Other Colloquial Names Mabel orchard orbweaver

Total Body Length Female 3.7–8.0 mm, Male 3.2–5.1 mm

Description The prosoma is light brown, with a dark median stripe. The opisthosoma is silver white and has a pattern of dark lines and yellow to orange spots. The dorsal opisthosoma has a dark central line and two lateral lines that run parallel the entire length. Some smaller diagonal lines connect the lateral lines to the central lines. There is often yellow shading within this area and, sometimes, yellow or orange spots near the posterior. The ventral opisthosoma is dark, with a copper-orange crescent-shaped marking (which in some cases is yellow). This marking leads many

female

people to the false conclusion that they are looking at an immature *Latrodectus* widow species. Yellow dots are near the spinnerets. The legs are green. A cluster of trichobothria (fine hairs) is on femur IV. These spiders often build their horizontal webs in humid environments.

L. argyrobapta is very similar to the other *Leucauge* species. The opisthosoma markings can help distinguish it from *Leucauge argyra*, but without looking at genitalia, it is morphologically very similar to *Leucauge venusta*, in which case, a fairly good assumption can be made based on location. If the specimen was seen in Florida or along the Gulf coast, it is likely *L. argyrobapta*. If it was seen anywhere north of Florida on the East Coast or in the Midwest, it is likely *L. venusta*.

Habitat These spiders are often found in shrubby meadows, forest edges, citrus groves, and suburban areas (including in landscaping plants).

Known records from FL, LA, TX

Leucauge venusta (Walckenaer, 1841)

Common Name Orchard orbweaver

Total Body Length Female 3.7–8.0 mm, Male 3.2–5.1 mm

Description The prosoma is light brown with a dark median stripe. The opisthosoma is silver white and has a pattern of dark lines and yellow to orange spots. The dorsal opisthosoma has a dark central line and two lateral lines that run parallel the entire length. Some smaller diagonal lines connect the lateral lines to the central lines. There is often yellow shading within this area and, sometimes, yellow or orange spots near the posterior. The ventral opisthosoma is dark, with a copper-orange crescent-shaped marking (which in some cases is yellow). This marking leads many

female

people to the false conclusion that they are looking at an immature *Latrodectus* widow species. Yellow dots are near the spinnerets. The legs are green. A cluster of trichobothria (fine hairs) is on femur IV. The spiders often build their horizontal webs in humid environments. This species is very similar to the other *Leucauge* species. The opisthosoma markings can help distinguish it from *Leucauge argyra*, but without looking at genitalia, it is morphologically identical to *Leucauge argyrobapta*, in which case, a fairly good assumption can be made based on location. If the specimen was seen in Florida or along the Gulf coast, it is likely *L. argyrobapta*. If it was seen anywhere north of Florida on the East Coast or in the Midwest, it is likely *L. venusta*.

Habitat These spiders often build their nearly horizontal webs in low vegetation in forested areas or in suburban and urban yards.

female

female

male

male

male ventral

juvenile ventral

juvenile

Known records from ON, QC (Canada); AL, AR, AZ, CA, CT, DC, GA, IA, IL, IN, KS, KY, LA, MA, MD, MI, MN, MO, MS, NC, NH, NJ, NY, OH, PA, RI, SC, TN, TX, VA, WI, WV (United States)

Meta ovalis (Gertsch, 1933)

Common Name Cave orbweaver

Other Colloquial Names Eastern cave long-jawed spider

Total Body Length Female 8.0–10.0 mm, Male 8.0–9.5 mm

Description The prosoma is orange to brown, with a dark median stripe and dark shading at the edges. The opisthosoma can vary in color from tan to deep brown; the markings can also be variable. Most spiders have a pale opisthosoma with an anterior fleur-de-lis pattern followed by a series of light and dark patterns that, at the posterior end, look like light-colored dots connected by horizontal lines. The ventral opisthosoma is black with yellow stripes down each side. The legs are banded.

Habitat The spiders will build their web in any orientation that the space allows, and they tend to prefer caves or rocky crevices. Suspended from cave walls or ceilings, the webs can be completely vertical; in other cases, they can be completely horizontal.

female

female

female ventral

197

female

female

Known records from NB, NL, NS, ON, QC (Canada); AR, CT, IA, IL, IN, KY, LA, ME, MN, MO, NC, NH, NY, OH, OK, PA, RI, TN, VA, VT, WI, WV (United States)

in web

in web

female

female ventral

Metellina mimetoides
Chamberlin & Ivie, 1941

Common Name This species does not have a common name.

Total Body Length Female 3.3–6.0 mm, Male 7.8–11.7 mm

Description The prosoma is yellow to white, with a darkly shaded median line that widens in the eye region. The opisthosoma has two humps about one-third of the way down the opisthosoma. They are usually pale white or yellow, with

paired dark markings just anterior of the humps. The ventral opisthosoma has a broad black area bordered by white lines. The first legs are long and usually banded.

Habitat The web often has a larger lower half, comprising thirty to fifty catching spirals, compared to fourteen to twenty on the upper half. Their webs are usually vertical, and they have an opening at the hub. The spiders tend to build their webs in woodpiles and low shrubs, on human structures, or in caves.

Known records from BC (Canada); AK, AZ, CA, CO, MT, NM, NV, OK, OR, TX, UT, WA (United States)

Pachygnatha autumnalis Marx, 1884

Common Name The genus *Pachygnatha* has the common name "thickjawed orbweavers."

Other Colloquial Names Big-eyed thick-jawed spider

Total Body Length Female 5.0–6.0 mm,
Male 4.0–4.5 mm

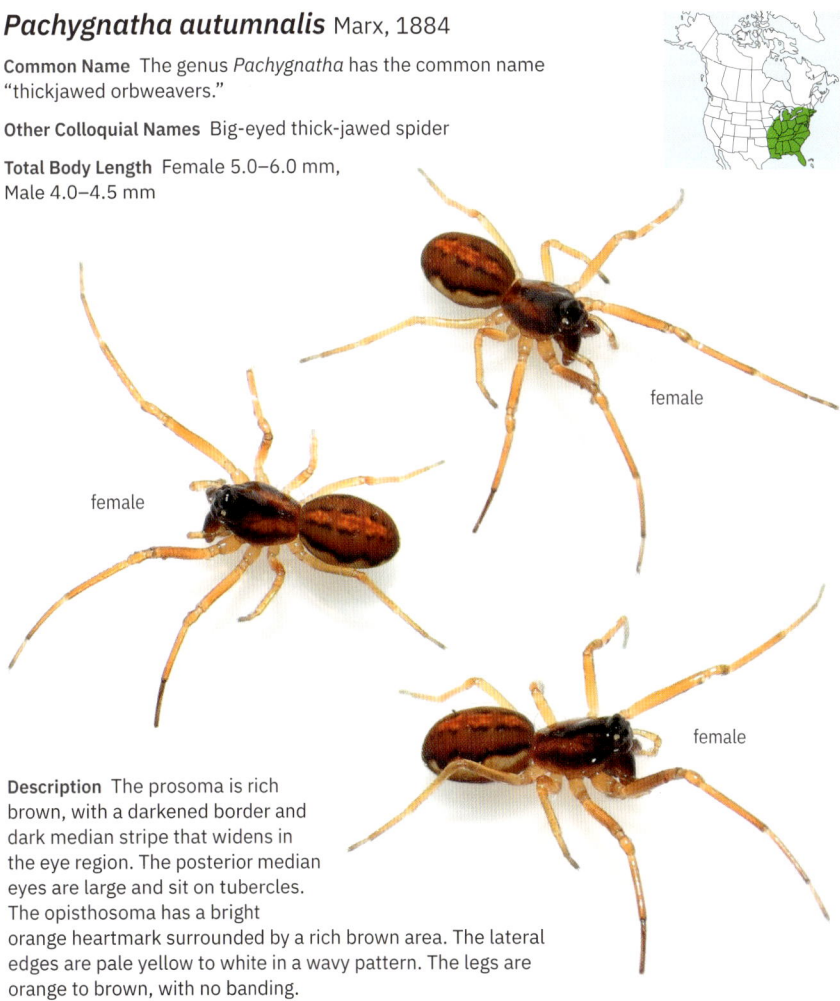

female

female

female

Description The prosoma is rich brown, with a darkened border and dark median stripe that widens in the eye region. The posterior median eyes are large and sit on tubercles. The opisthosoma has a bright orange heartmark surrounded by a rich brown area. The lateral edges are pale yellow to white in a wavy pattern. The legs are orange to brown, with no banding.

Habitat Juvenile spiders build an orb web within low vegetation close to the ground. As adults, they do not build webs, as they have actually lost the ability to do so. Adults hunt on foot, as members of the Ground Active Hunter Guild do. They can be found in wooded habitats and old fields.

Known records from ON, QC (Canada); AR, FL, IL, IN, MA, MD, ME, MI, MO, NC, NJ, NY, OH, PA, TX, VT (United States)

Tetragnatha elongata Walckenaer, 1841

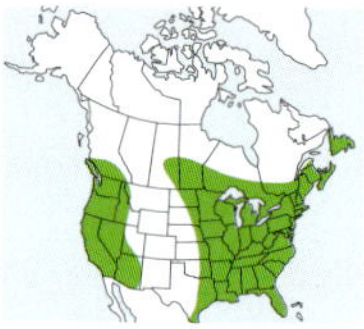

Common Name The genus *Tetragnatha* is known as "longjawed orbweavers."

Other Colloquial Names Elongate stilt spider

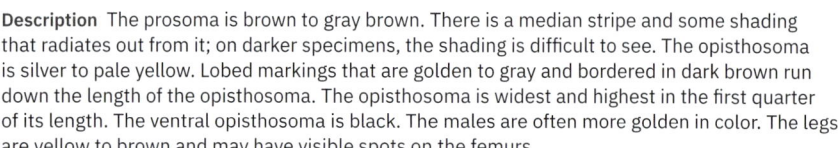

Total Body Length Female 8.2–13.2 mm, Male 7.8–11.7 mm

Description The prosoma is brown to gray brown. There is a median stripe and some shading that radiates out from it; on darker specimens, the shading is difficult to see. The opisthosoma is silver to pale yellow. Lobed markings that are golden to gray and bordered in dark brown run down the length of the opisthosoma. The opisthosoma is widest and highest in the first quarter of its length. The ventral opisthosoma is black. The males are often more golden in color. The legs are yellow to brown and may have visible spots on the femurs.

Habitat This species seems to be associated with water and tends to build its horizontal web near rivers, streams, or lakes.

female

Known records from AB, BC, MB, NL, NS, ON, PE, QC, SK (Canada); AL, AR, AZ, CT, DE, FL, GA, IA, ID, IL, IN, KS, KY, LA, MA, MD, ME, MI, MN, MO, MS, NC, NE, NH, NJ, NY, OK, OR, PA, RI, SC, SD, TN, TX, VA, VT, WA, WI, WV (United States)

female

female

female

male

male

male

male

female with egg sac

juvenile

female ventral

Tetragnatha guatemalensis O. Pickard-Cambridge, 1889

Common Name The genus *Tetragnatha* is known as "longjawed orbweavers."

Total Body Length Female 5.4–11.5 mm, Male 5.2–10.2 mm

Description The prosoma is brown to orange. Sometimes, faint shading can be seen. The opisthosoma is white with a scalloped pattern in tans and grays; sometimes, the males are orange to golden in color. The ventral opisthosoma is pale.

female

female

female with egg sac

female

juvenile

Habitat These spiders tend to build their horizontal webs in tall herbaceous vegetation along streams and lakes.

Known records from NB, NS, ON, PE, QC (Canada); AL, AR, AZ, CA, FL, GA, IL, LA, MA, ME, MO, MS, NC, NJ, NM, NY, OH, OK, SC, TN, TX, VA (United States)

male

male

Tetragnatha laboriosa Hentz, 1850

Common Name Silver longjawed orbweaver

Other Colloquial Names Silver long-jawed orbweaver

Total Body Length Female 5.2–9.0 mm,
Male 3.8–7.4 mm

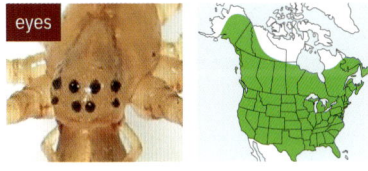

eyes

Description The prosoma is yellow to orange; some shading can be seen along the fovea and radiating outward. The opisthosoma is silver, with a dark band on the ventral side bordered in silver. The legs are yellow to orange.

female

female

female with egg sac

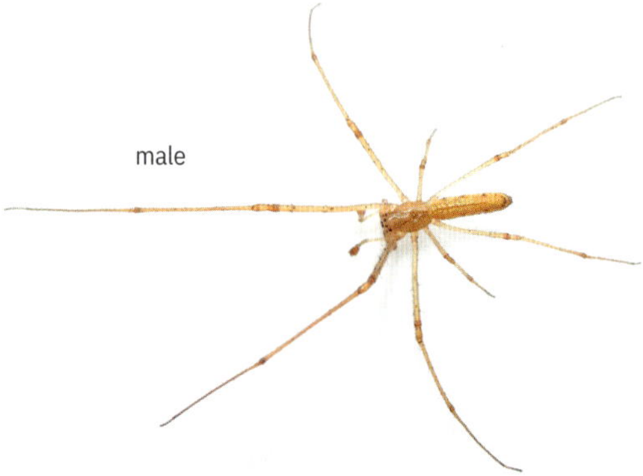

male

male

spiderling

male

female ventrolateral

Habitat This species can be found in many different habitats, including along roadsides and in suburban gardens, agricultural fields, meadows, and marshes. Their web tends to be horizontal in orientation.

Known records from AB, BC, MB, NB, NL, NS, ON, PE, QC, SK, YT (Canada); AK, AL, AR, AZ, CA, CO, CT, DE, FL, GA, IA, ID, IL, IN, KS, KY, LA, MA, MD, ME, MI, MN, MO, MS, MT, NC, ND, NE, NH, NJ, NM, NV, NY, OH, OK, OR, PA, RI, SC, SD, TN, TX, UT, VA, VT, WA, WI, WV, WY (United States)

Tetragnatha straminea Emerton, 1884

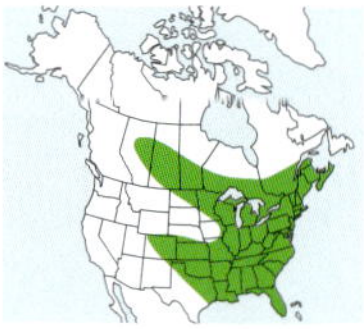

Common Name The genus *Tetragnatha* is known as "longjawed orbweavers."

Total Body Length Female 8.6–10.7 mm, Male 6.5–7.5 mm

Description The prosoma is orange. The opisthosoma is silver to golden in color dorsally; the lateral opisthosoma has a clear line of color change from the light dorsal color to a much darker shade. The ventral opisthosoma

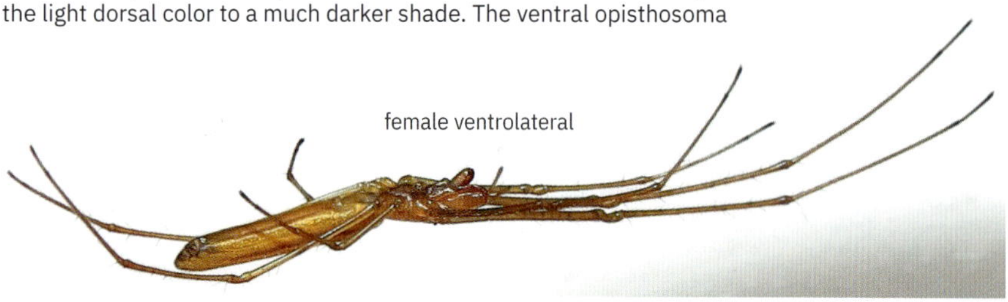

female

female

is black bordered in silver to golden. The stripe of color change visible on the lateral opisthosoma is helpful for identifying this species. The legs are pale orange.

Habitat These spiders tend to build their horizontal webs in moist habitats.

Known records from AB, MB, NB, NL, ON, QC, SK (Canada); AL, CT, FL, IL, IN, LA, MA, MD, ME, MT, NC, NE, NH, NY, OH, PA, SC, TX, UT, VA, VT, WV (United States)

female

male

male

egg sac

Tetragnatha versicolor Walckenaer, 1841

Common Name The genus *Tetragnatha* is known as "longjawed orbweavers."

Other Colloquial Names Versicolor long-jawed orbweaver

Total Body Length Female 5.4–13.3 mm, Male 4.3–9.2 mm

Description The prosoma is yellow to orange; some indistinct radial markings may be visible. The opisthosoma is silvery, with a darker lobed pattern. The ventral opisthosoma has a broad black stripe. The legs are yellow to orange.

Habitat The spiders tend to build their horizontal webs in moist habitats among low tree branches, tall herbaceous vegetation, or shrubs. They can also be found in upland woods and coniferous forest habitats.

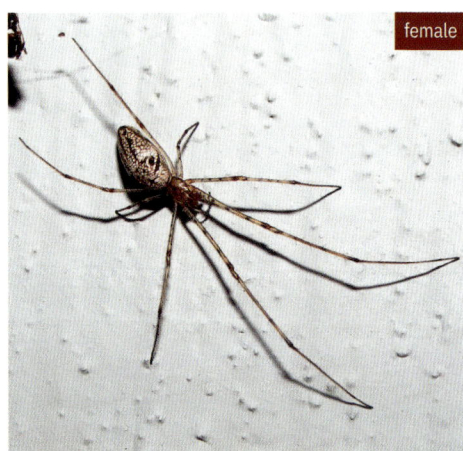

female

male

male

male

male

Known records from AB, BC, LB, MB, NB, NL, NS, NT, NU, ON, QC, SK, YT (Canada); AK, AL, AZ, CA, CO, CT, FL, GA, IA, ID, IL, IN, LA, MA, MD, ME, MI, MN, MO, MS, MT, NC, ND, NE, NH, NJ, NM, NV, NY, OH, OK, OR, PA, RI, SC, SD, TN, TX, UT, VA, VT, WA, WI, WV, WY (United States)

Tetragnatha viridis Walckenaer, 1841

Common Name Green longjawed orbweaver

Total Body Length Female 5.7–7.4 mm, Male 4.4–6.7 mm

Description The prosoma is green. The opisthosoma is green, often with a red marking on the anterior; occasionally, the entire ventral opisthosoma is red. Sometimes, a white midline is on the opisthosoma, and the dorsal surface is bordered in white. The legs are green. The ventral opisthosoma is green or red orange. This species is easy to identify by its unique color.

Habitat These spiders tend to build their horizontal webs in the low branches of coniferous trees, especially pines.

Known records from NB, NS, ON (Canada); AL, FL, GA, LA, MA, ME, NJ, NY, TX (United States)

female

male

male

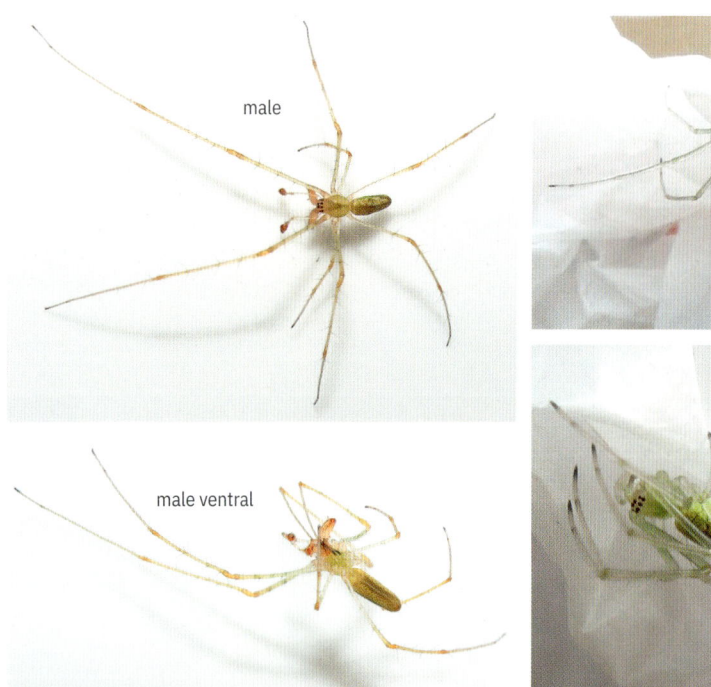

male

juvenile

juvenile

male ventral

THERIDIOSOMATIDAE > RAY ORBWEAVERS

Other Colloquial Names Slingshot spider

Genera in This Family *Parogulnius* (1) and *Theridiosoma* (2)

Total Body Length 1.3–2.4 mm

Description These small entelegyne ecribellate spiders have a globular opisthosoma that is usually dusky, with pale markings. They have pits in the prolateral area of the sternum. They have eight eyes in two rows. They have a pair of book lungs and a tracheal spiracle located near the spinnerets. They build an orb web that is pulled into a cone shape by a silk line from the center of the web. While resting, the spider maintains tension on this line until it detects prey contacting the web. Then the web is released to collapse onto the prey. The egg sacs are papery brown spheres that are suspended from a stiff stalk of silk.

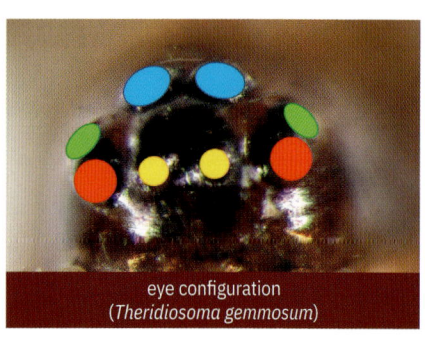

eye configuration
(*Theridiosoma gemmosum*)

Similar Families Anapidae, Araneidae, Linyphiidae, Mysmenidae, Symphytognathidae, Theridiidae

How to distinguish Theridiosomatidae from similar families:

Theridiosomatids build an orb web that is pulled into a cone shape; araneids build a flat (usually vertical) orb web; symphytognathids build a flat (usually horizontal) orb web; linyphiids build a sheet web; and theridiids and mysmenids build a tangle web. The sternal pits seen in theridiosomatids

sternal pits

Theridiosoma gemmosum

orb web pulled into a cone shape

Theridiosoma gemmosum egg sac

are also not found in any similar families. Anapid males have a scutum, and the females lack pedipalps, while theridiosomatids do not have scuta, and the females have pedipalps. They can be distinguished from mysmenids, as they lack the sclerotized spot on femur I that is found in mysmenids. The large male pedipalps seen in theridiosomatids also help to distinguish them from other similar families, as all males in similar families have proportionally smaller pedipalps.

Theridiosoma gemmosum (L. Koch, 1877)

Common Name Ray orbweaver

Other Colloquial Names Common eastern ray spider

Total Body Length Female 2.7 mm, Male 1.6 mm

Description The prosoma is brown, and sometimes a few dark radiating marks are visible. The opisthosoma is very globular and mottled with brown, white, and tan spots. The males' modified pedipalps are very large, considering the size of the spider.

female

female

female ventral

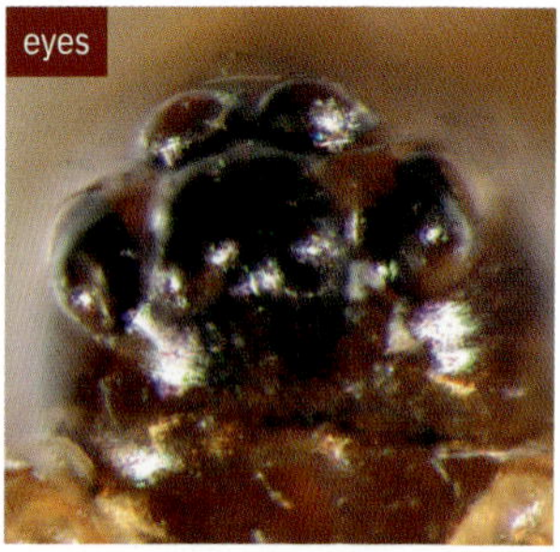

eyes

egg sac

web

Habitat Most often, the spider is not seen, and identification can be made by observing either its unusual web or the distinctive egg sacs. The web is a traditional orb shape oriented vertically that the spider then pulls into a cone shape by extending a line from the hub. The spider holds the line to maintain the shape until prey is detected near the web. Then the spider releases the line, collapsing the web on the prey. The egg sacs are round, approximately 3 mm in diameter, and look like they are covered in tan or orange paper. They are attached to the vegetation by a stiff stalk of silk and hang down from the vegetation 2.5 cm or more. These spiders tend to build their webs in humid habitats.

Known records from NB, NL, NS, ON, QC (Canada); AL, CT, FL, GA, IA, IL, IN, KY, MA, MD, ME, MI, MN, MO, NC, NH, NY, OH, PA, RI, SC, TN, VA, WI, WV (United States)

male

male

ULOBORIDAE ▶ HACKLED ORBWEAVERS

Other Colloquial Names Cribellate orbweaver, venomless orbweaver

Genera in This Family *Hyptiotes* (4), *Miagrammopes* (1), *Octonoba* (1), *Philoponella* (3), *Siratoba* (1), *Uloborus* (4), and *Zosis* (1)

Total Body Length 4.0–8.0 mm

Description These unique-looking spiders are entelegyne. They are cribellate, and the calamistrum is made up of a single row of setae. The femurs of legs II, III, and IV have a row of long trichobothria. The webs are the traditional orb-style, a triangle web (similar to taking a wedge out of an orb web), or a simple web of only a few capture lines. Members of this family of spiders lack venom glands and use silk to subdue their prey.

Similar Families Araneidae

How to distinguish Uloboridae from other similar families:

Uloborids are cribellate, whereas araneids are ecribellate, so the presence of a cribellum and calamistrum distinguishes the uloborids. Uloborids have elongated endites, usually one and a half to two times the width, whereas araneids have endites that are squarish.

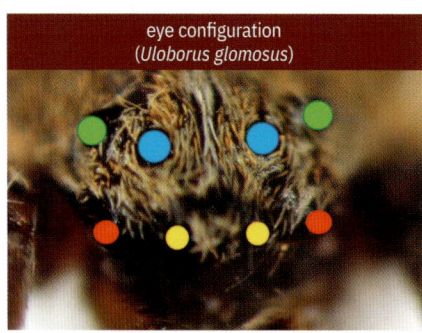

eye configuration
(*Uloborus glomosus*)

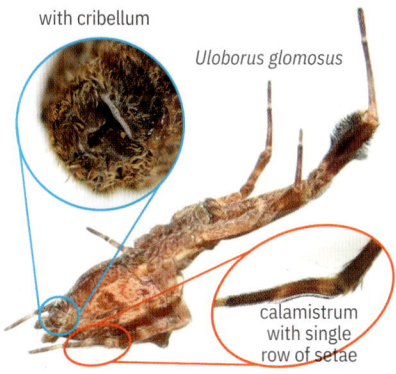

with cribellum

Uloborus glomosus

calamistrum with single row of setae

Hyptiotes cavatus (Hentz, 1847)

Common Name Triangle weaver

Other Colloquial Names Triangle web spider, mouse spider

Total Body Length Female 2.3–3.8 mm, Male 2.0–2.6 mm

Description The prosoma is brown and covered with hairs in shades of tans, browns, black, and orange. It is uniquely trapezoidal, and the posterior

female

female ventral

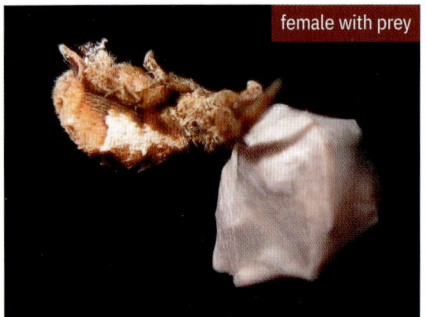

female with prey

female with prey

female with prey

female with web deployed

male

web with spider

lateral eyes project off the sides on tubercles. The anterior lateral eyes are tiny and can easily be overlooked. The opisthosoma is round and has a series of humps that look like tufts of hair. The markings are mottled in shades that provide camouflage on tree branches. The ventral opisthosoma is similar in color. The legs are also covered with hair and in similar shades. The first pair of legs is robust.

Habitat This spider can easily be identified by its web, which looks like a pie piece that has been removed from a traditional orb web. There are four radial lines, with cribellate silk spun between them, similar to the sticky spiral of an orb web. The radials connect to a single line that extends to where the spider sits, holding the web taut. When prey touches the web, the spider releases the tension, allowing the web to collapse over the prey. These spiders are the first nonhumans proven to use an external device (their web) for power amplification (Han et al. 2019).

Known records from ON, QC (Canada); AL, CT, FL, GA, IL, IN, KY, MA, MD, ME, MI, MO, NC, NH, NJ, NY, OH, PA, RI, TN, TX, VA, VT, WI, WV (United States)

Octonoba sinensis (Simon, 1880)

Common Name This species does not have a common name.

Total Body Length Female 3.0–5.8 mm, Male 2.8–3.3 mm

Description The prosoma is grayish tan, with a pale tan median stripe and pale stripes on the edge. The opisthosoma is tan to pale yellow. There is a pale brown median stripe that has an irregular margin. Four pairs of small tubercles are on either side of the median stripe, usually tipped in white to cream; the tubercles are less apparent in the males. The legs are brown, with white bands on the distal femurs, proximal and median tibiae, and proximal metatarsi. The first tibiae lack the feather-like hairs that are seen in *Uloborus* species. The sides are often mottled with black or brown.

Habitat These spiders are often found in and around human structures.

Known records from AL, KS, MD, MO, OH, SC, TN, TX, VA
This is an introduced species.

female

female

female ventral with egg sac

female with egg sac

Uloborus glomosus (Walckenaer, 1841)

Common Name Featherlegged orbweaver

Total Body Length Female 2.8–4.3 mm,
Male 2.3–3.2 mm

eyes

Description The prosoma is pale golden to brown.
The opisthosoma has four humps on the anterior and
is covered in a blanket of hairs. The first pair of legs has
a very conspicuous tuft of hairs with a feather-like appearance.

Habitat The web is often horizontal in orientation and is built in low vegetation or on human
structures. There may be a stabilimentum in a loose spiral near the hub. The egg sacs are
shaped like a multipointed star and strung in a line in the web or secured nearby.

Known records from ON, QC (Canada); AL, AR, CO, CT, FL, GA, IL, IN, KS, KY, LA, MA, MD, ME, MO,
MS, NC, NE, NH, NJ, NM, NY, OH, OK, PA, RI, SC, TN, TX, VA, WI (United States)

female

female ventral

female in web

female with egg sacs

juvenile in web

male

male

male

male ventral

Zosis geniculata (Olivier, 1789)

Common Name This species does not have a common name.

Other Colloquial Names Ninjastar ceiling spider

Total Body Length Female 5.2–8.3 mm, Male 4.0–6.3 mm

Description The prosoma is yellow to grayish brown. The eyes are ringed in black. There is a pale median stripe. The opisthosoma is tan to pale reddish brown. The heartmark is brown, with a cream to white stripe on either side of it. There is a hump on the anterior region of the opisthosoma. Paired white to cream spots are on either side of the midline. The legs are pale yellow. The femurs have a median white band followed by a broad dark distal band. The patellae are dark at each end. The tibiae have a median white band, a thin dark band is toward the proximal end, and the distal ends have a broad dark band. Faint dusky bands may be on the metatarsi.

Habitat These spiders are often found in and around human structures.

Known records from ON (Canada); FL, MS (United States)

female with egg sac

THE SPACE WEB WEAVER GUILD

The spiders of the Space Web Weaver Guild construct a web that takes up three-dimensional space and often does not seem to have any sort of organized structure; the web has an overall messy appearance. This would include those webs commonly referred to as cobwebs or tangle webs. All the spiders in this guild are Araneomorphae. This guild includes the following families: Dictynidae, Diguetidae, Hypochilidae, Leptonetidae, Mysmenidae, Nesticidae, Pholcidae, Theridiidae, Titanoecidae, and Trogloraptoridae.

KEY TO FAMILY

1a.	With cribellum --- 2
1b.	Without cribellum -- 5

Emblyna sublata   *Theridion glaucescens*

2a.	Cribellum divided -- 3
2b.	Cribellum entire or appears entire --- 4

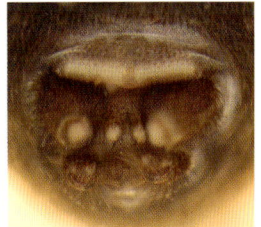
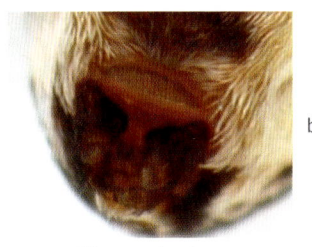

Titanoeca americana   *Dictyna volucripes*

3a. Robust spiders (4.0–8.0 mm total body length), eight eyes in two rows, usually found in dry habitats -- **Titanoecidae** (page 294)

3b. Small spiders (1.1–1.8 mm total body length), long slender legs, six eyes, found only in CA, found in damp habitats ------------------------ **Some Leptonetidae** (*Archoleptoneta*), not covered in this guide

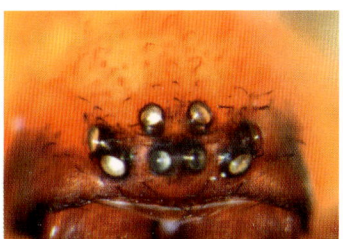

Titanoeca americana

4a. Two pairs of book lungs, dorsoventrally flattened, long legs, builds lampshade-shaped web, 10.0–20.0 mm total body length ---------- **Hypochilidae** (page 228)

4b. One pair of book lungs, small (1.2–8.0 mm total body length), web not lampshade shaped --------------------------------------- **Some Dictynidae** (page 219)

a

b

Hypochilus pococki

Emblyna sublata

Hypochilus pococki

5a. Subsegmented raptorial tarsi, six eyes, found in caves and old-growth forests of CA and OR, 6.9–9.6 mm total body length ------ **Trogloraptoridae** (page 295)

5b. Tarsi not subsegmented --- **6**

a

Trogloraptor marchingtoni

6a. Leg IV with tarsal comb -- **7**

6b. Leg IV without tarsal comb -- **8**

a

b

Steatoda grossa

Pholcus phalangoides

7a. Labium not rebordered, endites converging, opisthosoma usually spherical or globular, sometimes lives in webs of other spiders as kleptoparasites, 0.8–12.0 mm total body length --- **Theridiidae** (page 242)

7b. Labium rebordered, endites subparallel, tends to be pale, opisthosoma usually globular, lives in caves or cave-like habitats (cool and moist), 1.8–10.0 mm total body length -- **Nesticidae** (page 234)

ABOVE: *Steatoda grossa*

Eidmannella pallida

LEFT: *Argyrodes elevatus* in *Neoscona* orb web

8a. Six eyes in three diads, lives in desert scrub habitats, web with vertical retreat, 5.0–10.0 mm total body length ---------------------------------- **Diguetidae** (page 226)

8b. Eyes otherwise --- **9**

Diguetia imperiosa

Diguetia canities

9a. Usually with six eyes with median eyes posteriorly placed, 1.0–3.0 mm total body length ------------------------------------ **Some Leptonetidae** (page 229)

9b. Eyes otherwise --- **10**

Calileptoneta helferi

10a. Femur I with sclerotized spot on ventral surface, 0.5–2.0 mm total body length, eight eyes in two rows -------------------------------------- **Mysmenidae** (page 232)

10b. Femur I without sclerotized spot -- **11**

10a

Microdipoena guttata

11a. Long legs, six to eight eyes with six eyes in two triads (if anterior median eyes are present, they are located between triads), eye region usually elevated, 1.0–10.0 mm total body length ---------------------------------- **Pholcidae** (page 236)

11b. Eight eyes in two rows, found only in salt crust or intertidal habitats of FL and CA, 3.2–5.9 mm total body length, widely spaced spinnerets --------------------------------- **Some Dictynidae, not covered in this guide**

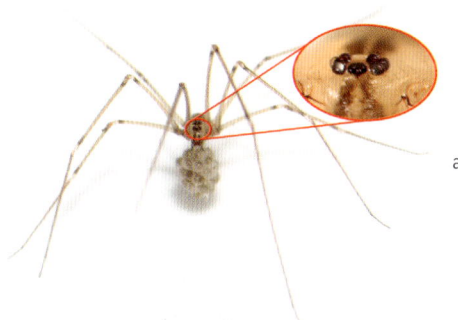

a

Pholcus manueli

DICTYNIDAE ▶ MESHWEAVERS

Other Colloquial Names Mesh web weavers

Genera in This Family *Arctella* (1), *Argenna* (2), *Argennina* (1), *Brigittea* (1), *Brommella* (3), *Dictyna* (44), *Emblyna* (64), *Hackmania* (2), *Iviella* (3), *Lathys* (10), *Mallos* (7), *Mexitlia* (1), *Nigma* (1), *Paratheuma* (1), *Phantyna* (9), *Saltonia* (1), *Thallumetus* (1), *Tivyna* (4), and *Tricholathys* (9)

Total Body Length 1.2–8.0 mm

Description These spiders are entelegyne. This family contains both cribellate and ecribellate species. They usually have six or eight eyes.

eye configuration
(*Dictyna volucripes*)

The ecribellate species tend to live close to or below ground and are found only in intertidal zones in Florida and desert salt crust habitat in the Southwest. The cribellate species choose much more varied habitats, ranging from leaf litter to high in tree canopies.

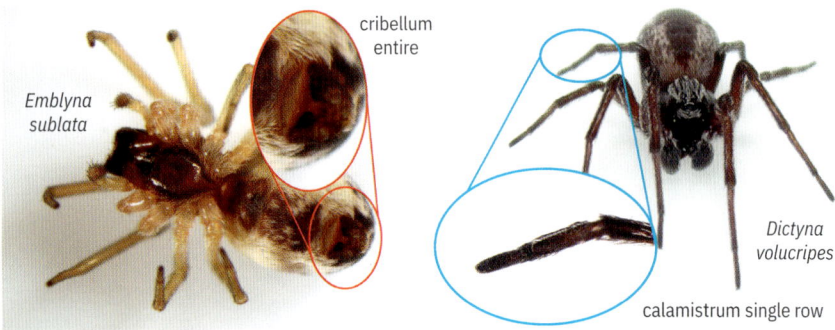

Emblyna sublata

cribellum entire

Dictyna volucripes

calamistrum single row

Similar Families Amaurobiidae, Agelenidae, Cybaeidae, Titanoecidae

How to distinguish Dictynidae from other similar families:

Some dictynids are ecribellate, which would distinguish them from amaurobiids and titanoecids. The cribellum of those that are cribellate is entire (or at least it appears entire), whereas it is divided in amaurobiids and titanoecids. Dictynids can be distinguished from the agelenids by the lack of elongated posterior lateral spinnerets, which are present in agelenids. Some dictynids are cribellate, which would distinguish them from the agelenids and cybaeids. Those that are ecribellate are found only in desert salt crust habitats of the Southwest and intertidal zones in Florida, which would not be ideal habitat for most of the agelenids or cybaeids.

Dictyna calcarata Banks, 1904

Common Name This species does not have a common name.

Total Body Length Female 2.5–4.5 mm, Male 3.0 mm

Description The prosoma is orange to brown, with rows of thick white hairs from the fovea to the eye region. The opisthosoma is yellow orange, with a dark heartmark and a pair of dark spots toward the posterior. The opisthosoma is covered in thick hairs. The female epigynum is very broad, with openings laterally on the opisthosoma. The legs are banded in yellow and brown. Each of the male pedipalps has a remarkable tibial apophysis (a tall erect spur that points up and forward from the pedipalps). Other species have similar tibial apophyses, but they are most pronounced in *D. calcarata*.

Habitat These spiders frequently make their webs on human structures.

Known records from BC (Canada); AZ, CA, CO, ID, IL, MT, NM, NV, OK, OR, TX, UT, WA (United States)

female

female with egg sac

male

male

Dictyna minuta Emerton, 1888

Common Name This species does not have a common name.

Total Body Length Female 1.8–2.0 mm, Male 1.5–1.6 mm

Description The prosoma is dusky to dark brown. The median stripe is made up of stripes of white hairs. The opisthosoma is covered with white to cream hairs. The heartmark is dark brown, and a series of dark brown chevrons (that may be fused) is toward the posterior end. The sides are often spotted with dark brown. The legs are pale brown, with distinct cream to white banding, especially on the tibiae.

Habitat These spiders are often found on low vegetation.

Known records from AB, BC, MB, NB, NL, NS, ON, QC, SK (Canada); AK, CT, IA, ID, IL, IN, MA, ME, MI, MN, ND, NE, NH, NJ, NY, OH, PA, RI, UT, WA, WI (United States)

female

female

Dictyna volucripes Keyserling, 1881

Common Name This species does not have a common name.

Total Body Length Female 2.5–4.2 mm, Male 2.7–3.3 mm

Description The prosoma is brown, with rows of thick white hairs from the fovea to the eye region in a broad band. The opisthosoma is dark brown with white hairs. Usually, a dark band runs down the middle, and light bands are on either side. The overall markings can vary. The legs are brown, and sometimes faint banding is visible.

eyes

egg sac

female

female ventral

female

male

male

male

Habitat These spiders often build their webs in low vegetation or on walls and fences.

Known records from AB, MB, NB, ON, PE, QC, SK (Canada); AL, AR, CO, CT, FL, GA, IL, KS, KY, LA, MA, MD, ME, MI, MN, MO, MS, MT, NC, NE, NH, NJ, NM, NY, OH, OK, SC, TN, TX, UT, WI (United States)

Emblyna annulipes (Blackwall, 1846)

Common Name This species does not have a common name.

Total Body Length Female 1.7–4.5 mm, Male 2.4–3.8 mm

Description The prosoma is red brown, with faint radial bands; there are white hairs from the fovea to the eye region. The opisthosoma is orange brown, and thick hairs give it a variegated appearance. There is a dark heartmark, and a series of dark chevrons can sometimes be seen. The legs are brown with pale banding; sometimes the males' legs are unbanded.

female

Habitat These spiders often build their webs in herbaceous vegetation.

Known records from AB, BC, LB, MB, NB, NL, NS, NT, ON, PE, QC, SK, YT (Canada); AK, AR, CO, CT, FL, IA, ID, IL, IN, MA, MD, ME, MI, MN, MO, MS, MT, ND, NE, NH, NJ, NY, OH, OR, PA, SC, SD, TX, VA, VT, WA, WI, WY (United States)

Emblyna sublata (Hentz, 1850)

Common Name This species does not have a common name.

Total Body Length Female 2.0–3.7 mm, Male 2.0–2.5 mm

Description The female prosoma is dark brown to almost black, with white hairs from the fovea to the eye region. The opisthosoma is a rich red-brown color and can have variable markings. In some females, the entire dorsal opisthosoma is the deep red-brown color, and there are pale hairs on the sides. In others, the dorsal surface is covered in various yellows, whites, and browns. The legs are yellow and do not

female

female

female ventral

female with egg sac

male

male

male

have any visible bands. The male prosoma is orange red and elevated in the eye region. The opisthosoma is red orange, with a slightly purple iridescence. The pedipalps are black. The legs are unbanded and red orange and sometimes somewhat darker at the distal ends.

Habitat These spiders often build their webs in low herbaceous vegetation.

Known records from AB, BC, MB, NB, NS, ON, QC, SK (Canada); AR, CT, FL, GA, IA, IL, IN, KS, MA, MD, ME, MI, MN, MO, MS, NC, NE, NH, NJ, NY, OH, OK, PA, SC, SD, TN, TX, VA, VT, WI (United States)

Nigma linsdalei (Chamberlin & Gertsch, 1958)

Common Name This species does not have a common name.

Total Body Length Female 3.0 mm, Male 2.8 mm

Description The female prosoma is pale green, with white hairs from the fovea to the eye region, although they are less obvious on this pale spider. The opisthosoma is pale green, with paler spots. The legs are green to yellow with faint banding. The males have an orange-brown prosoma. Pale bands are on the edges. From the fovea to the eye region is a widening orange-brown band that has white hairs. The opisthosoma is pale green on the dorsal surface, with white to yellow-brown spots. The lateral edge is encircled with a dark band. The legs are pale yellow to green and usually unbanded.

Habitat These spiders seem to have an affinity for oak trees and often build their webs at the tips of the branches.

Known records from CA

Phantyna terranea (Ivie, 1947)

Common Name This species does not have a common name.

Total Body Length Female 2.2–2.4 mm, Male 2.0–2.1 mm

Description The prosoma is dark brown, slightly lighter in the eye region. The median stripe is made up of stripes of white hairs. At the base of the chelicerae, the males have a tooth that comes to a sharp point. The opisthosoma is yellow brown. The heartmark is dark brown. At the posterior end, a short transverse band makes a T shape. There are many white hairs in the median stripe. The median stripe is bordered in dark brown. The legs are yellow brown, with dusky bands at the distal end of the femurs and the proximal tibiae.

male

male

male ventral

Habitat These spiders usually build their webs close to the ground under debris.

Known records from AB, BC, ON, SK, YT (Canada); CO, ID, OR, UT, WA, WY (United States)

Tivyna moaba (Ivie, 1947)

Common Name This species does not have a common name.

Total Body Length Female 1.8–1.9 mm, Male 1.7–1.8 mm

Description The prosoma is orange red, with a median stripe made up of white hairs. The opisthosoma is deep red to almost purple. It is carpeted with white and dark hairs. The legs are orange to yellow and unbanded.

Habitat These spiders are often found in low vegetation.

Known records from AZ, CA, UT

female

male

DIGUETIDAE ▶ DESERTSHRUB SPIDERS

Genera in This Family *Diguetia* (7)

Total Body Length 5.0–10.0 mm

Description These spiders are haplogyne and ecribellate. They have six eyes arranged in three diads. Their chelicerae are fused at the base. They have a pair of book lungs and lack a tracheal spiracle. They are found in desert scrub habitat in the Southwest, where they build their extensive webs in low vegetation. Their web is a tangle web with a vertical tubelike retreat.

Similar Families Plectreuridae, Scytodidae, Sicariidae

How to distinguish Diguetidae from other similar families:

Diguetids can be distinguished from plectreurids by their eye configuration: plectreurid eyes are in two rows that occupy more than half the width of the prosoma, whereas diguetids have six eyes in three diads. Also, diguetids build an extensive space web

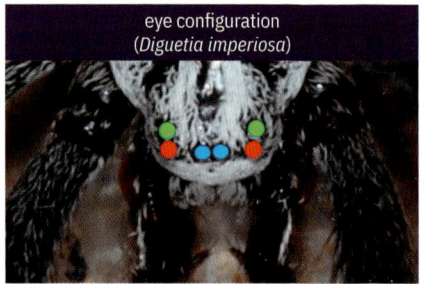

eye configuration
(*Diguetia imperiosa*)

vertical retreat
(*Diguetia canities*)

for prey capture, whereas plectreurids are ground-hunting spiders. With six eyes in three diads, it could be easy to confuse diguetids with scytodids and sicariids. Sicariids are plainly marked spiders that lack leg spines; diguetids are usually brown with white markings, and at least a few spines are present on the legs. Diguetids also tend to have at least some banding on the legs; sicariids do not. Scytodids have a very high prosoma, whereas diguetids do not.

Diguetia canities (McCook, 1890)

Common Name This species does not have a common name.

Total Body Length Female 5.0–10.0 mm, Male 5.0–7.0 mm

Description The prosoma is brown, with thick white hairs. There are six eyes in three pairs. The opisthosoma is golden brown, with thick white hairs. A scalloped cream-colored pattern runs down the dorsal midline. The legs are brown to golden brown. Faint banding may be seen at the patella and medially and distally on the tibiae.

Habitat These spiders' somewhat large webs can be found on cacti or low shrubs. A tubular retreat is suspended within the web that the spiders decorate with various pieces of debris. The females place their egg sacs within the retreat.

Known records from AZ, CA, NM, NV, OK, TX

female

female

female

female entering retreat

HYPOCHILIDAE ▶ LAMPSHADE WEAVERS

Other Colloquial Names
Lampshade spiders

Genera in This Family
Hypochilus (10)

Total Body Length
10.0–20.0 mm

Description The bodies of these cribellate spiders are dorsoventrally flattened.

lampshade web
(*Hypochilus pococki*)

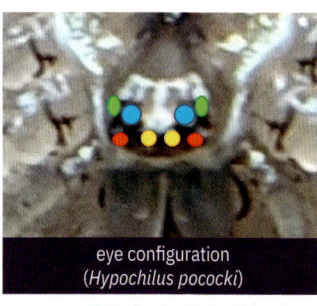

eye configuration
(*Hypochilus pococki*)

The spiders are often yellow to gray, with dusky black markings and sometimes an iridescent sheen. They have eight eyes; the anterior eye row is slightly procurved, and the posterior eye row is recurved. They have very long legs (up to 150 mm in length). They have two pairs of book lungs, which is unique for an Araneomorphae. Their webs are constructed in a unique lampshade shape, usually on bare surfaces of rocks, cliffs, caves, and sometimes human structures.

dorsoventrally flat, long legs (*Hypochilus pococki*)

Similar Families Pholcidae

How to distinguish Hypochilids from other similar families:

Hypochilids can be distinguished from pholcids, as they are dorsoventrally flattened and have two pairs of book lungs; pholcids have one pair of book lungs and are cylindrical or globular.

Hypochilus pococki Platnick, 1987

Common Name This species does not have a common name.

Other Colloquial Names Pococki's lampshade-web spider

Total Body Length Female 11.0–13.0 mm, Male 10.0–11.0 mm

Description The prosoma is tan to gray. There is a pale irregular band around the edges, a light marking at the fovea, and dark purple-brown stripes radiating from the fovea. The opisthosoma is mottled with shades of white, black, gray, and brown. The spiders are somewhat flattened in appearance. The legs are banded in gray and dark brown. Femur I is about three times the length of the prosoma in females and about five times the length of the prosoma in males.

Habitat These spiders build their unique lampshade-shaped webs on the surfaces of walls, overhangs, and rocks.

Known records from GA, NC, SC, TN, VA

female

female with prey

female with egg sac

web

LEPTONETIDAE ▸ TINY CAVE SPIDERS

Genera in This Family *Appaleptoneta* (9), *Archoleptoneta* (2), *Calileptoneta* (9), *Chisoneta* (1), *Darkoneta* (1), *Montanineta* (1), *Ozarkia* (9), and *Tayshaneta* (19)

Total Body Length 1.0–3.0 mm

Description These small haplogyne spiders can be cribellate or ecribellate. They have six eyes, and in most species, the medial eyes are posteriorly displaced. There are eyeless species and species whose six eyes cluster together anteriorly (but this cluster never occupies more than half the width of the prosoma). The spiders tend to be pale, and

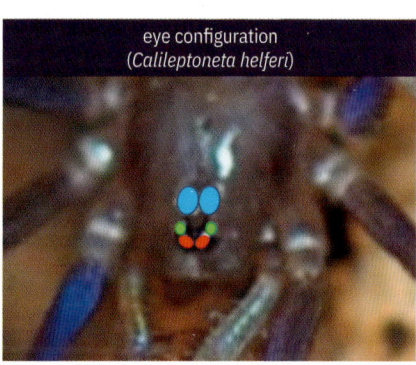
eye configuration
(*Calileptoneta helferi*)

the legs often have an iridescent sheen. They have long slender legs, and autospasy occurs at the patella-tibia joint. They have a pair of book lungs and a single tracheal spiracle near the spinnerets. They build a small tangle web, under which they hang inverted. When disturbed, they will drop from the web. They are often found in humid areas, such as under rocks, in leaf litter, or in caves.

Similar Families Ochyroceratidae, Linyphiidae, Pholcidae, Telemidae, Trogloraptoridae

How to distinguish Leptonetidae from other similar families:

Many leptonetids can be distinguished from other similar-looking families by the unique eye arrangement: usually, the posterior medians are set back from the other eyes. (Some leptonetids are eyeless, and a few have contiguous eyes that occupy less than half the width of the prosoma.) Pholcids' eyes are arranged in a pair of triads, and the anterior median eyes, if present, are located between and just below the triads. Linyphiids usually have eight eyes in two rows. Trogloraptorids have six eyes, and the lateral eyes are adjacent to one another, and the posterior median eyes are between them. Leptonetids can be distinguished from ochyroceratids, as they lack an onychium (an extension of the tarsus). Also, only one ochyroceratid species has been recorded in North America, and it has been found only in Florida. Leptonetids have a pair of book lungs and a single posteriorly placed tracheal spiracle, whereas telemids usually have two pairs of tracheal spiracles: one pair anterior of the epigastric furrow, the other pair between the epigastric furrow and the spinnerets.

Calileptoneta helferi (Gertsch, 1974)

Common Name This species does not have a common name.

Total Body Length Female 2.2–2.3 mm, Male 2.3 mm

Description The prosoma is tan to brown and darker around the border. The posterior median eyes are located posterior to the other eyes and thinly ringed in black. The anterior eye cluster is surrounded by black. The opisthosoma is golden brown, with thin, pale, iridescent, longitudinal bands. The legs are tan to brown, sometimes with an iridescent look to them.

female

male

male

Habitat These spiders like cool moist habitats, such as in caves, in leaf litter, or under rocks.

Known records from CA

Darkoneta garza (Gertsch, 1974)

Common Name This species does not have a common name.

Total Body Length Female 1.1 mm (The male of this species has not been described.)

Description The prosoma is white to pale yellow. The eyes are located in a cluster; the anterior median eyes are missing. The eye cluster occupies less than half the width of the prosoma and is surrounded in black. The opisthosoma is off-white to pale yellow, with a slight iridescent sheen. The heartmark may be slightly darker. The first pair of legs is the longest. The legs are pale yellow, often with an iridescent sheen that is most prominent on the femurs. The specimen shown here is likely an undescribed species and was found in Arizona but is similar in outward appearance to *D. garza*.

Habitat These spiders are usually found in caves or similar types of habitats.

Known records from TX

Ozarkia apachea (Gertsch, 1974)

Common Name This species does not have a common name.

Total Body Length Male 1.5 mm (The female of this species has not been described.)

Description The prosoma is light tan to pale orange and slightly dusky around the edge. The posterior median eyes are located posterior to the other eyes and thinly ringed in black. The anterior eye cluster is surrounded by black. The opisthosoma is tan to orange, with no visible markings and covered in thin dusky hair. The legs are pale, and dusky hairs and a few

spines are present. Often, the femurs have an iridescent sheen. The specimen shown here was not definitively identified but matches the outward appearance of *O. apachea*.

male

male

Habitat These spiders can be found in leaf litter or other moist habitats.

Known records from AZ

MYSMENIDAE ▸ DWARF COBWEB WEAVERS

Genera in This Family *Maymena* (1), *Microdipoena* (1), *Mysmena* (2), *Mysmenopsis* (1), and *Trogloneta* (1)

Total Body Length 0.5–2.0 mm

Description These small entelegyne and ecribellate spiders are globular and often pointed posteriorly. The prosoma is often high, and the eyes are located on a tubercle. The spiders have eight eyes in two rows. They are usually dark, with pale dots or markings. The legs may be uniformly colored or banded. They have a sclerotized spot on the apical ventral surface of femur I. They tend to build their small tangle webs in damp areas, such as in leaf litter, under debris, or in other similar types of habitats. These spiders are extremely difficult to identify, even to genus, without viewing a mature specimen under a microscope; this is especially true with female specimens.

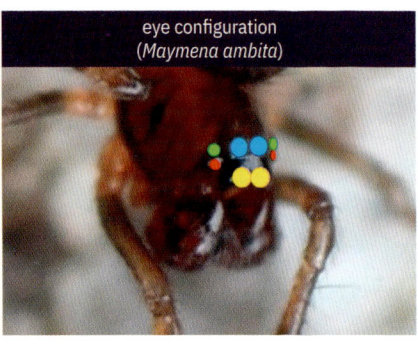

eye configuration
(*Maymena ambita*)

Similar Families Anapidae, Araneidae, Linyphiidae, Symphytognathidae, Theridiidae, Theridiosomatidae

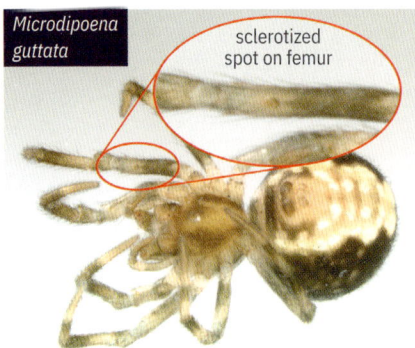

Microdipoena guttata

sclerotized spot on femur

globular opisthosoma
(*Trogloneta paradoxa*)

How to distinguish Mysmenidae from other similar families:

Mysmenids can be distinguished from anapids by the presence of pedipalps in the females (which anapids lack) and the lack of a scutum on the male opisthosoma (which anapids have). They can be distinguished from symphytognathids by the presence of eight eyes in two rows; symphytognathids have only four eyes located in two pairs. Mysmenids can be distinguished from all similar families by the presence of the sclerotized spot on femur I, which anapids, araneids, linyphiids, symphytognathids, theridiids, and theridiosomatids lack. They can be distinguished from theridiosomatids by the lack of the prolateral sternal pits that are seen only in theridiosomatids. Additionally, the web structure of mysmenids is a tangle web; in contrast, anapids, araneids, symphytognathids, and theridiosomatids build orb webs, and linyphiids build a sheet web.

Maymena ambita (Barrows, 1940)

Common Name This species does not have a common name.

Total Body Length Female 2.0 mm, Male 1.4 mm

Description The prosoma is brown, and a few dark radial lines are visible. It is elevated in the eye region, and the clypeus is high. The anterior eyes are dark, and the other eyes appear white. The opisthosoma is dark brown, and paired white spots ringed in golden brown run on either side of the midline down the dorsal surface. Additional white spots ringed in golden brown may also be on the sides. The legs are banded in dark and golden brown.

Habitat These spiders tend to live in caves and other dark moist areas.

Known records from AL, FL, KY, OH, TN, VA

Trogloneta paradoxa Gertsch, 1960

Common Name This species does not have a common name.

Total Body Length Female 1.0 mm, Male 0.9–1.0 mm

Description The prosoma is yellow brown, with dark markings just behind the eyes. The eyes are elevated on a tubercle that is extremely tall in the males. There may be faint radial lines. The opisthosoma is higher than it is long and shaped to overhang the prosoma. The opisthosoma is brown, with pale markings that are most pronounced on the posterior dorsal area. The legs are brown, with yellow bands.

female

female with egg sac

female with egg sac

male

Habitat The spiders tend to build their webs under logs or within debris near ground level; webs have also been found in woodrat middens.

Known records from AB, BC (Canada); AZ, CA, OR, UT (United States)

NESTICIDAE ▸ CAVE COBWEB SPIDERS

Other Colloquial Names Scaffold web spiders, comb-footed cellar spiders

Genera in This Family *Eidmannella* (7), *Gaucelmus* (1), and *Nesticus* (30)

Total Body Length 1.8–10.0 mm

Description These entelegyne ecribellate spiders are known troglophiles (like to live in caves). They have six or eight eyes in two rows. The labium is rebordered, and the endites are subparallel. They have a globular opisthosoma, and many are pale. They have a tarsal comb on leg IV. They tend to build their tangle webs in cool moist areas, such as in caves, beneath rocks, in leaf litter, or in crevices.

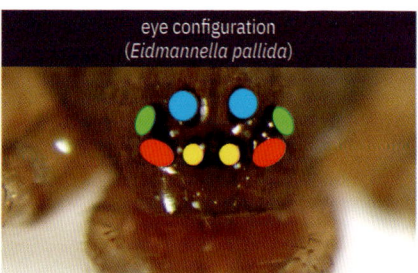

eye configuration
(*Eidmannella pallida*)

Similar Families Linyphiidae, Theridiidae

How to distinguish Nesticidae from other similar families:

Linyphiids lack the tarsal comb that is seen in nesticids and theridiids. Linyphiids also build a sheet web, rather than the tangle web built by nesticids. Theridiids lack the rebordered labium and have converging endites, whereas nesticids have a rebordered labium and subparallel endites.

labium rebordered, endites subparalell

Eidmannella pallida

tarsal comb

pale color

Eidmannella pallida

Eidmannella pallida
(Emerton, 1875)

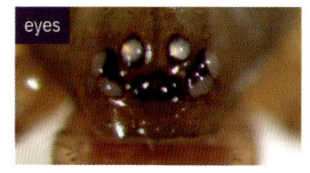

eyes

Common Name This species does not have a common name.

Total Body Length Female 3.2–4.0 mm, Male 2.2–2.8 mm

Description The prosoma is yellow to golden orange. The opisthosoma is yellow to dusky brown. When the opisthosoma is darker, paler chevrons may be just visible in a line down the dorsal surface. The legs are yellow to golden in color and may have some faint banding.

female

female ventral

Habitat These spiders like to build their webs in moist cool habitats and can be found in caves or cave-like habitats, in leaf litter, under rocks and boulders, or in crevices. Occasionally, they can be found in and around human structures.

Known records from NB, NL, NS, ON, QC (Canada); AL, AR, AZ, CA, FL, GA, IA, IL, IN, KS, KY, LA, MA, MD, MS, NC, NJ, NM, OH, OK, OR, PA, SC, TN, TX, UT, VA, WI, WV (United States)

female

juvenile

male

male

PHOLCIDAE > CELLAR SPIDERS or DADDYLONGLEG SPIDERS

Other Colloquial Names Daddy long-legs spider, granddaddy long-legs spider, carpenter spider, daddy long-legger, vibrating spider, gyrating spider, long daddy

Genera in This Family
Artema (1), *Chisosa* (1), *Crossopriza* (1), *Holocnemus* (1), *Metagonia* (1), *Micropholcus* (1), *Modisimus* (2), *Pholcophora* (2), *Pholcus* (13), *Physocyclus* (5), *Psilochorus* (19), *Smeringopus* (1), and *Spermophora* (1)

Total Body Length 1.0–10.0 mm

Description These spiders are usually pale, yellow to light brown or gray (often with darker

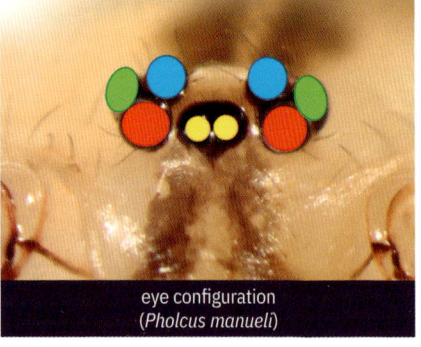
eye configuration
(*Pholcus manueli*)

markings), haplogyne, and ecribellate. They have six or eight eyes, which are placed in two triads, and if the anterior median eyes are present, they are located between and just anterior of the triads. The opisthosoma is cylindrical to globular. The spiders have one pair of book lungs, and they often have a posteriorly placed tracheal spiracle near the spinnerets. The legs are often very long. The females carry their egg sac in their chelicerae. They tend to live in dark spaces, under rocks, in leaf litter, and in caves, but also readily adapt to basements and cellars. When disturbed, they often vibrate rapidly or shake in the web.

Similar Families Hypochilidae, Leptonetidae, Ochyroceratidae, Plectreuridae, Scytodidae, Telemidae, Trogloraptoridae

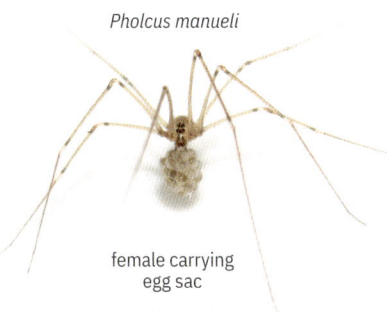

Pholcus manueli

female carrying
egg sac

How to distinguish Pholcids from other similar families:

Hypochilids are dorsoventrally flattened and have two pairs of book lungs, whereas pholcids have one pair of book lungs and are not dorsoventrally flattened. Leptonetids have a very distinctive eye arrangement: usually, the posterior medians are set back from the other eyes. (Some are eyeless, and a few have contiguous eyes that occupy less than half the width of the prosoma.) These eye configurations are not similar to those of pholcids, whose eyes are arranged in a pair of triads, and the anterior median eyes, if present, are located between and just below the triads. Ochyroceratids are very small spiders and are easily distinguished if one can get a view of the tarsal claws, which are on an onychium (an extension of the tarsus). Pholcids lack an onychium. Also, only one known ochyroceratid species has been introduced to North America, and it can be found only in Florida. Pholcids can be distinguished from plectreurids by their very long slender legs, as plectreurids tend to have shorter, stockier legs. Also, pholcids build a web for prey capture, and plectreurids are ground-hunting spiders (although they do build a small web near their retreat). Pholcids can be distinguished from scytodids by the shape of the prosoma: scytodids have a highly domed prosoma, which is distinctly different from pholcids. Telemids can be distinguished from pholcids by the lack of book lungs, as pholcids have a pair of book lungs. Telemids also have a sclerotized zigzag just above the pedicle, which is lacking in pholcids. Trogloraptorids can be distinguished from pholcids by the presence of their unique subsegmented raptorial tarsi. Trogloraptorids can also be distinguished from pholcids by their eye configuration. Trogloraptorids have six eyes. (The anterior median eyes are absent.) The lateral eyes are located adjacent to one another, and the posterior median eyes are between them. There is also confusion regarding pholcids and opilionids (harvestmen), as the common name "daddy-longlegs" and variations of it are often associated with both groups. Opilionids are arachnids, but they are not spiders. They can be distinguished from spiders by their fused body segments; in comparison, spiders have a clearly separated prosoma and opisthosoma. For more information, see the descriptions of the extant arachnid orders in the introduction.

Crossopriza lyoni (Blackwall, 1867)

Common Name This species does not have a common name.

Other Colloquial Names Tailed cellar spider, tailed daddy longlegs spider, box spider

Total Body Length Female 5.0–9.0 mm, Male 4.0–8.0 mm

Description The prosoma is circular, and the eyes are clustered. It is tan, with a dark median stripe that is slightly wider in the eye region, and the fovea is deep. The opisthosoma is gray, with pale lines, and comes to a point at the posterior. Viewed laterally, it appears very square. The legs are tan, with a pale band distally on each femur, dark patellae, and then a dark band distally on the tibiae, followed by a pale band.

Habitat The females carry a loosely spun egg sac in their chelicerae. This species is synanthropic in North America and usually found in buildings or other human structures.

Known records from AZ, FL, IN, KS, TX
This is an introduced species.

female with
egg sac

female

female

female

Holocnemus pluchei
(Scopoli, 1763)

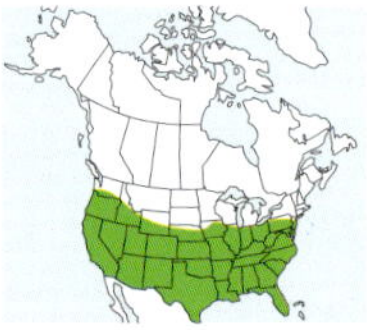

Common Name
Marbled cellar
spider

Total Body Length
Female 5.0–9.0 mm, Male 5.0–7.0 mm

Description The prosoma is tan, with a
dark median stripe. The fovea is fairly
deep. The opisthosoma is tan, with a
marbled pattern of grays, browns, and
tans. The ventral opisthosoma is dark.
The legs are yellow, with some banding
near the joints. Females have swollen
pedipalps and can easily be mistaken
for immature males.

male approaching female in retreat

Habitat These spiders build their webs in shrubs or on the outside of buildings, and they are found more often outside than inside. The females build an interesting dome-shaped web, and they sit in it when they have egg sacs and newly emerged young.

Known records from AZ, CA, NM
This is an introduced species.

Pholcus manueli Gertsch, 1937

Common Name This species does not have a common name.

Total Body Length Female 3.5–5.0 mm, Male 3.5–4.1 mm

Description The prosoma is pale cream to gray, and two slightly irregular dark markings are on either side of the midline. The clypeus has two dark lines that are joined near the eye region. There are two clusters of three eyes, and the anterior median eyes are between them. The opisthosoma is pale yellow to gray, with a slightly darker heartmark. The legs are pale yellow to gray and may have faint banding at the joints.

eyes

female with egg sac

female with egg sac

female with egg sac

Habitat The females carry their loosely spun egg sacs in their chelicerae. This species is synanthropic and usually found in or around human structures. When these spiders are disturbed in their web, they shake violently. They are known to take down prey much larger than themselves, including other spiders.

Known records from ON (Canada); CT, DE, IA, IL, IN, KS, KY, MA, MD, ME, MI, MN, NC, NE, NJ, NY, OH, PA, VA, WA (United States)
This is an introduced species.

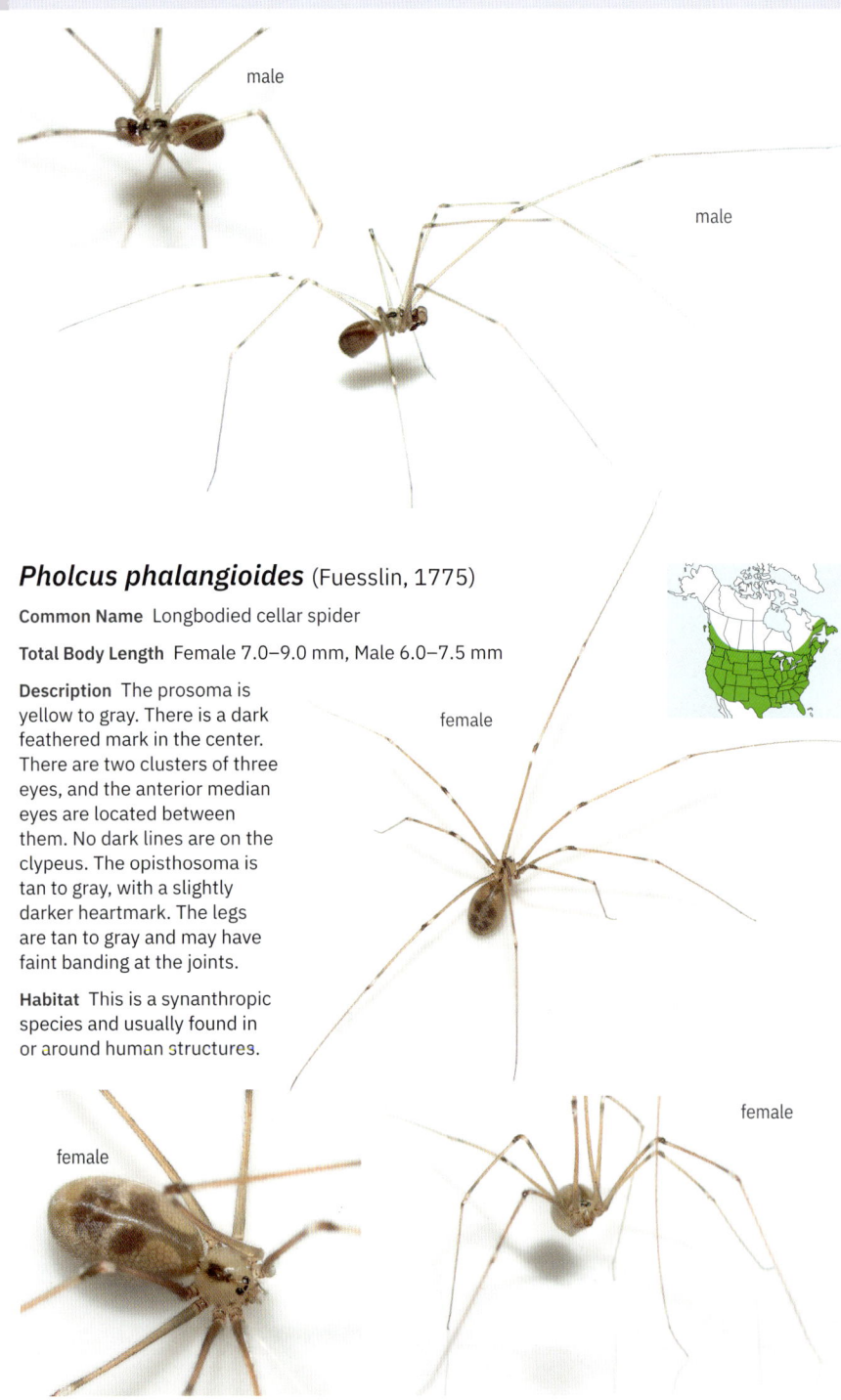

male

male

Pholcus phalangioides (Fuesslin, 1775)

Common Name Longbodied cellar spider

Total Body Length Female 7.0–9.0 mm, Male 6.0–7.5 mm

Description The prosoma is yellow to gray. There is a dark feathered mark in the center. There are two clusters of three eyes, and the anterior median eyes are located between them. No dark lines are on the clypeus. The opisthosoma is tan to gray, with a slightly darker heartmark. The legs are tan to gray and may have faint banding at the joints.

Habitat This is a synanthropic species and usually found in or around human structures.

female

female

female

female with egg sac

male

male

male

male

When these spiders are disturbed in their web, they shake violently. They are known to take down prey much larger than themselves, including other spiders.

Known records from AB, BC, NL, NS, ON, QC, SK (Canada); AK, AL, AR, AZ, CA, CO, CT, DE, FL, IL, IN, MA, ME, MI, MT, NY, OH, OK, OR, MA, TX, VA, WA, WI, WV (United States)
This is an introduced species.

Pholcus zichyi Kulczyński, 1901

Common Name This species does not have a common name.

Total Body Length Female 3.6 mm, Male 3.8–3.9 mm

Description The prosoma is light gray to pale orange. A dark marking branches from the fovea to the pedicel. Two pairs of dark markings are on the edge of the prosoma. The eyes are elevated. An X-shaped markings covers the anterior median eyes and sits between the other eyes. The opisthosoma is gray to pale orange, and a series of triangular shapes in brown runs down the midline. There may also be spotting on either side of this marking. The legs are tan, and faint banding may be visible, especially on the femurs.

Habitat These spiders tend to build their webs between rocks or on human structures.

Known records from DE
This is an introduced species. So far, the only confirmed records are from DE, but similar-looking specimens from GA, KS, MD, MO, and NC have been reported on social media.

male

male

male

male

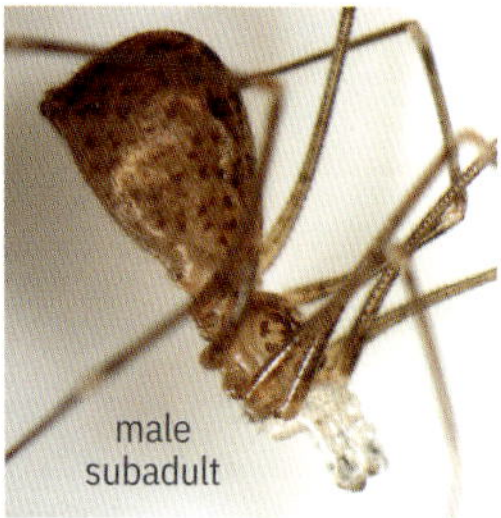

male
subadult

THERIDIIDAE ▷ COBWEB WEAVERS

Other Colloquial Names Tangle-web spider, cobweb spider, comb-footed spider

Genera in This Family *Anelosimus* (3), *Argyrodes* (3), *Asagena* (4), *Canalidion* (1), *Chrosiothes* (7), *Chrysso* (3), *Coleosoma* (3), *Crustulina* (2), *Cryptachaea* (9), *Dipoena* (15), *Emertonella* (2), *Enoplognatha* (10), *Episinus* (1), *Euryopis* (18), *Faiditus* (8), *Hentziectypus* (5), *Lasaeola* (2), *Latrodectus* (5), *Meotipa* (1), *Neopisinus* (1), *Neospintharus* (3), *Neottiura* (1), *Nesticodes* (1), *Ohlertidion* (1), *Parasteatoda* (2), *Paratheridula* (1), *Pholcomma* (4), *Phoroncidia* (1), *Phycosoma* (1), *Phylloneta* (2), *Platnickina* (5), *Rhomphaea* (2), *Robertus* (17), *Rugathodes* (2), *Simitidion* (1), *Spintharus* (1), *Steatoda* (21), *Stemmops* (2), *Styposis* (2), *Tekellina* (1), *Theonoe* (1), *Theridion* (56), *Theridula* (3), *Thymoites* (13), *Tidarren* (2), *Wamba* (2), and *Yunohamella* (1)

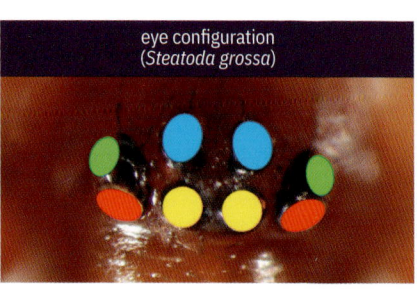

eye configuration
(*Steatoda grossa*)

Total Body Length 0.8–12.0 mm

kleptoparasite
(*Argyrodes elevatus*)
in *Neoscona* orb web

tarsal comb
(*Steatoda grossa*)

endites converge, labium not
rebordered (*Steatoda grossa*)

Description These entelegyne ecribellate spiders tend to be spherical or globular but in some cases can be elongated. They come in a variety of colors. They have eight eyes. They have a pair of book lungs, and a tracheal spiracle is near the spinnerets. They usually lack leg spines. A tarsal comb is on leg IV. They do not have a rebordered labium, and their endites converge distally. Some species are known to live in the webs of other spiders, where they may feed on the host, its prey, and its eggs. Most species build a three-dimensional tangle web that seems to lack any organized structure. They occupy a variety of habitats, and some species are known to be highly synanthropic and live in our homes.

Similar Families Anapidae, Linyphiidae, Mimetidae, Mysmenidae, Nesticidae, Tetragnathidae, Theridiosomatidae

How to distinguish Theridiidae from other similar families:

Anapids build an orb web, the males have a scutum on the dorsal opisthosoma, and the females lack pedipalps. Theridiids build a tangle web, the males do not have a scutum, and the females, unless they have suffered an injury, have pedipalps. Linyphiids build a sheet web, lack the tarsal comb, and have leg autospasy at the patella-tibia joint. Autospasy in theridiids occurs at the coxa-trochanter joint. Theridiids lack the diagnostic leg spination of legs I and II that is seen in mimetids. Mysmenids have a sclerotized spot on femur I that theridiids lack. Nesticiids tend to be pale and are troglophilic (like to live in caves). Their labium is rebordered, and the endites are subparallel; in theridiids, the labium is not rebordered, and the endites converge distally. Tetragnathids build an orb web (or, in a couple cases, are ground hunters), whereas theridiids build a tangle web. Theridiosomatids build an orb web that is pulled into a cone. They also have pits in the prolateral area of the sternum; pits are not seen in theridiids.

Argyrodes elevatus
Taczanowski, 1873

Common Name The genus *Argyrodes* has the common name "dewdrop spiders."

Other Colloquial Names American dewdrop spider

Total Body Length Female 3.4–5.2 mm, Male 3.3–4.0 mm

Description The prosoma is dark brown to black and shiny. The opisthosoma is an erect triangle shape and ranges in color from blacks, oranges, and reds to shiny silver, with a black midline. The legs are dark proximally and have paler distal segments.

Habitat This species resides in the webs of other spiders, usually Araneidae (orbweavers). They are kleptoparasites, stealing food from the host's web; sometimes this is small prey that is ignored by the large orbweaver. Occasionally, they have been observed feeding

female

female

female ventral

female

egg sac

female with egg sac

on the host spider. The egg sacs are suspended on a thick stalk of silk and appear as if they are covered in tan paper. The egg sac is teardrop shaped, and an extension that looks like a tube projects out the bottom.

Known records from AL, CA, FL, GA, LA, MO, MS, NC, OH, OK, SC, TX, VA

Asagena americana Emerton, 1882

Common Name Twospotted cobweb spider

Total Body Length Female 3.5–4.7 mm, Male 3.2–4.4 mm

Description The prosoma is reddish brown and slightly darker in the eye region. The opisthosoma is dark red, almost purple, with two pale white to yellow dashes just over halfway down toward the lateral edges. Gravid females often lack these marks. The males have many stout cusps on their legs and clasping spurs on leg II, and the prosoma is edged in what look like gear cogs. These are likely used in stridulation during courtship.

Habitat These spiders tend to build their webs close to the ground and are often found under rocks and debris. This species is considered Imperiled (S2) in Ontario, Canada.

Known records from AB, BC, MB, NB, ON, QC, SK (Canada); AL, AZ, CO, CT, DC, FL, GA, ID, IL, IN, MA, MD, ME, MI, MN, MO, MS, NC, NE, NH, NJ, NM, NY, OH, OK, OR, PA, RI, SC, TN, TX, UT, VA, VT, WA, WI, WV (United States)

female

female

male

male

Crustulina sticta (O. Pickard-Cambridge, 1861)

Common Name This species does not have a common name.

Total Body Length Female 2.2–2.7 mm, Male 1.9–2.7 mm

Description The prosoma is dark brown and covered in granulations. The opisthosoma is yellow to orange, with a few dark blotches. The ventral opisthosoma has a broad red band. The legs are dark orange and lack banding. The male pedipalps are black and bulbous at the ends of each distal joint, giving them a unique appearance.

Habitat These spiders tend to build their webs among leaf litter and other ground debris.

Known records from AB, BC, MB, NB, NL, NS, NT, ON, PE, QC, SK (Canada); AK, AZ, CA, CO, CT, IA, ID, IL, MA, ME, MI, MN, MT, NE, NH, NJ, NM, NY, OH, OR, TX, UT, VA, WA, WI, WV, WY (United States)

male
male
male
male ventral

Cryptachaea porteri (Banks, 1896)

Common Name This species does not have a common name.

Total Body Length Female 2.2–4.9 mm, Male 1.6–2.8 mm

Description In the females, the prosoma is orange to dark brown; in the males, it is orange, with a dark stripe from the fovea that widens toward the eye region. The opisthosoma is mottled with black, yellows, oranges, and browns. A small tubercle may or may not be on the dorsal surface toward the posterior end, although this may not be obvious in gravid or well-fed specimens. The legs are banded. Species of *Cryptachaea* are difficult to identify without looking at the genitalia with a microscope.

Habitat These spiders tend to build their webs in humid habitats, such as in caves, under woody debris, or under rocks.

Known records from AL, AR, CO, FL, GA, IL, IN, KS, KY, LA, MO, NC, NM, NY, OH, SC, TN, TX, VA

male

male

Cryptachaea rupicola
(Emerton, 1882)

Common Name
This species does not have a common name.

Total Body Length
Female 1.8–2.7 mm, Male 1.4–2.2 mm

Description The prosoma is dusky yellow to black. The opisthosoma is a variegated mix of black, browns, yellows, and white. A small tubercle is on the dorsal surface toward the posterior end, although this may not be obvious in gravid or well-fed specimens. The legs are banded. Species of *Cryptachaea* are difficult to identify without looking at the genitalia with a microscope.

Habitat These spiders tend to build their webs under ledges or large rock slabs; debris is often incorporated into the webs. The egg sacs are gray papery-looking spheres.

female with
egg sac

female

female

Known records from ON, QC (Canada); AL, CT, GA, IL, IN, KY, LA, MA, MD, ME, MI, MS, NC, NH, NJ, NY, OH, PA, SC, TN, VA, WI (United States)

Dipoena nigra (Emerton, 1882)

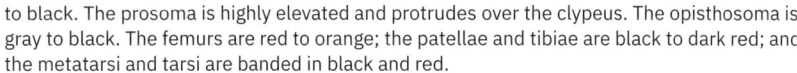

female eyes

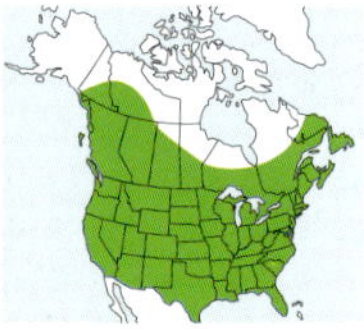

Common Name This species does not have a common name.

Total Body Length Female 2.1–4.3 mm, Male 1.3–2.0 mm

Description The prosoma is dusky yellow to black. The prosoma is highly elevated and protrudes over the clypeus. The opisthosoma is gray to black. The femurs are red to orange; the patellae and tibiae are black to dark red; and the metatarsi and tarsi are banded in black and red.

Habitat These spiders are often found in leaf litter, and they are known to feed on ants. The egg sacs are spheres covered in fluffy golden silk.

Known records from BC, MB, NB, NL, NS, NT, ON, QC, SK, YT (Canada); AR, AZ, CA, CO, CT, DC, FL, GA, IA, ID, IL, MA, ME, MI, MN, MO, MS, MT, NC, NH, NJ, NM, NY, OH, OK, OR, PA, SC, SD, TN, TX, UT, VA, VT, WA, WI, WV, WY (United States)

female

female

female with egg sac

egg sac

female ventral

Emertonella taczanowskii (Keyserling, 1886)

Common Name This species does not have a common name.

Total Body Length Female 1.9–3.0 mm, Male 2.0–2.5 mm

Description The prosoma is dark orange to dark brown. The opisthosoma is cream to off-white, sometimes silvery. The markings on the opisthosoma vary. When the prosoma is dark, the opisthosoma has a single contiguous dark brown marking that is widest anteriorly and undulates slightly but is basically triangular. When the prosoma is lighter in color, the opisthosoma marking has a pale heartmark, and there is a constriction just posterior to where the pale lateral color almost touches in the middle. Then a series of diamond-shaped markings runs down the midline, and scattered dark markings are on the lateral edges. The legs are mostly dark, except for the proximal femurs, or yellow with irregular banding.

Habitat These spiders are often found in and around human structures.

Known records from AZ, CO, FL, MA, NM, TX, UT

Enoplognatha caricis (Fickert, 1876)

Common Name This species does not have a common name.

Total Body Length Female 4.3–9.0 mm, Male 4.5–5.3 mm

Description The prosoma is dark yellow brown and darker in the eye region and at the edges. The opisthosoma is a variegated pattern of whites, browns, blacks, and tans. The dark heartmark is usually bordered in white or cream. The markings may be very faint in gravid females, but the white marking around the heartmark is usually visible. The legs are dusky yellow to orange, with slightly darker rings at the distal end of each segment. The literature states that the opisthosoma pattern may be similar in both *E. caricis* and *Enoplognatha marmorata*, so without looking at genitalia (and in the case of the female, dissecting the specimen), species-level identification is very difficult.

female

female

female with egg sac

female with young

Habitat These spiders like to build their webs under debris, such as stones or wood, or in leaf litter. The egg sacs are a fluffy cream-colored ball.

Known records from AB, MB, NL, ON, QC, SK (Canada); AK, CO, CT, IA, IL, IN, MA, MD, ME, MI, MN, NJ, NY, OH, PA, RI, TX, VA, WA, WI (United States)
This is an introduced species.

juvenile

male

male

Enoplognatha marmorata (Hentz, 1850)

Common Name Marbled cobweb spider

Total Body Length Female 3.9–7.0 mm, Male 4.0–6.4 mm

Description The prosoma is dark yellow brown and darker in the eye region and at the edges. The opisthosoma is a variegated pattern of whites, browns, blacks, and tans. The dark heartmark is usually bordered in white or cream. The markings may be very faint in gravid females, but the white marking around the heartmark is usually visible. The legs are dusky yellow to orange, with slightly darker rings at the distal end of each segment. The literature states that the opisthosoma pattern may be similar in both *Enoplognatha caricis* and *E. marmorata*, so without looking at genitalia (and in the case of the female, dissecting the specimen), species-level identification is very difficult.

Habitat These species like to build their webs under debris, such as stones or wood, or in leaf litter. The egg sacs are a fluffy cream-colored ball.

Known records from AB, BC, MB, NB, NL, NS, ON, QC, SK (Canada); AL, AZ, CA, CO, CT, GA, IA, IL, MO, MD, ME, MI, MN, MS, MT, NC, ND, NH, NJ, NM, NY, OH, OR, RI, TN, TX, UT, VA, WA, WI, WY (United States)

female

female

male

male

female with egg sac

Enoplognatha ovata (Clerck, 1757)

Common Name This species does not have a common name.

Other Colloquial Names Common candy-striped spider, candy stripe spider, polymorphic spider

Total Body Length Female 4.3–7.2 mm, Male 3.5–5.2 mm

Description The prosoma is pale yellow, with a dusky median stripe and a narrow dark edge. The sternum is yellow, with a dark border and dark median stripe. The opisthosoma can have a variety of coloration morphs. The *lineata* form has a pale yellow opisthosoma with paired dark spots. The *redimita* form has a cream-colored opisthosoma with a pair of red to pink stripes down the length. The *ovata* form has a cream-yellow opisthosoma with a broad red to pink band that covers the dorsal surface. The ventral opisthosoma has a broad black line. The legs are yellow, and a dark band is at the distal end of the tibiae. The males have elongated chelicerae and fangs.

female

female

Habitat These spiders tend to build their webs in shrubby or herbaceous vegetation.

Known records from BC, NB, NL, NS, ON, PE, QC, SK (Canada); CA, CO, IL, IN, MA, ME, NH, NY, OH, RI, UT, VT, WA, WI (United States)

female ventral

female

male

male ventral

Episinus amoenus
Banks, 1911

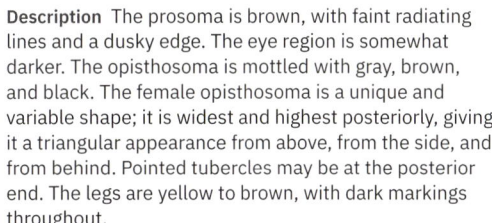

Common Name This species does not have a common name.

Total Body Length
Female 3.0–4.5 mm, Male 3.0 mm

Description The prosoma is brown, with faint radiating lines and a dusky edge. The eye region is somewhat darker. The opisthosoma is mottled with gray, brown, and black. The female opisthosoma is a unique and variable shape; it is widest and highest posteriorly, giving it a triangular appearance from above, from the side, and from behind. Pointed tubercles may be at the posterior end. The legs are yellow to brown, with dark markings throughout.

Habitat These spiders are often found in leaf litter near logs in forested habitats. They do not build the traditional space-filling web; rather, they create a frame of two main lines that reach from the ground to a low branch. While hanging, the spider holds these two lines until unsuspecting prey brushes against them. They have been documented preying upon ants.

Known records from AL, DC, FL, GA, MD, MS, NC, OH, TN, VA

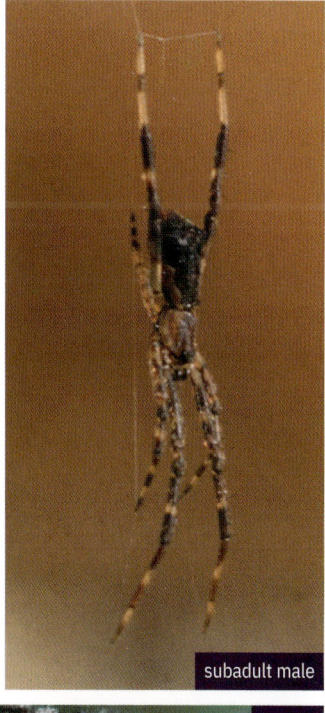

subadult male

female

Euryopis funebris (Hentz, 1850)

Common Name This species does not have a common name.

Total Body Length Female 3.0–4.7 mm, Male 2.3–3.0 mm

Description The prosoma is dark brown. The opisthosoma is teardrop shaped and dark brown, with a broad silver band with a wavy border on the lateral and posterior edges. The legs are banded in black and cream.

Habitat These spiders tend to be found in low vegetation, under leaves, or on the bark of trees. They do not build the traditional cobweb; rather, they hang head down from a few strands and capture ants that wander past. The egg sacs are a spikey dome attached to the surface. These egg sacs are often confused for *Latrodectus geometricus* egg sacs, but *L. geometricus* hangs its spherical egg sacs within the web. *Euryopis* egg sacs are not spherical and are attached completely to the surface.

female

female

female

Known records from BC, MB, NS, ON, PE, QC (Canada); AL, AR, CO, CT, FL, GA, IA, IL, IN, LA, MA, MD, ME, MI, MS, NC, ND, NH, NJ, NY, OH, OK, PA, SC, TN, VA, WI, WV (United States)

Euryopis scriptipes Banks, 1908

Common Name This species does not have a common name.

Total Body Length Female 4.0–6.0 mm, Male 3.0–4.0 mm

Description The prosoma is pale yellow to light brown and darker in the eye region. A black band across the clypeus is as wide as the eye region. The opisthosoma is cream to silvery on the lateral edges. Its markings can vary somewhat, but it usually has a dark triangular area with undulating edges; this may be either pale at the anterior or dark throughout. The legs are yellow with scattered spots.

Habitat These spiders tend to be found in low vegetation, under leaves, or on the bark of trees. They do not build the traditional cobweb; rather, they hang head down from a few strands and capture ants that wander past. The egg sacs are a spikey ball attached to the surface of the vegetation or other structures. These egg sacs are often confused for *Latrodectus geometricus* egg sacs, but *L. geometricus* hangs its spherical egg sacs within the web. *Euryopis* egg sacs are not completely spherical and are attached completely to a surface.

Known records from AB, BC, SK (Canada); AZ, CO, MT, NE, NM, SD, UT, WY (United States)

female with ant prey

Faiditus cancellatus female (see below)

Faiditus cancellatus
(Hentz, 1850)

Common Name This species does not have a common name.

Total Body Length Female 2.8–3.0 mm, Male 2.8–3.0 mm

Description The prosoma is dark brown to black and shiny. The opisthosoma is dark gray, with speckled white or silver spots. This species can change the shape of the opisthosoma. At times, the end can appear rounded and smooth, while at other times, there is a pair of tubercles at the posterior end and a pair on each lateral edge. The legs are pale brown proximally, with the distal ends of the femur and the distal segments darker.

female with egg sac

female with wasp cocoon

male

male

male

male

Habitat These spiders are often found inhabiting the webs of other spiders, such as Araneidae, Agelenidae, Linyphiidae, and other theridiids. The egg sac is suspended on a stiff stalk of silk; the sac is oval, and an extension that looks like a tube projects out the bottom.

Known records from ON (Canada); AL, AR, CT, DC, FL, KY, LA, MO, MS, NC, NH, NY, OH, PA, SC, TN, TX, VA (United States)

Hentziectypus globosus (Hentz, 1850)

Common Name This species does not have a common name.

Total Body Length Female 1.6–2.0 mm, Male 1.2–1.8 mm

Description The prosoma is dark brown, with a darker median line that widens in the eye region. The female opisthosoma is highly elevated. The

female

female

female

female with egg sac

dorsal surface is cream white, with a central black spot. The ventral surface is dark. When viewed from the side, a line clearly defines light and dark on the lateral edges. The males have a much less elevated opisthosoma that tends to be more orange. The males lack the large cream-white surface but still have the central black spot. The legs are dusky brown and unbanded.

Habitat These spiders tend to build their small cobwebs in leaf litter, hollow logs, or other humid environments. The egg sacs taper at both ends and look like they are covered in cream paper.

Known records from MB, NS, ON, QC (Canada); AL, AR, CT, DC, FL, GA, IA, IL, IN, KY, LA, MA, MD, ME, MI, MN, MO, MS, NC, NH, NJ, NY, OH, PA, SC, TN, TX, VA, WI (United States)

Latrodectus bishopi Kaston, 1938

Common Name Red widow

Total Body Length Female 7.8–9.0 mm, Male 3.0–3.8 mm

Description The prosoma is red to orange. The lateral eyes are distinctly separated from each other. The female opisthosoma is globular. It is shiny black, with red-orange spots outlined in white. A line of larger spots on the dorsal surface usually runs the length of the opisthosoma; smaller spots appear laterally. Sometimes the central spots are heart shaped or triangular. The ventral opisthosoma has one or two red spots or markings. The book lung covers are dark. The male opisthosoma is lighter, orange to yellow, with white markings.

Habitat These spiders seem to be found exclusively in inland dune and sand-pine habitats. They tend to make their webs up in the vegetation, and they usually have a conical retreat, which they create by rolling a leaf of the vegetation. The egg sacs are smooth and cream white to gray. This species is considered Imperiled (G2G3).

Known records from BC (Canada); FL, IN, KS, MA, MT, NH, UT, VA, WA, WI (United States)

female

Envenomation by mature females of this species is considered of medical concern. If you are bitten, please seek medical attention.

Latrodectus geometricus C. L. Koch, 1841

Common Name Brown widow

Other Colloquial Names Geometric button spider, brown button spider, button spider, gray widow spider

Total Body Length Female 5.0–16.0 mm, Male 3.0–8.0 mm

Description The prosoma is typically dark brown. The lateral eyes are distinctly separated from each other. The opisthosoma is globular and tan to brown, with a series of spots down the midline and smaller spots laterally. The spots are red to orange, ringed with white, and bordered with black. The lateral spots have a black spot toward the midline. Some individuals are completely dark, almost black, creating difficulty distinguishing them from other *Latrodectus* species. The hourglass is fairly consistent in shape but can vary from bright yellow to brilliant red. The book lung covers are pale on the pale specimens and darker on the dark specimens. The legs are yellow to orange, with dark patellae and distal tibiae.

female ventral

female

female

female

female in retreat with egg sacs

female with egg sac

Habitat These spiders seem to like human structures and prefer to be outside; they are rarely found inside structures. They like to build their webs under mobile homes or some sort of overhang, such as fence supports or the recessed handles of trash bins; they are rare in agricultural settings. The egg sacs are spherical and covered with spikes. The egg sacs are usually suspended within the web. Envenomation by *L. geometricus* is considered less severe than envenomation by other *Latrodectus* species in North America, typically causing mild localized symptoms, although a few rare cases of more severe symptoms do exist in the literature (Vetter et al. 2012).

Known records from AZ, CA, FL, OH, TX
This is an introduced species.

Latrodectus hesperus Chamberlin & Ivie, 1935

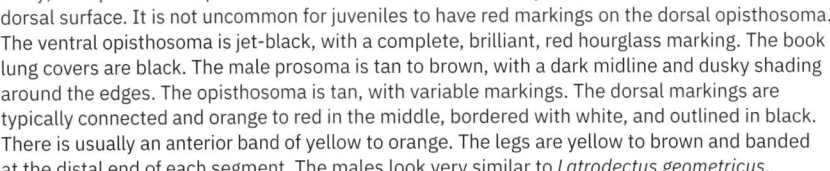

Common Name Western black widow

Total Body Length Female 8.0–16.0 mm, Male 3.0–6.5 mm

Description The female prosoma is black and shiny. The lateral eyes are distinctly separated from each other. The female opisthosoma is black and shiny; red spots or red spots bordered in white are occasionally on the dorsal surface. It is not uncommon for juveniles to have red markings on the dorsal opisthosoma. The ventral opisthosoma is jet-black, with a complete, brilliant, red hourglass marking. The book lung covers are black. The male prosoma is tan to brown, with a dark midline and dusky shading around the edges. The opisthosoma is tan, with variable markings. The dorsal markings are typically connected and orange to red in the middle, bordered with white, and outlined in black. There is usually an anterior band of yellow to orange. The legs are yellow to brown and banded at the distal end of each segment. The males look very similar to *Latrodectus geometricus*.

Habitat This species tends to prefer xeric habitats and can be quite common in agricultural settings.

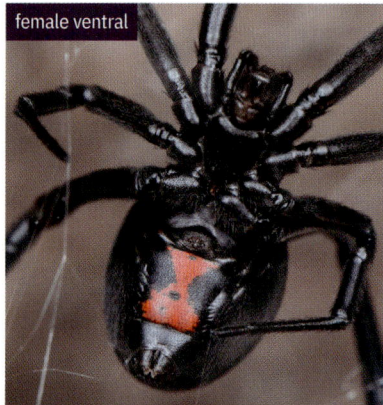
female ventral

female with prey (male in background)

male

Envenomation by mature females of this species is considered of medical concern. If you are bitten, please seek medical attention.

Known records from AB, BC, SK (Canada); AR, AZ, CA, CO, IA, KS, MN, NE, NM, NV, OK, OR, TX, UT, WA, WY (United States)

Latrodectus mactans (Fabricius, 1775)

Common Name Southern black widow

Other Colloquial Names Shoe button spider

Total Body Length Female 8.0–13.5 mm, Male 2.9–4.0 mm

Description The female prosoma is shiny black. The lateral eyes are distinctly separated from each other. The female opisthosoma is globular. It is shiny black, sometimes with red markings on the dorsal surface. The ventral opisthosoma is black, usually with a complete, brilliant, red hourglass that is wider toward the posterior. The legs are black. The male prosoma is black to dusky brown, sometimes with a visible midline. The male opisthosoma is black to tan brown, with a line of white to red spots down the dorsal surface. The legs have wide bands at the joints.

female

female

Envenomation by mature females of this species is considered of medical concern. If you are bitten, please seek medical attention.

female ventral

male

male

male

juvenile

Habitat This species is often found in human-disturbed areas and can be found in woodpiles, on buildings, and in basements or crawl spaces.

Known records from AL, AR, CA, CT, FL, GA, IL, IN, KS, LA, MN, MO, MS, NC, NJ, NY, OH, OK, SC, TN, TX, VA

Latrodectus variolus Walckenaer, 1837

Common Name Northern black widow

Total Body Length Female 9.2–11.0 mm, Male 5.5–6.7 mm

Description The prosoma is black and shiny. The lateral eyes are distinctly separated from each other. The female opisthosoma is globular. It is black and shiny and often has a series of red markings (sometimes outlined in white) down the dorsal midline. The ventral opisthosoma is black, with two separated red triangles or trapezoids. The book lung covers are dark. The legs are black. The males look similar, and white dashes are often on the lateral opisthosoma. The male legs can be red with dark banding at the joints.

Envenomation by mature females of this species is considered of medical concern. If you are bitten, please seek medical attention.

female ventral

female

Habitat This species likes to live in forested and other undisturbed habitats. They often have a funnel-like retreat that is surrounded by curled vegetation. The egg sacs are smooth spheres and can range in color from white to light brown.

Known records from AB, BC, ON (Canada); AR, CA, FL, GA, IL, KS, MA, MI, MO, OH, SC, TX, UT, VT, WA, WI, WV (United States)

female

male

male

spiderlings

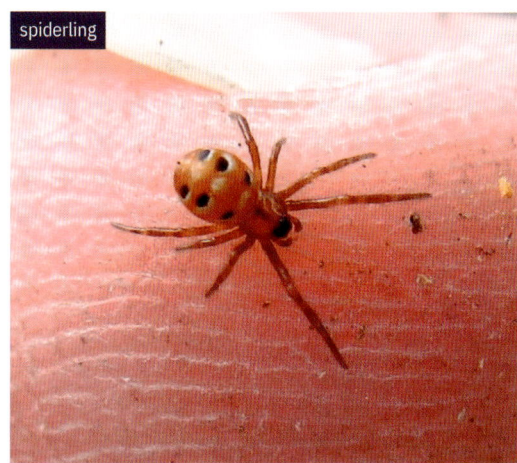

spiderling

Neopisinus cognatus (O. Pickard-Cambridge, 1893)

Common Name This species does not have a common name.

Total Body Length Female 4.5–6.0 mm, Male 3.9–4.3 mm

Description The prosoma is tan, with a dark brown median stripe. There are irregular dark brown submarginal bands. White hairs are scattered over the entire prosoma. The opisthosoma is gray tan, with a dark heartmark. There are scattered brown, gray, and black markings. The opisthosoma is uniquely shaped: it is tallest just over halfway down, where there is a pair of tubercles. The overall shape is very angular; viewed from the side, it makes almost a perfect triangle. The ventral opisthosoma has a wide black median stripe. The male opisthosoma is less dramatically shaped and may completely lack the tubercles and hump. The male pedipalps are very long.

Habitat These spiders are often found on twigs and tree branches.

Known records from AZ, TX

female

female

Neospintharus trigonum (Hentz, 1850)

Common Name This species does not have a common name.

Other Colloquial Names Horned parasitic cobweaver

Total Body Length Female 3.7–4.2 mm, Male 2.5–3.3 mm

Description The prosoma is orange. The opisthosoma is triangular, and the spider can change the shape to be more rounded or pointed at the posterior end. The color can range from pale golden to brown.

female

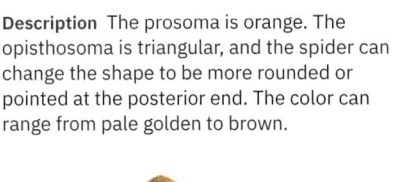

female

female

female with egg sac

juvenile

male

male

male

Some underlying iridescence is usually visible. The legs are orange and unbanded. The male prosoma has two horns: one projecting from just above the eyes, the other projecting from the clypeus.

Habitat This species can often be found in Araneidae (orbweavers) webs, where they steal small prey from the host's web and sometimes kill the host spider, although they do occasionally build their own cobweb. The egg sacs are suspended from a stiff silk stalk and are teardrop shaped; a tubular extension projects from the bottom.

Known records from NB, NS, ON, PE, QC (Canada); AL, AR, CT, FL, GA, IL, KY, MA, MD, ME, MI, MO, MS, NC, NH, NY, OH, PA, SC, TN, TX, VA, WI, WV (United States)

Nesticodes rufipes (Lucas, 1846)

Common Name This species does not have a common name.

Other Colloquial Names Red house spider

Total Body Length Female 4.2–6.0 mm, Male 2.8–4.0 mm

Description The prosoma is yellow to red. The opisthosoma is globular and grayish red to black. There is a white marking on the posterior end along the midline. The color of the ventral opisthosoma is similar to that of the dorsal surface. The spinnerets are ringed with a darker color. One to two pairs of white dashes may also radiate from the midline toward the posterior.

Habitat This species appears to be synanthropic and is usually found within human structures.

Known records from FL, TX, WA

female with egg sac

female with egg sac

juvenile

Parasteatoda tabulata (Levi, 1980)

Common Name This species does not have a common name.

Total Body Length Female 3.2–5.9 mm, Male 2.9–4.1 mm

Description The prosoma is orange to dark brown. A dark midline may be visible on the paler specimens and often on the males. The opisthosoma is highly variable in coloration, ranging from tans and oranges to black and cream. The pattern, although abstract, forms transverse lines across the opisthosoma. A pale crescent marking is just over halfway back on the dorsal surface. The legs are banded, usually to a heavier degree than is seen in the very similar-looking *Parasteatoda tepidariorum*, with which it is often confused.

female

female

female

female with egg sac

male

male

P. tabulata is likely more prevalent in North America than is known. To confirm a species-level identification, one would need to look at the genitalia under a microscope.

Habitat This species is synanthropic and often found in and around human structures. A debris-covered retreat is often in the middle of the web. The egg sacs are spherical to teardrop shaped and are usually within the retreat.

Known records from NB, NL, NS, ON, QC, SK (Canada); IL, IN, MA, ME, OH (United States) This is an introduced species.

Parasteatoda tepidariorum (C. L. Koch, 1841)

Common Name Common house spider

Total Body Length Female 3.5–7.5 mm, Male 3.5–6.5 mm

Description The prosoma is tan to dark brown. The opisthosoma is highly variable in colors and markings. Some are very pale tan, with scattered markings of white and brown. Others are almost black, with white and tan speckling. A faint white marking that resembles square brackets is sometimes on the dorsal opisthosoma. Crescent-shaped, dark brown to black marking is just over halfway down the opisthosoma. The legs are banded, to a lesser degree in the males. This species looks very similar to *Parasteatoda tabulata* and easily misidentified as such. To confirm a species-level identification, one would need to look at the genitalia under a microscope.

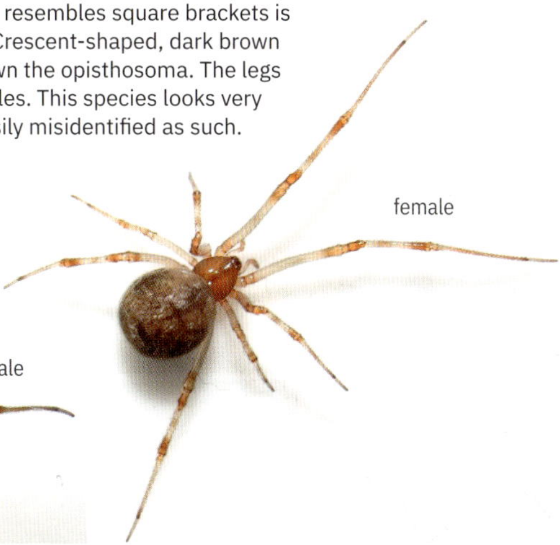

female

female

female with prey

female with egg sacs (one with young emerging)

female with egg sac

juvenile

male

male

Habitat This species is synanthropic and usually found in or around human structures. The egg sacs are spherical to teardrop shaped and are suspended within the web.

Known records from AB, BC, MB, NL, NS, ON, QC, SK (Canada); AL, AR, AZ, CA, CO, CT, FL, GA, IA, IL, IN, KS, KY, LA, MA, MD, ME, MI, MN, MO, MS, NC, NH, NJ, NM, NY, OH, OK, OR, PA, SC, TN, TX, VA, VT, WA, WI, WV (United States) This is an introduced species.

Phylloneta pictipes (Keyserling, 1884)

Common Name This species does not have a common name.

Total Body Length Female 2.8–5.0 mm, Male 2.5–3.5 mm

Description The prosoma is cream to yellow, with a dark median line that widens in the eye region. The edge is black. The legs are pale yellow, with dark bands at the middle and both ends of each segment. The opisthosoma is white to yellow, with black patches. The ventral opisthosoma is white, with a broad black band from the epigastric furrow to the spinnerets. This species looks very similar to *Theridion frondeum* and *Theridion albidum*, but *P. pictipes* has banding at both ends and the middle of each leg segment and a dark line on the edge of the prosoma. These features distinguish them from the *Theridion* species.

Habitat These spiders tend to build their webs in forested habitats and are often found under leaves.

Known records from AL, FL, GA, OH, SC, VA

male

male

male

Platnickina alabamensis (Gertsch & Archer, 1942)

Common Name This species does not have a common name.

Total Body Length Female 1.9–3.7 mm, Male 1.8–2.7 mm

Description The prosoma is yellow to yellow orange, with a dark median stripe that widens in the eye region. The opisthosoma is dark brown, with faint white markings. A white to yellow marking is just above the spinnerets, which are surrounded by a dusky ring. The legs are yellow to yellow orange.

male

male

male

male

Habitat These spiders are often found in humid habitats, including under bark, in hollow tree logs, or on the ground.

Known records from NB, ON, QC (Canada); AL, CA, CT, FL, GA, IL, LA, MA, MD, ME, MS, NC, NJ, NM, NY, OH, PA, RI, TX, WI (United States)

Rhomphaea fictilium (Hentz, 1850)

Common Name This species does not have a common name.

Other Colloquial Names Lizard spider

Total Body Length Female 5.0–10.5 mm, Male 4.0–7.0 mm

female

egg sac with young emerging

female

male

Description The prosoma is pale yellow to golden in color. The opisthosoma is uniquely shaped: wormlike, it possesses a long, flexible, tubular extension that can be extended and curled. It is usually yellow to pale brown, with some silver spots. The legs are pale and very slender. The male pedipalps are very long and thin.

Habitat These spiders are often found in understory vegetation. The egg sacs are an elongate tube with a latticelike pattern.

Known records from BC, MB, NS, ON, QC (Canada); AL, AZ, CA, CO, CT, FL, GA, LA, MA, ME, MI, MN, MO, MS, NC, NY, OH, OR, SC, TN, TX, WA (United States)

Rugathodes sexpunctatus (Emerton, 1882)

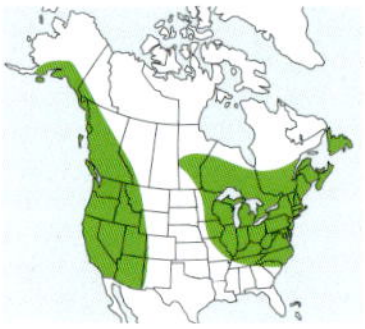

Common Name This species does not have a common name.

Total Body Length Female 1.5–2.6 mm, Male 1.8–2.4 mm

Description The prosoma is yellow, with a dark midline that widens slightly in the eye region. There is a dark band around the edge. The chelicerae are elongated in the male. The opisthosoma is yellow, with four paired black shapes, each of which contains a large pale dot. There are some white flecks throughout. The most anterior of the shapes tends to be triangular, while the others are somewhat rectangular. These markings range from very distinct to somewhat mottled. The ventral opisthosoma is yellow, and a pair of black bands wrap around from the dorsal but do not meet. The legs are yellow and banded in brown.

Habitat These spiders tend to be found in coniferous forests.

Known records from AB, BC, MB, NB, NL, ON, QC, SK, YT (Canada); AK, AZ, CA, CO, FL, ID, MA, MD, ME, MI, NC, NH, NY, OH, OR, PA, TN, UT, VT, WA, WI, WV, WY (United States)

female

female ventral

male

male

Spintharus flavidus Hentz, 1850

Common Name This species does not have a common name.

Total Body Length Female 4.0–4.5 mm, Male 2.8–3.1 mm

Description The prosoma is yellow to cream white. The posterior lateral eyes are in line with the anterior eye row, giving this spider the appearance of six eyes in the anterior row and only two posterior eyes. The opisthosoma

female

female

female

is teardrop shaped. The coloration can be quite variable. The lateral edges are often white with a red dorsal surface. Large yellow spots can be within the red, and white spots are in the middle of each yellow spot. The ventral opisthosoma is cream yellow. The legs are yellow, with red banding at the distal tibiae of legs I and IV.

Habitat These spiders are often found on the undersides of leaves. Their web is delicate and nearly invisible.

Known records from AL, AR, CT, DC, FL, GA, IN, KY, MA, MD, MS, NC, NJ, NY, OH, OK, SC, TN, TX, VA, WV

Steatoda albomaculata (De Geer, 1778)

Common Name The genus *Steatoda* has the common name "false widows."

Other Colloquial Names White-spotted false widow

Total Body Length Female 4.0–8.0 mm, Male 3.3–6.8 mm

male

male

male

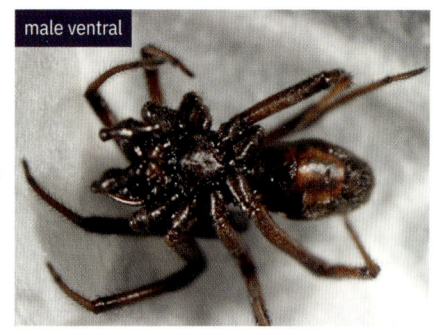

male ventral

Description The prosoma is very dark brown. The opisthosoma is dark brown, with a white line along the anterior edge. The markings on the opisthosoma can vary. Some are almost completely dark brown, with no evidence of any markings, while others have paired white spots on either side of the opisthosoma. The legs are dark brown, and some banding may be visible.

Habitat These spiders tend to build their webs close to the ground and can be found under rocks, logs, and other large debris.

Known records from AB, BC, MB, NB, NT, ON, PE, QC, SK, YT (Canada); AZ, CA, CO, CT, IA, ID, IL, IN, MA, MI, MN, MT, NE, NH, NM, NY, OH, OR, SD, UT, WA, WI, WV, WY (United States)

Steatoda borealis (Hentz, 1850)

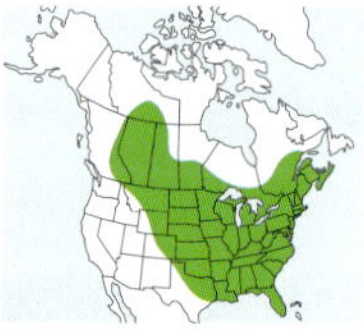

Common Name Northern cobweb weaver

Other Colloquial Names Boreal combfoot, false widow

Total Body Length Female 3.8–7.0 mm, Male 4.3–6.0 mm

female

Description The prosoma is dark brown. The opisthosoma is dark brown to deep purple. A white band on the anterior edge joins a white midline that can vary in length and, in some specimens, runs the entire length of the opisthosoma (although it may or may not be contiguous). The ventral opisthosoma is brown, with a dark V between the spinnerets and the epigastric furrow, the base of which encircles the spinnerets. The legs are brown and banded in dark brown. The male pedipalps look quite hairy, and the structures can look almost like claws.

Habitat These spiders are often found in webs close to the ground under debris. They are also often found on and around human

female

female ventral

female

structures. The egg sacs are fluffy white spheres and are suspended in the web or attached to the surfaces. Sometimes, the eggs appear to have a pink hue.

Known records from AB, BC, MB, NB, NS, NT, ON, QC, SK, YT (Canada); AK, AZ, CO, CT, DE, FL, IA, IL, IN, KS, KY, LA, MA, MD, ME, MI, MN, MO, MS, MT, NC, ND, NE, NH, NJ, NY, OH, PA, RI, SC, SD, TX, UT, VA, VT, WI, WV (United States)

female

female

female with egg sac

male

male

male

Steatoda grandis Banks, 1901

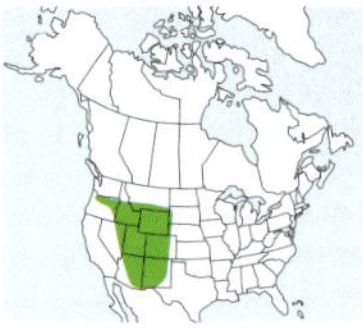

Common Name The genus *Steatoda* has the common name "false widows."

Total Body Length Female 4.9–9.0 mm, Male 4.3–7.4 mm

Description The prosoma is reddish brown. The opisthosoma is reddish brown, with a cream-colored anterior band that extends around the lateral edges. A pale cream marking that is shaped like an irregular lobe is sometimes on the dorsal opisthosoma. The females can look very similar to *Steatoda borealis*, but location and size should help distinguish the two. The legs are orange to brown. The male pedipalps appear hairy.

female

male

male

male

Habitat These spiders are often found under rocks or on rocky ledges. They are usually found at lower elevations.

Known records from AZ, CO, MT, NM, NV, OR, SD, UT, WY

Steatoda grossa (C. L. Koch, 1838)

Common Name False black widow

Other Colloquial Names False widow

Total Body Length Female 5.9–10.5 mm, Male 4.1–10.0 mm

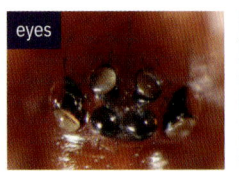

eyes

Description The female prosoma is dark brown to almost black. The opisthosoma is more dorsoventrally flattened than on a *Latrodectus* species. The opisthosoma is purplish to dark brown to almost black. A series of pale triangles often runs in a line down the midline, but in some darker specimens, these are not visible. The ventral opisthosoma is dark. In contrast to most *Latrodectus* species, which have dark brown to black book lung covers, the book lung covers

female

female

female

female

female

are light brown. The male prosoma is more orange. The marking of the opisthosoma of the male is similar to that of the female, but the male opisthosoma is much smaller. The legs are brown to black. These spiders have been documented preying upon *Latrodectus* species.

female ventral

egg sac with young emerging

juvenile

male

Habitat This species is usually found in or around human structures. The egg sacs are fluffy white spheres that are usually suspended within the web.

Known records from BC, NS, ON (Canada); AL, AZ, CA, CO, CT, FL, GA, IL, IN, LA, MA, MS, NM, OH, OR, RI, SC, WA (United States)
This is an introduced species and, from what is known in the scientific literature, seems to be expanding its range.

Steatoda nobilis (Thorell, 1875)

Common Name The genus *Steatoda* has the common name "false widows."

Other Colloquial Names Noble false widow

Total Body Length Female 7.8–10.6 mm, Male 9.5–14.0 mm

Description The prosoma is dark brown. The opisthosoma is dark brown, with a large, rectangular/diamond-shaped, cream to white-colored marking that is sometimes said to resemble a simplistic drawing of a house with windows or a doorway. A cream to white anterior band continues down the lateral edges almost to the spinnerets. The legs are yellow brown to brown, and some banding is usually visible.

female

female

female female

Habitat This species appears to be synanthropic and is often found in and around human structures. The spiders create a web with a funnel-like retreat that could easily be mistaken as an Agelenidae web.

Known records from CA
This is an introduced species and likely expanding its range.

Steatoda triangulosa (Walckenaer, 1802)

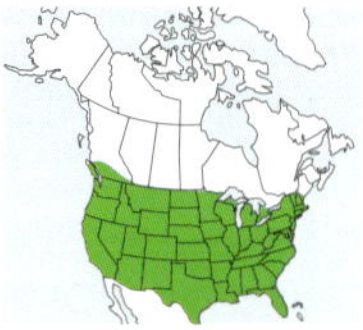

Common Name Checkered cobweb weaver

Other Colloquial Names False widow, triangulate combfoot, triangulate cobweb spider

Total Body Length Female 3.6–8.6 mm, Male 3.5–5.0 mm

female

female

female

female

female with egg sac

male

male

Description The prosoma is dark orange to brown. Faint radial lines may be visible. The opisthosoma is brown, with a row of cream-white diamonds down the midline. A pattern of darker and lighter markings is often on the lateral edges in a checkered pattern. The legs are dark yellow to orange, with faint banding.

Habitat These spiders are often found in and around human structures. The egg sacs are fluffy white spheres that are usually suspended within the web.

Known records from ON, QC (Canada); AL, AR, CA, CO, CT, DC, DE, FL, GA, IA, ID, IL, IN, KS, LA, MA, MD, ME, MI, MN, MO, MS, MT, NC, NE, NJ, NM, NY, OH, OK, OR, PA, SC, TX, UT, VA, WI, WV, WY (United States)
This is an introduced species.

Theridion albidum Banks, 1895

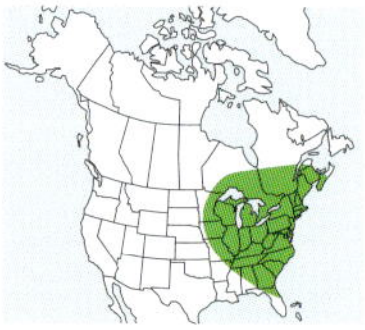

Common Name This species does not have a common name.

Total Body Length Female 2.2–3.3 mm, Male 2.2–2.8 mm

Description The prosoma is white to yellow, with a dark midline. The opisthosoma is gray to yellow and sometimes has black spots on the dorsum. The male tends to be slightly darker than the female. The distal end of each segment of the legs has some dark banding, although it may be faint. Some specimens have a small band in the middle of tibia I. The males have a small mastidion (tooth) on the chelicerae just below the clypeus.

Habitat This species is often found on the underside of leaves or branches.

Known records from NS, ON, QC (Canada); AL, CT, DE, IA, IL, IN, KY, LA, MA, MD, ME, MI, MN, MO, NC, NJ, NS, NY, OH, PA, TN, UT, VA, WI, WV (United States)

female

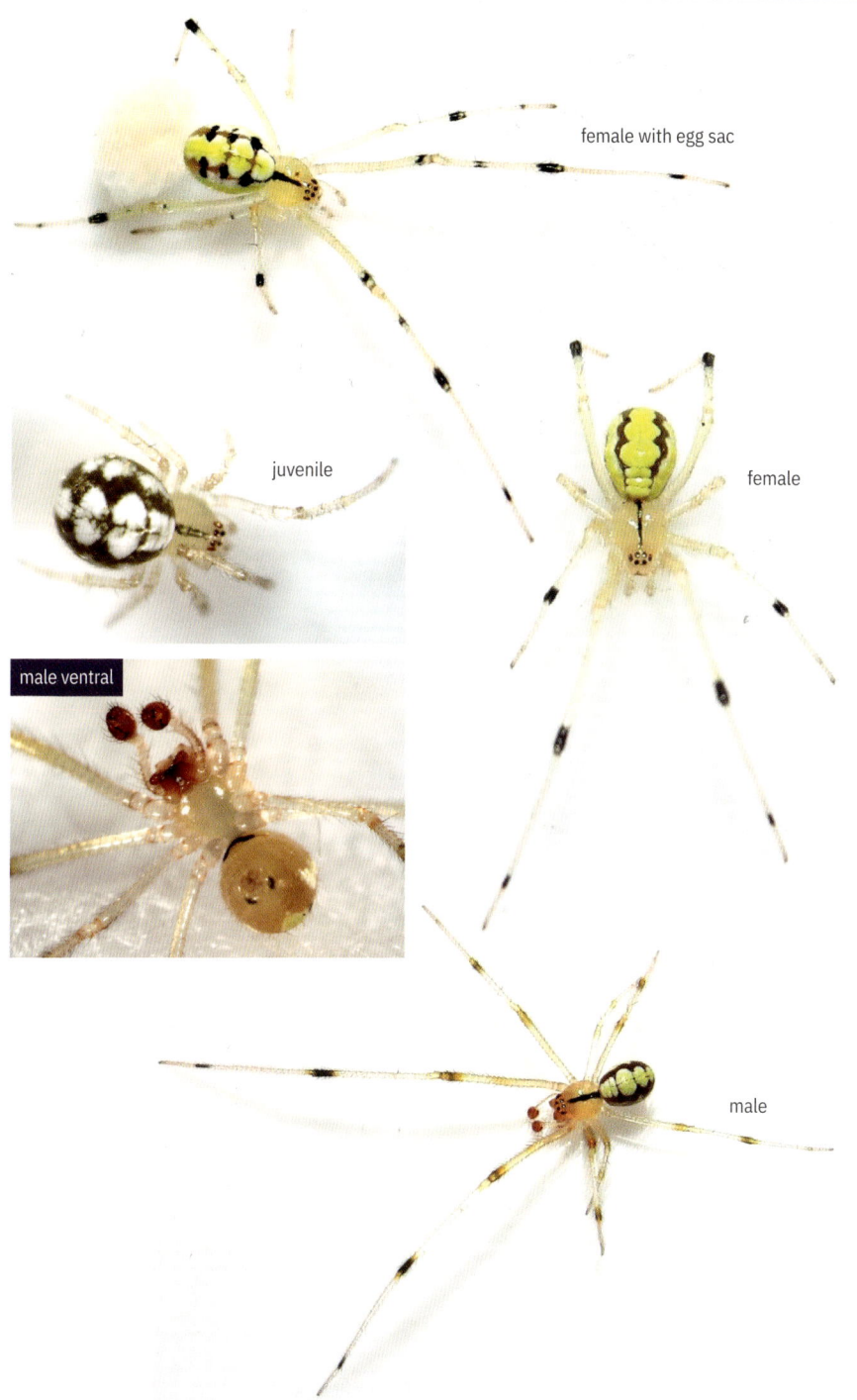

female with egg sac

juvenile

female

male ventral

male

Theridion australe Banks, 1899

Common Name This species does not have a common name.

Total Body Length Female 2.0–3.0 mm, Male 1.9–2.3 mm

Description The prosoma is yellow to deep orange red. A dark marking usually starts at the fovea and widens anteriorly to the eye region. The males may have a dark marking just around the eyes. The opisthosoma is orange to dark brown. There is a lobed cream median stripe. At the tips of each lobe is a yellow spot. The lateral edges are cream and appear in an undulating pattern. A pair of black spots is just above the spinnerets; this characteristic is one of the first markings visible on young spiderlings. The ventral opisthosoma is orange to orange brown. The opisthosoma markings on the males may be greatly reduced; they sometimes are mottled with white and orange, with dark spots. The legs are yellow and banded in brown.

female

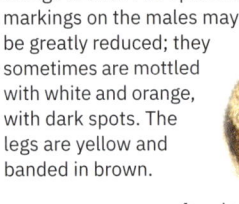
female

juvenile

egg sac and spiderling

male

male

Habitat The spiders are often found in grassy habitats.

Known records from AR, DE, FL, LA, MD, MS, NC, NJ, TX, UT

Theridion differens Emerton, 1882

Common Name This species does not have a common name.

Total Body Length Female 1.8–3.5 mm, Male 1.6–2.5 mm

Description The prosoma is orange red, with a dark midline that widens in the eye region. The opisthosoma is brown, with a pair of wavy white to cream lines that run down the length. The coloration of the space between these lines is variable. In some specimens, there is a reddish tone anteriorly, yellow in the middle, and red posteriorly. Other specimens are tan to gray in this area. The markings are much less obvious in the male, whose opisthosoma may be uniformly dark orange. The legs are pale yellow, with faint banding. These spiders can easily be confused with other *Theridion* species. In many cases, looking at the genitalia with a microscope is the only way to confirm the identification.

female

subadult male

female

Habitat These spiders are usually found in low vegetation. The egg sacs are white to tan spheres and are attached within the web.

Known records from AB, BC, MB, NB, NL, NS, NT, ON, PE, QC, SK (Canada); AK, AL, AR, CA, CO, CT, FL, GA, IA, ID, IL, IN, KS, KY, MA, MD, ME, MI, MN, MO, MS, MT, NC, NE, NH, NJ, NY, OH, OK, OR, PA, SC, SD, TN, TX, UT, VA, VT, WA, WI, WY (United States)

male

male

Theridion frondeum Hentz, 1850

Common Name This species does not have a common name.

Other Colloquial Names Eastern long-legged cobweaver

Total Body Length Female 3.0–4.2 mm, Male 3.0–3.9 mm

Description The prosoma is white to yellow, with a single or paired median stripe. The opisthosoma is white to yellow and sometimes has black spots and other times is almost completely black. These spiders can be quite variable in their appearance. Two black spots are above the spinnerets. The males have a large mastidion (tooth) on the chelicerae just below the clypeus. The legs are pale yellow; banding at the distal end of each segment may be very faint. This species looks very similar to *Phylloneta pictipes* and *Theridion albidum*. The lack of mid-segment banding on the legs differentiates this species from *P. pictipes*, and the overall size can be used to distinguish it from *T. albidum*. (There is some overlap in the sizes, so viewing the genitalia under a microscope would be needed to confirm an identification.)

female

female

female

female

Habitat These spiders are usually found in shrubs and herbaceous vegetation.

Known records from AB, BC, MB, NB, NS, ON, PE, QC, SK (Canada); AL, AR, AZ, CA, CO, CT, FL, IA, IL, IN, LA, MA, MD, ME, MI, MN, MS, NC, ND, NH, NJ, NY, OH, PA, SC, SK, TN, TX, VA, VT, WA, WI, WV (United States)

female with egg sac

male

male

Theridion glaucescens Becker, 1879

Common Name This species does not have a common name.

Total Body Length Female 1.6–3.0 mm, Male 1.4–2.5 mm

Description The prosoma is yellow to orange, with a dark median band that widens in the eye region. There is a dusky border to the prosoma. The opisthosoma is yellow brown, with a scalloped white median band. The males are usually darker, and the median band may be dark yellow. Faint striping or mottling may be on the lateral edges. The ventral opisthosoma is yellow to orange. The legs are yellow with banding, which may be faint in some individuals. These spiders can easily be confused with other *Theridion* species. In many cases, looking at the genitalia with a microscope is the only way to confirm the identification.

female

female

female ventral

female with egg sac

male

male

Habitat These spiders are often found on the underside of leaves of low vegetation. The egg sacs are a papery white to gray sphere that is usually suspended within the web.

Known records from MB, NB, NL, NS, ON, PE, QC (Canada); AL, CT, FL, GA, IL, IN, KY, LA, MA, MD, ME, MI, MS, NC, NE, NH, NJ, NY, OH, PA, SC, TN, TX, VA, VT, WI, WV (United States)

Theridion goodnightorum Levi, 1957

Common Name This species does not have a common name.

Total Body Length Female 3.0–4.0 mm, Male 2.7 mm

Description The prosoma is red orange. A median stripe starts at the fovea and widens slightly as it approaches the eye region. The edges are dusky. The males often have a completely dusky-colored prosoma. The opisthosoma is mottled with cream white and yellow. There is the faint appearance of a pair of white to yellow median stripes that zigzag down the dorsal surface. A pair of dark brown spots is just above the spinnerets. The ventral opisthosoma is dark brown. The legs are white to cream, often with yellow banding.

Habitat These spiders are usually found near the ground, where they build their webs under debris or grasses.

Known records from AZ, CA, CO, NM, TX, UT, WY

female ventral

female

female

Theridion murarium
Emerton, 1882

Common Name This species does not have a common name.

Other Colloquial Names Fence long-legged cobweaver

Total Body Length Female 2.8–4.3 mm, Male 2.1–3.2 mm

Description The prosoma is yellow, with a dark median line that widens in the eye region. There is a thin dark line around the edge. The opisthosoma is brown or orange, with a pair of scalloped white to cream lines that are sometimes bordered in dark brown. Anteriorly and

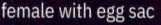

female with egg sac

female

female

male

juvenile

posteriorly
between the lines
is a bold white to
yellow marking;
otherwise, the area
between the lines is usually
red to orange. The lines may be
very faint in the males, but the bold yellow to white marking
is usually very clear. The legs are pale with banding.
T. murarium can easily be confused with other *Theridion*
species. In many cases, looking at the genitalia with a
microscope is the only way to confirm the identification.

juvenile

Habitat These spiders are often found on low vegetation
and human structures. The egg sacs are a papery gray
sphere and are often suspended within the web.

Known records from AB, BC, MB, NB, NL, NS, ON, PE, QC,
SK (Canada); AL, AR, AZ, CA, CO, CT, FL, GA, IA, IL, IN, LA,
MA, MD, ME, MI, MN, MS, MT, NC, ND, NE, NH, NJ, NM, NY,
OH, OK, OR, PA, RI, SC, SD, TN, TX, UT, VA, VT, WA, WI
(United States)

Theridion pennsylvanicum Emerton, 1913

Common Name This species does not have a common name.

Total Body Length Female 2.4–3.3 mm, Male 1.8–2.6 mm

Description The prosoma is yellow to orange, with a thin dark median line
and a dark submarginal band. The opisthosoma is yellow to red, and a pair

male

male

of white lines run the entire length. Two pairs of transverse dashes intersect the median lines about one-third and two-thirds of the way down. There are small black spots throughout. The legs are yellow to orange, with some banding.

Habitat These spiders tend to be found in forested habitats under the leaves of trees and shrubs.

Known records from ON (Canada); AL, CT, FL, IL, IN, MA, MD, MO, NC, NJ, NY, OH, SC, TN (United States)

Theridion pictum (Walckenaer, 1802)

Common Name This species does not have a common name.

Other Colloquial Names Painted cobweb weaver, red arrow cobweb spider

Total Body Length Female 2.9–4.7 mm, Male 3.1 mm

Description The prosoma is yellow to pale orange. A median stripe widens anterior of the fovea to the eye region. The edges are dusky. The opisthosoma is dark brown. The heartmark is white. A white to pink median stripe in the shape of connected arrows points

female

toward the anterior; this shape is filled with dark pink to red. A pair of dark spots may be just above the spinnerets. The sides have a cream to white edge. The ventral opisthosoma is gray. The legs are yellow to orange, with dusky banding.

Habitat These spiders are most often found in low vegetation in coniferous forests and other moist habitats. They sometimes construct a tent of debris to protect the egg sacs and newly emerged young.

Known records from AB, BC, LB, MB, NB, NL, NS, NT, ON, QC, SK (Canada); CO, ID, IL, MA, ME, MI, MN, MT, NE, NH, NY, OH, SD, UT, WA, WI, WY (United States)

Theridula emertoni
Levi, 1954

Common Name This species does not have a common name.

Other Colloquial Names Emerton's bitubercled cobweaver

Total Body Length Female 2.3–2.8 mm, Male 2.0–2.3 mm

Description The prosoma can be yellow or orange with a black median stripe or completely black. The female opisthosoma is a unique shape: it is wider than it is long and seems almost to come to a point at each lateral extension.

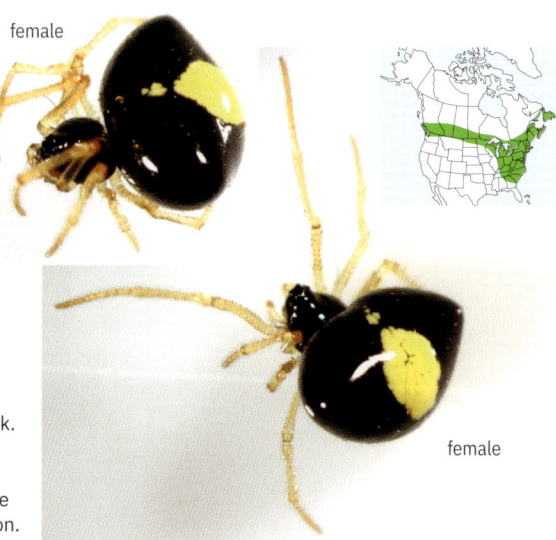

female

female

The color can vary from black to bright red, with a central yellow spot. When the opisthosoma is red, the points on the lateral extensions darken toward the points. The males have an ovoid opisthosoma that is orange, sometimes with a white or yellow central spot. The legs are yellow and unbanded.

female ventral

Habitat These spiders are often found in low vegetation. The egg sacs are a white sphere, and the females have been observed carrying them.

Known records from AB, BC, MB, NB, NL, NS, ON, QC, SK (Canada); AR, CT, DE, IL, MA, ME, MI, MN, MS, NC, NH, NY, OH, SC, TN, WI, WV (United States)

Thymoites pallidus (Emerton, 1913)

Common Name This species does not have a common name.

Total Body Length Female 1.5–2.9 mm, Male 1.3–1.8 mm

Description The prosoma is orange, with a dark median band from the fovea to the eye region. The opisthosoma is yellow to orange, with a dashed dark median line. A thin dark band is around the lateral edges. Paired dark dashes are on either side of the median line. The ventral opisthosoma has a dark band that runs from the epigastric furrow and rings the spinnerets. The legs are orange and unbanded.

female

female

female

female ventral

Habitat These spiders can be found close to the ground, often under debris.

Known records from CA, CO, FL, GA, MA, MS, NC, NM, NY, OH, OK, RI, TN, TX, UT

female ventral

subadult male

Thymoites unimaculatus (Emerton, 1882)

Common Name This species does not have a common name.

Other Colloquial Names Spotted cobweaver

Total Body Length Female 1.2–2.3 mm, Male 1.4–2.0 mm

Description The prosoma is orange red, with a dark median line from the fovea to the eye region. The opisthosoma is tan to red orange, and a black dot is often toward the posterior end. The ventral opisthosoma is tan to red orange, with a black ring around the spinnerets. The legs are yellow, with brown to black shading. The color and size of this species could lead one to misidentify it as one of the Linyphiidae, but the presence of the tarsal comb on the fourth tarsi identifies it as a theridiid.

female

female

female ventral

female with egg sac

male

subadult male

Habitat These spiders are often found in low vegetation. The egg sacs are a slightly fluffy white sphere.

Known records from MB, NB, NS, ON, PE, QC, SK (Canada); AL, CT, FL, GA, IA, IL, IN, LA, MA, MD, ME, MI, MN, MS, NC, NJ, NY, OH, PA, RI, SC, TN, TX, WI (United States)

Tidarren sisyphoides (Walckenaer, 1841)

Common Name This species does not have a common name.

Other Colloquial Names Tent cobweb weaver

Total Body Length Female 5.8–8.6 mm, Male 1.2–1.4 mm

Description The prosoma is light brown, and a dark median stripe widens in the eye region. There is a dark marginal band. The opisthosoma is pale

female and male

female and male

female and male mating

male

brown to tan, with dark speckling and somewhat of an orange hue. A white band extends up from the spinnerets. Paired white lines are often on the sides about halfway down the opisthosoma. The legs are pale brown and banded. The male removes one palp at the penultimate stage, and when he matures, his one remaining palp is large.

Habitat These spiders are often found around human structures or in rocky areas. The web often has a curled leaf for a retreat; the female hides her egg sacs within this retreat.

Known records from AL, AR, AZ, CA, FL, IL, KS, KY, LA, MS, NC, NM, SC, TX

Wamba crispulus
(Simon, 1895)

female

Common Name This species does not have a common name.

Total Body Length Female 1.4–2.6 mm, Male 1.2–1.6 mm

Description The prosoma is white to yellow, with a dark median stripe that widens in the eye region. Anteriorly, the opisthosoma has a black to brown triangle that is surrounded by cream white. A pair of black spots are on the posterior end. The ventral opisthosoma is dark. The legs are yellow, with banding and spots.

Habitat These spiders are often found on shrubby vegetation.

Known records from NB, NS, ON, PE, QC (Canada); AL, AR, CA, CO, FL, GA, IN, LA, MA, MD, ME, MS, NC, NH, NY, OH, OK, OR, SC, TN, TX, VA (United States)

male

Yunohamella lyrica
female (see below)

Yunohamella lyrica
(Walckenaer, 1841)

Common Name
This species does not have a common name.

Other Colloquial Names
Lyric cobweaver

Total Body Length
Female 2.5–3.5 mm,
Male 2.1–2.8 mm

female with egg sac

Description The prosoma is yellow, with a dusky submarginal band. The opisthosoma has a white to yellow median line that is often veined with red. The rest of the opisthosoma is mottled with cream and black. Anteriorly, the median line is bordered in black. The legs are heavily banded.

male

male

Habitat These spiders are often found in low vegetation or on fences and occasionally within human structures. The egg sac is a papery tan to brown sphere that is suspended within the web. The females have been documented carrying them in their chelicerae at times.

Known records from NB, NS, ON, PE, QC (Canada); AL, AR, CT, FL, GA, IA, IL, IN, KY, LA, MA, ME, MI, MS, NC, NH, NJ, NY, OH, PA, SC, TN, TX, VA, WI (United States)

TITANOECIDAE ▶ ROCK WEAVERS

Genera in This Family *Titanoeca* (4)

Total Body Length 4.0–8.0 mm

Description These are medium-sized to large entelegyne and cribellate spiders. The cribellum is divided, and the calamistrum is long, usually more than 70 percent of the length of metatarsus IV. They have eight eyes in two rows; both rows are nearly straight. They are usually found in xeric microhabitats near ground level.

Similar Families Amaurobiidae, Desidae, Dictynidae, Zoropsidae

How to distinguish Titanoecidae from other similar families:

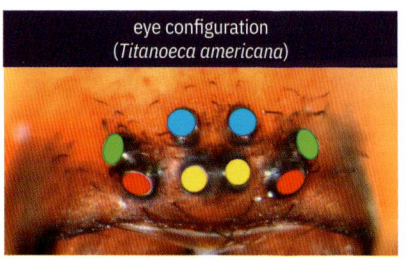

eye configuration
(*Titanoeca americana*)

divided cribellum

long calamistrum

Titanoeca americana

Titanoecids can be distinguished from amaurobiids and desids by their long calamistrum: in titanoecids, the calamistrum is over 70 percent of the length of metatarsus IV, while in amaurobiids it is less than 50 percent of the length and in desids it is about 40–50 percent of the length. Titanoecids lack markings on the opisthosoma, whereas visible markings are usually on the opisthosoma of amaurobiids. The enlarged anterior median eyes of *Badumna longinqua* (Desidae) also distinguish them from titanoecids. The cribellate dictynid species tend to have a complete cribellum (or at least it appears complete), whereas titanoecids have a divided cribellum. Also, most of the cribellate dictynids are small (less than 4 mm total body length). Most of the zoropsids are ecribellate. Those zoropsids with a cribellum (*Zoropsis* and *Zorocrates* species) can be distinguished by the posterior eyes: they are recurved in *Zoropsis* but nearly straight in titanoecids. *Zorocrates* (Zoropsidae) are found in Texas and Arizona (and areas south of there), while most of the titanoecids are found in more northerly locations.

Titanoeca americana Emerton, 1888

Common Name This species does not have a common name.

Other Colloquial Names American rock weaver

Total Body Length Female 4.4–7.4 mm, Male 5.7–7.4 mm

Description The prosoma is red, with a dark eye region covered with hairs. The opisthosoma is dark brown to black and covered with velvety hair. That there are no visible markings on the opisthosoma helps to distinguish these spiders from the similar-looking species in the family Amaurobiidae. The legs are dark brown to black and appear quite fuzzy.

Habitat These spiders are usually found in leaf litter or under rocks and seem to prefer dry habitats.

Known records from ON, QC, YT (Canada); AR, CO, CT, IA, IL, IN, KS, MA, ME, MI, MO, NE, NH, NJ, NM, NY, OK, PA, SD, TX, VA, WI, WY (United States)

male

subadult male

female ventral

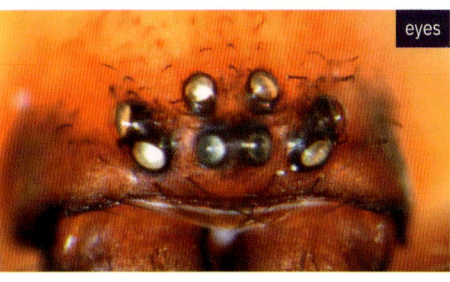
eyes

TROGLORAPTORIDAE ▷ LONGCLAWED CAVE SPIDERS

Other Colloquial Names Bigfoot spiders, Sasquatch spiders, trogloraptors, cave robbers, cave hunters

Genera in This Family *Trogloraptor* (1)

Total Body Length 6.9–9.6 mm

Description These spiders are haplogyne and ecribellate. They have six eyes. (The anterior median eyes are absent.) The lateral eyes are located adjacent to one another, and the posterior median eyes are between them. Their eyes occupy more than half the width of the prosoma. They have very unique subsegmented tarsi: the tarsi appear to have an additional joint. As subsegmented tarsi are on all their legs, they are most likely not used for prey

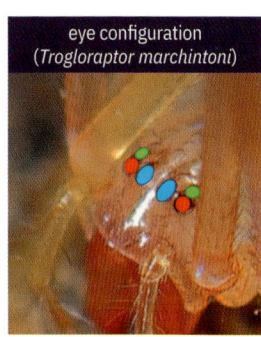
eye configuration (*Trogloraptor marchintoni*)

Trogloraptor marchintoni

subsegmented tarsi

capture but may be used to grasp the walls of the caves they inhabit (Murphy and Roberts 2015). The spiders have a pair of book lungs and a broad tracheal spiracle located anterior of the spinnerets. One described species has been found only in caves in Oregon, and one undescribed species was found in the understory of old-growth forests in California.

Similar Families Caponiidae, Dysderidae, Leptonetidae, Ochyroceratidae, Oonopidae, Pholcidae, Telemidae

How to distinguish Trogloraptoridae from other similar families:

Trogloraptorids can be distinguished from all listed similar families by their unique subsegmented raptorial tarsi. Trogloraptorids can be distinguished from all listed similar families by their eye configuration. Trogloraptorids have six eyes. (The anterior median eyes are absent.) The lateral eyes are located adjacent to one another, and the posterior median eyes are between them. Their eyes occupy more than half the width of the prosoma. Caponiids have either two eyes or eight eyes in a cluster. Dysderids have their six eyes clustered together; they occupy less than a third of the width of the prosoma. Leptonetids usually have their posterior medians set back from the other eyes. (Some are eyeless, and a few have contiguous eyes that occupy less than half the width of the prosoma.) Ochyroceratids have six eyes arranged in three diads. Oonopids have six eyes in a cluster. Pholcids have either eight eyes or six eyes that are in two triads and usually elevated. Telemids have six eyes in three diads.

Trogloraptor marchingtoni
Griswold, Audisio & Ledford, 2012

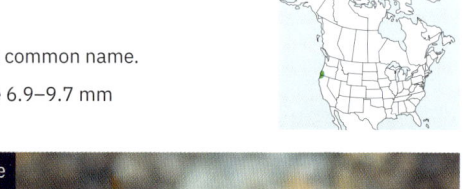

Common Name This species does not have a common name.

Total Body Length Female 8.3–9.6 mm, Male 6.9–9.7 mm

Description The prosoma is pear-shaped and yellow to pale orange, and a V-shaped brown marking is on the prosoma near the fovea. The anterior median eyes are absent. The opisthosoma is orange to purple brown, and pale chevrons are at the posterior end. The legs are orange, and all the tarsi are subsegmented and have raptorial claws.

female

Habitat This species has been found only within caves, where they suspend their webs from the ceiling.

Known records from CA, OR

The spiders of the Ambush Hunter Guild are sit-and-wait predators; they do not rely on silk for prey capture. They often resemble the substrate on which they sit; unsuspecting prey will walk right up to them. All the spiders in this guild are Araneomorphae. This guild includes the following families: Deinopidae, Selenopidae, Sicariidae, and Thomisidae.

KEY TO FAMILY

1a. Six eyes in three diads, long slender legs that lack spines,
5.0–13.0 mm total body length -- **Sicariidae** (page 302)
1b. Eight eyes -- **2**

Loxosceles reclusa

a   b

Misumessus oblongus

2a. Posterior median eyes extremely large, all other eyes very small,
10.0–14.0 mm total body length -- **Deinopidae** (page 298)
2b. Eyes not as above -- **3**

a

Deinopis spinosa

3a. Legs I and II much more robust and longer than legs III and IV, overall
crab-like appearance, 1.5–11.3 mm total body length ------------------ **Thomisidae** (page 307)
3b. All legs of equal or near equal build and length, dorsoventrally flattened,
7.5–13.0 mm total body length -- **Selenopidae** (page 300)

a   b

Xysticus aucificus   *Misumenoides formosipes*   *Selenops* sp.

DEINOPIDAE ▶ NETCASTING SPIDERS

Other Colloquial Names Ogrefaced spiders

Genera in This Family *Deinopis* (1)

Total Body Length 10.0–14.0 mm

Description These unique-looking spiders are entelegyne and cribellate. They have eight eyes. The posterior median eyes are extremely large, and all other eyes are very small. The spiders have a pair of book lungs and a single tracheal spiracle anterior of the spinnerets. They are nocturnal hunters. During the day, they appear sticklike and blend easily into their environment. They build a netlike web that is suspended from the first two pairs of legs and used to scoop up prey.

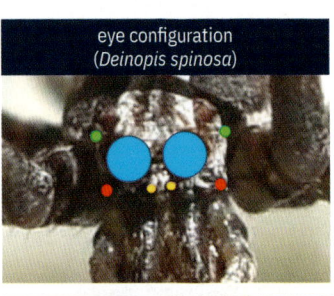
eye configuration
(*Deinopis spinosa*)

web in anterior legs waiting for prey

web held with anterior legs (*Deinopis spinosa*)

Similar Families Pisauridae, Tetragnathidae

How to distinguish Deinopidae from other similar families:

When at rest, deinopids can look similar to some tetragnathids and pisaurids. Their extremely large posterior median eyes, the fact that they are cribellate, and their unusual method of prey capture distinguish them from pisaurids and tetragnathids. Both pisaurids and tetragnathids are ecribellate. There may be some size differences in the eyes of pisaurids and tetragnathids, but not to the extent that is seen in deinopids. Pisaurids do not build any web structure for prey capture, and tetragnathids usually build a traditional orb web.

Deinopis spinosa Marx, 1889

Common Name This species does not have a common name.

Other Colloquial Names Ogre-faced spider, net-casting spider

Total Body Length Female 12.0–17.0 mm, Male 10.0–14.0 mm

Description The prosoma is mottled with gray tan to light brown. The posterior median eyes are extremely large, taking up most of the frontal view. The opisthosoma is mottled with gray tan to light brown, with two humps approximately halfway down the opisthosoma.

female

Two darker brown circles are posterior to the humps. The legs are extremely long and mottled with gray tan to light brown. When at rest, this species looks like a twig or branch and is very difficult to see.

Habitat These spiders tend to be found in forested habitats. They are nocturnal. At night, they suspend themselves from a framework of silk. They create a net of silk that they hold outstretched between the four anterior legs. When prey wanders or flies past, the webbing is used to scoop up the prey. During the day, these spiders are excellent stick mimics and can be difficult to find.

Known records from AL, FL, IL

female with web

female resting (looking very twig-like)

male

female with web ready to capture prey

male

SELENOPIDAE ▶ FLATTIES

Other Colloquial Names Crescent-eyed spiders, flatties, wall crab spiders

Genera in This Family *Selenops* (7)

Total Body Length 7.5–13.0 mm

Description These spiders are entelegyne and ecribellate. They are dorsoventrally flattened and are sometimes referred to as "flatties" due to this compressed body style. They have eight eyes; the posterior eye row is so strongly procurved that it appears that there are six eyes in the anterior row and only two eyes in the posterior row. They have a pair of book lungs and a single tracheal spiracle located near the spinnerets. Their legs are laterigrade, allowing them to fit into narrow cracks and crevices. The posterior sternum is notched. These are extremely fast-moving spiders.

Similar Families Philodromidae, Sparassidae, Zoropsidae

How to distinguish Selenopidae from other similar families:

The extremely dorsoventrally flattened appearance of selenopids helps to distinguish them from other similar families. *Lauricius* (Zoropsidae) species have laterigrade legs, which could cause them to be confused with selenopids. The eye configuration in selenopids, with the posterior eye row so strongly procurved that it appears that there are six eyes in the anterior row and only two eyes in the posterior row, is unique to selenopids. In contrast, philodromids, sparassids, and zoropsids have their eight eyes in two rows of four eyes.

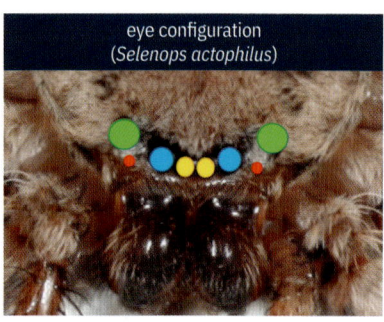
eye configuration
(*Selenops actophilus*)

dorsoventrally flattened, legs laterigrade
(*Selenops* sp.)

Selenops actophilus Chamberlin, 1924

Common Name This species does not have a common name.

Other Colloquial Names Crescent-eyed spider, flattie, wall crab spider

Total Body Length Female 8.7–13.0 mm, Male 8.0–9.5 mm

Description The prosoma is mottled with shades of browns, tans, creams, and red brown. An indistinct dark marking starts at the fovea and splits and runs to the posterior lateral eyes in the shape of a U. The edges are sometimes slightly darker. The opisthosoma is mottled with brown gray, with a slightly darker heartmark and median stripe. The opisthosoma markings can be somewhat variable. The legs are banded in dark brown and orange to tan. These spiders can move extremely fast, and their flattened bodies allow them to fit into small cracks and crevices, making them difficult to photograph or collect.

Habitat These spiders are often found under rocks.

female

female

female

female

Known records from AZ, CA, NM, TX

Selenops submaculosus
Bryant, 1940

female

Common Name This species does not have a common name.

Other Colloquial Names
Crescent-eyed spider,
flattie, wall crab spider

Total Body Length Female 9.5–12.5 mm, Male 8.5–11.0 mm

Description The prosoma is tan to brown. The opisthosoma is gray tan to brown; the heartmark is sometimes dark brown. The posterior edge is usually quite dark, almost black. The legs are banded in tan and brown and darken distally. The bands join into a longitudinal stripe on the anterioventral surface of the femurs. These spiders can move extremely fast, and their flattened bodies allow them to fit into small cracks and crevices, making them difficult to photograph or collect.

Habitat These spiders are often found on human structures.

Known records from FL

female

female

SICARIIDAE ▶ SIXEYED BROWN SPIDERS

lacks leg spines
(*Loxosceles reclusa*)

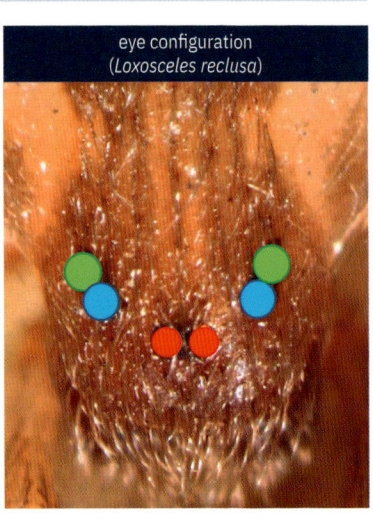

eye configuration
(*Loxosceles reclusa*)

Other Colloquial Names Brown spiders, sixeyed sicariid spiders, violin spiders

Genera in This Family *Loxosceles* (13)

Total Body Length 5.0–13.0 mm

Description These infamous haplogyne spiders are ecribellate. They have six eyes arranged in three diads. They have a pair of book lungs and a single tracheal spiracle located near the spinnerets. They tend to be somewhat plain in appearance. The only markings seen are usually on the prosoma (sometimes in the shape of a violin). Their slender legs lack large spines. They typically hide in crevices, and some species can be found in homes.

Similar Families Diguetidae, Scytodidae

How to distinguish Sicariidae from other similar families:

The eye configurations of diguetids, scytodids, and sicariids are similar: their six eyes are in three diads. Sicariids are plainly marked spiders that lack leg spines. Diguetids are usually brown with white markings, and at least a few spines are present on the legs. Sicariids, fairly flat spiders, do not have the highly domed prosoma that is seen in scytodids.

Loxosceles arizonica Gertsch & Mulaik, 1940

Common Name Arizona recluse

Total Body Length Female 6.0–10.0 mm, Male 6.0–8.0 mm

Description The prosoma is yellow to orange brown. There are six eyes arranged in three pairs. Starting at the fovea is a brown marking that widens toward the eye region, making a violin-shaped pattern. Faint dusky spots may be near each coxa. The opisthosoma is tan to brown, sometimes with a slightly darker heartmark and no other markings. The legs are thin and lack any banding or prominent spines. Leg II is the longest. The species of *Loxosceles* look very similar, but location is helpful in narrowing the possibilities. To confirm an identification, you would need to examine a mature specimen under a microscope.

Habitat These spiders are often found under rocks or other debris in arid environments.

Known records from AZ, CA, NM

Envenomation by this species is possibly of medical concern. If you are bitten, please seek medical attention.

female

male

male

Loxosceles deserta Gertsch, 1973

Common Name Desert recluse

Total Body Length Female 6.0–9.0 mm, Male 6.0–8.0 mm

Description The prosoma is pale yellow to tan. There are six eyes arranged in three pairs. Starting at the fovea, there may be a faint marking that widens toward the eye region, making a violin-shaped pattern. The opisthosoma is yellow to tan, sometimes with a slightly darker heartmark but no other markings. The legs are thin and lack any

female

Envenomation by this species is possibly of medical concern. If you are bitten, please seek medical attention.

banding or prominent spines. Leg II is the longest. The species of *Loxosceles* look very similar, but location is helpful in narrowing the possibilities. To confirm an identification, you would need to examine a mature specimen under a microscope.

Habitat These spiders are often found under rocks or other debris in arid environments. This species is rarely found inside human structures.

Known records from AZ, CA, NM, NV, UT

Loxosceles reclusa Gertsch & Mulaik, 1940

Common Name Brown recluse

Other Colloquial Names Violin spider

Total Body Length Female 7.0–12.0 mm, Male 7.0–12.0 mm

Description The prosoma is yellow to orange brown. Six eyes are arranged in three pairs. A brown marking starts at the fovea and widens toward the eye region, making a violin-shaped pattern. Dusky spots are near each coxa. The opisthosoma is tan to brown, with a slightly darker heartmark but no other markings. The legs are thin and lack any banding or prominent spines. Leg II is the longest. The species of *Loxosceles* look very similar, but location is helpful in narrowing the possibilities. To confirm an identification, you would need to examine a mature specimen under a microscope.

> **Envenomation by this species is considered of medical concern.**
> **If you are bitten, please seek medical attention.**

female

female

female

female ventral

male

male

male

Habitat These spiders are often found in human structures and seem to have some affinity for cardboard. People can move them unintentionally to areas outside their range. At these locations, populations can be established in the structure to which they were introduced, but this species does not disperse well (they do not balloon and seem averse to being in exposed areas), so they are not expanding their range.

Known records from AL, AR, GA, IA, IL, IN, KS, KY, LA, MO, MS, NE, NM, OH, OK, TN, TX

Loxosceles rufescens (Dufour, 1820)

Common Name Mediterranean recluse

Total Body Length Female 7.0–7.5 mm, Male 7.0–7.5 mm

Description The prosoma is dull orange to light brown. Six eyes are arranged in three pairs. The eyes are smaller than those in the native species. A brown marking starts at the fovea and widens toward the eye region, making a violin-shaped pattern; this tends to be slightly more angular in this species than in the native *Loxosceles*. Faint radial lines may be visible, and the edge of the prosoma is often slightly dusky. The opisthosoma is orange to light brown, sometimes with a slightly darker heartmark but no other markings. Fine black hairs usually cover the opisthosoma. The legs are thin and lack any banding or prominent spines. Leg II is the longest. The species of *Loxosceles* look very similar. To confirm an identification, you would need to examine a mature specimen under a microscope.

> **Envenomation by this species is possibly of medical concern.**
> **If you are bitten, please seek medical attention.**

female

female

Habitat These spiders can become established in the human structure to which they were introduced, but they do not disperse well enough to establish a true population.

Known records from CA, CO, FL, GA, IL, LA, MA, MD, MI, MO, NC, NE, OH, OK, PA, TN, TX, UT
This is an introduced species and does not have a "range" in North America; each location listed is based on an introduction by human transport.

female

Loxosceles sabina Gertsch & Ennik, 1983

Common Name Tucson recluse

Total Body Length Female 8.5 mm total body length
(The male is undescribed for this species.)

Description The prosoma is yellow to pale orange. Six eyes are arranged in three pairs. A very faint marking starts at the fovea and widens toward the eye region, making a violin-shaped pattern; this may be completely lacking. The edge of the prosoma may be slightly lighter. The opisthosoma is gray to brown, sometimes with a slightly darker heartmark but no other markings. The legs are thin and lack any banding or prominent spines. Leg II is the longest. The species of *Loxosceles* look very similar, but location is helpful in narrowing the possibilities. To confirm an identification, you would need to examine a mature specimen under a microscope.

Envenomation by this species is possibly of medical concern. If you are bitten, please seek medical attention.

female with prey

Habitat These spiders are often found under rocks or other debris in arid environments.

Known records from AZ

THOMISIDAE > CRAB SPIDERS

eye configuration
(*Mecaphesa asperta*)

Other Colloquial Names Flower spiders

Genera in This Family *Bassaniana* (3), *Bucranium* (1), *Coriarachne* (1), *Diaea* (2), *Isaloides* (1), *Mecaphesa* (19), *Misumena* (2), *Misumenoides* (1), *Misumenops* (3), *Misumessus* (4), *Modysticus* (3), *Ozyptila* (19), *Psammitis* (5), *Spiracme* (4), *Synema* (3), *Thomisus* (5), *Tmarus* (6), and *Xysticus* (63)

Total Body Length 1.5–11.3 mm

Description These crab-like spiders are entelegyne and ecribellate. They have eight eyes in two rows. They have a pair of book

occupying flower waiting for pollinator (*Misumenoides formosipes*)

lungs and a single tracheal spiracle near the spinnerets. Their legs are laterigrade, and the first two pairs are longer and more robust than the others. Some are well known for inhabiting flowers, where they wait to ambush pollinators and have been documented shifting in color to match the environment. Others live in cracks and crevices on bark or in leaf litter or herbaceous vegetation.

Similar Families
Philodromidae, Salticidae

first two pairs of legs more robust than others (*Xysticus ferox*)

How to distinguish Thomisidae from other similar families:

Both thomisids and philodromids have a crab-like appearance with laterigrade legs, but thomisids have robust legs I and II, and legs III and IV are small in comparison. In contrast, in philodromids, the second pair of legs is the longest (in some cases, extremely so), and legs III and IV are of similar build as the first two pairs. Salticids have greatly enlarged anterior median eyes; the configuration is unlike that of thomisids.

Bassaniana versicolor
(Keyserling, 1880)

Common Name The genus *Bassaniana* is known as "bark crab spiders."

Other Colloquial Names Multicolored bark crab spider

Total Body Length Female 4.4–7.7 mm, Male 3.9–5.7 mm

Description The prosoma is dark brown to jet-black. There is a yellow border in the females, but the males lack this feature. The eye region is slightly paler. The opisthosoma is variegated in shades of tans, browns, yellows, and black, and lighter in the female than in the male. The legs are tan to light brown, with dark brown to black banding and spotting.

female

female

female with egg sac

juvenile

male

male

male

Habitat These spiders are often found under bark or in small cracks and crevices in dead wood or rocks.

Known records from ON (Canada); AL, AR, AZ, CO, CT, FL, IA, IL, IN, KS, KY, LA, MA, MD, MI, MO, MS, NC, NE, NM, NY, OH, OK, OR, PA, RI, TN, TX, VA, WI, WV, WY (United States)

Diaea livens Simon, 1876

Common Name This species does not have a common name.

Total Body Length Female 6.0–6.5 mm, Male 4.0–5.0 mm

Description The prosoma is green but somewhat yellow around the eyes. The opisthosoma is yellow to ivory on the lateral edges and red on the dorsal surface. The red marking fades closer to the midline, changing to pink or even white. The legs are green, with black spots. The male has red banding at the distal end of each leg segment.

male and female

Habitat These spiders are usually found in coastal woodlands and seem to be associated with oak.

Known records from CA

Isaloides yollotl Jiménez, 1992

Common Name This species does not have a common name.

Total Body Length Female 5.6–6.2 mm, Male 3.8–4.9 mm

Description The female prosoma is reddish brown, with a pale median stripe that widens as it approaches the eye region. There is a tan submarginal band. The opisthosoma is red. There is a dark brown median stripe in the form of several chevron shapes of differing sizes. Five pairs of red dots are outlined in white on the tips of the chevron shapes. The first two pairs of legs are brown; the other legs are cream to off-white. The male prosoma is reddish brown, with a pale median stripe and pale edges, and is overall shinier than the female. The male opisthosoma is yellow, with a reddish-brown median stripe in the form of several chevron shapes. Five pairs of red dots are on the tips of the chevron shapes. The first two pairs of legs are reddish brown with indistinct banding. The other legs are pale orange and spotted.

female

male

Habitat These spiders are often found on herbaceous vegetation.

Known records from AZ

Mecaphesa asperata (Hentz, 1847)

Common Name Northern crab spider

Total Body Length Female 4.4–6.0 mm, Male 3.0–4.0 mm

Description The prosoma is yellow to tan and occasionally looks pale green, with a pair of broad brown lines on either side of the midline. There are numerous spines on the surface, giving it a somewhat hairy appearance. The opisthosoma is yellow, with red to brown spots. A V-shaped marking in pale white to yellow starts at the anterior and opens toward the posterior. The legs are pale yellow and have lots of small dark spots. The legs are unbanded in the female, while the males have dark red rings at each joint. The legs are proportionally much longer in the males than the females.

Habitat These spiders are often found in and around blooming vegetation.

Known records from AB, BC, MB, ON, QC, SK (Canada); AL, AR, CA, CO, CT, DC, FL, GA, ID, IL, IN, KS, KY, LA, MA, MD, ME, MI, MN, MO, MT, NC, NE, NH, NJ, NM, NY, OH, OK, PA, RI, TN, TX, UT, VA, WI, WY (United States)

female

juvenile

female

juvenile

male

male

male

Mecaphesa celer (Hentz, 1847)

Common Name Celer crab spider

Other Colloquial Names Swift crab spider

Total Body Length Female 5.0–6.7 mm, Male 2.5–3.2 mm

Description The prosoma is pale green and has an off-white X marking near the fovea. There are darker submarginal bands, which can be a little darker than the coloration of the rest of the prosoma or brown. The eyes are on white tubercles. The male prosoma has a red line around the edge. Both the prosoma and opisthosoma have many erect dark spines. The opisthosoma is cream to pale

female

female

female

yellow. There is an indistinct tan-colored heartmark, and tan stripes run down the lateral edges. Paired blotches of dark brown to black are on the posterior half of the opisthosoma; these often shift to brown or reddish brown the more posterior they are. The legs are pale green. The females lack any markings on the legs. The males have red spots where there are spines and red bands on the distal tibiae of legs I and II.

Habitat These spiders are often found in agricultural fields and meadows or on low tree branches.

Known records from AB, BC, SK (Canada); AL, AZ, CA, CO, FL, GA, ID, IL, IN, KS, LA, MA, ME, MI, MN, MO, MS, MT, NC, NE, NM, NV, NY, OH, OK, OR, TX, UT, VA, WA, WI, WY (United States)

Misumena vatia (Clerck, 1757)

Common Name Goldenrod crab spider

Other Colloquial Names Flower spider

Total Body Length Female 6.0–9.0 mm, Male 2.9–4.0 mm

Description This species has been documented to shift in color, usually to match the flowers on which it is hunting. The female prosoma can be white to yellow, sometimes with broad bands of green to brown. The eye region is usually yellow to orange. The female opisthosoma is yellow to white. When yellow, a pair of green bands may start near the pedicel and run along the lateral edge to about halfway down. If the opisthosoma is white, pink to red bands may be located in this area. The legs are white to yellow. The male prosoma is purple brown, sometimes with a broad pale median band. The eye region is white to pink. The opisthosoma is cream white to yellow, with brown spots on either side of the midline toward the posterior; there is a dark ring around the lateral edges of brown to red. The first two pairs of legs are purple brown, and sometimes there is pale banding on the distal segments. The legs are proportionally much longer in the males than the females. The other legs are yellow. The spiders use the spines on the first two pairs of legs to help them grab their prey. Note this species lacks the transverse ridge (carina) that is found on the clypeus of *Misumenoides formosipes*.

female

Habitat These spiders can often be found in or on flowers, where they sit and wait to capture pollinators.

Known records from AB, BC, LB, MB, NB, NL, NS, NT, ON, PE, QC, SK, YT (Canada); AK, AL, AR, AZ, CA, CO, DC, FL, IA, ID, IL, IN, KS, MA, MD, ME, MI, MN, MT, NC, NH, NJ, NM, NY, OR, PA, SD, TN, TX, UT, VA, VT, WA, WI, WY (United States)

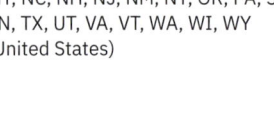

female

Misumenoides formosipes (Walckenaer, 1837)

Common Name Whitebanded crab spider

Other Colloquial Names Red banded crab spider, ridge-faced flower spider

Total Body Length Female 5.0–11.3 mm, Male 2.5–3.2 mm

Description The female prosoma is white to yellow and sometimes has a pair of submarginal bands that vary from green to brown. Along the clypeus, a carina (transverse ridge) that is white to yellow extends posteriorly. The opisthosoma varies from white to yellow, occasionally pink to lavender. Brown to red spots may be on the dorsal surface in a V shape that opens toward the posterior. A pair of brown bands may also be on the lateral edge, starting at the pedicel and ending about halfway down the length. The legs are white to yellow and can be unbanded or have dark brown bands at the patella and distal tibia, while the remaining segments are dark brown. The male prosoma is green to orange. The carina is present and usually easy to see on the darker-colored prosoma. The opisthosoma ranges from red to yellow. The first two pairs of legs are very dark, almost black. The other legs are green to yellow. The legs are proportionally much longer in the males than the females. These spiders use the

female

female

female

male

female in Jewelweed flower

spines on the first two pairs of legs to help them grab their prey. This species also has been reported to change color to match the environment.

Habitat These spiders can often be found in or on flowering vegetation, where they wait to capture pollinators.

Known records from AB, BC, ON, QC (Canada); AL, AR, AZ, CA, CT, DC, DE, FL, GA, IA, IL, IN, KS, KY, LA, MA, MD, ME, MI, MN, MO, MT, NC, NE, NH, NJ, NM, NY, OH, OK, PA, RI, SD, TN, TX, VA, VT, WI, WV (United States)

Misumessus lappi Edwards, 2017

Common Name This species does not have a common name.

Total Body Length Female 6.2–7.9 mm, Male 3.3–4.1 mm

Description The female prosoma is off-white to pale yellow. The opisthosoma is pale green to off-white, with a yellow anterior band. The legs are off-white to pale yellow. The male prosoma is pale cream to orange yellow. The opisthosoma is pale green, and paired red spots are on either side of the midline. The first two pairs of legs are pale green to off-white, and red bands are on the distal femurs and proximal and distal tibiae. The other legs are off-white to pale yellow. In both the males and the females, the eyes are situated on a white to yellow or pale pink mask. There is a projection between the median eyes that is unique to this species.

Habitat These spiders are often found on trees or understory vegetation.

Known records from CO, OK, TX

female

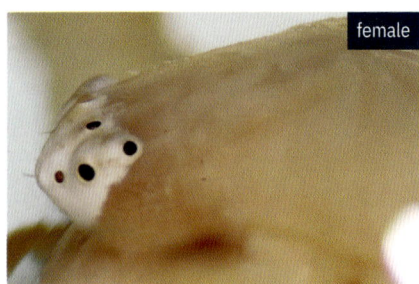
female

male

female and male

male

Misumessus oblongus (Keyserling, 1880)

Common Name This species does not have a common name.

Other Colloquial Names American green crab spider

Total Body Length Female 4.9–6.2 mm, Male 1.5–2.6 mm

Description The female prosoma is pale yellow to white to pale green. The eye region is white to pink. The opisthosoma is white to yellow green, and two broad pink stripes may start at the pedicle and make a V shape that opens toward the posterior. The prosoma and opisthosoma lack any stiff hairs. The legs are white to pale yellow green and unbanded. The male prosoma is yellow to green, with a thin dark line on the edge. There are many stiff hairs, resembling *Mecaphesa* species.

female

female

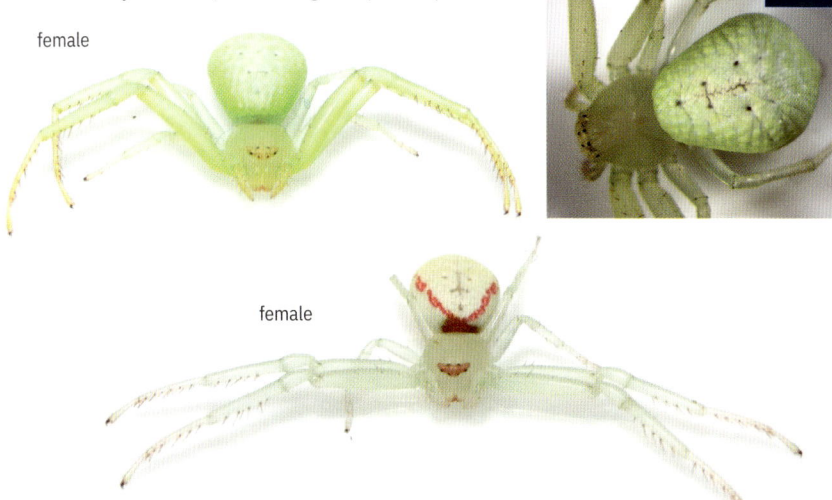

female

female with egg sac

juvenile

The opisthosoma is green to yellow, also with stiff hairs. The first two pairs of legs have green femurs and are banded in red and white distally. The other legs are green to yellow, and faint banding may be seen. The legs are proportionally much longer in the males than the females.

Habitat The spiders are often found in fields, prairies, and understory vegetation.

Known records from ON (Canada); AL, AR, AZ, CA, CO, CT, DC, FL, GA, IA, IL, IN, KY, LA, MA, MD, MI, MN, MO, MS, NC, NE, NH, NJ, NM, NY, OH, OK, PA, TN, TX, UT, VA (United States)

juvenile

male

male

male

Ozyptila pacifica Banks, 1895

Common Name This species does not have a common name.

Total Body Length Female 4.0 mm, Male 3.0–3.5 mm

Description The prosoma is reddish brown. There is a pale brown median stripe. Within the median stripe is a brown marking that starts at the fovea and widens to the eye region. Brown stripes are on either side of the median stripe, and some dark dotting is around the edge. The opisthosoma is brown, with scattered white and black markings. The legs are brown, and the femurs and patellae have scattered black spots. The males tend to be somewhat darker than the females.

Habitat These spiders are often found in moss, under bark, or in the litter of hemlock and cedar-hemlock forests.

Known records from BC (Canada); AK, CO, OR, WA (United States)

female

male

Synema parvulum (Hentz, 1847)

Common Name This species does not have a common name.

Other Colloquial Names Black-banded crab spider

Total Body Length Female 2.5–2.8 mm, Male 2.5 mm

Description The prosoma is yellow to orange, usually darker in the males. The eye region is usually paler, sometimes with white rings around the eyes. The opisthosoma is yellow to orange, and a broad black band covers all of the posterior third of the opisthosoma except the posterior tip, which is yellow orange. The first two pairs of legs are dark yellow to black (usually darkest on the males), while the other legs are pale yellow.

Habitat These spiders are often found on low vegetation. On one occasion, a specimen was observed working its way up and down the leaf of some forest understory vegetation; surprisingly, it was always walking backward. Then, when a small fly landed behind the spider and started interacting with it, the spider held

female

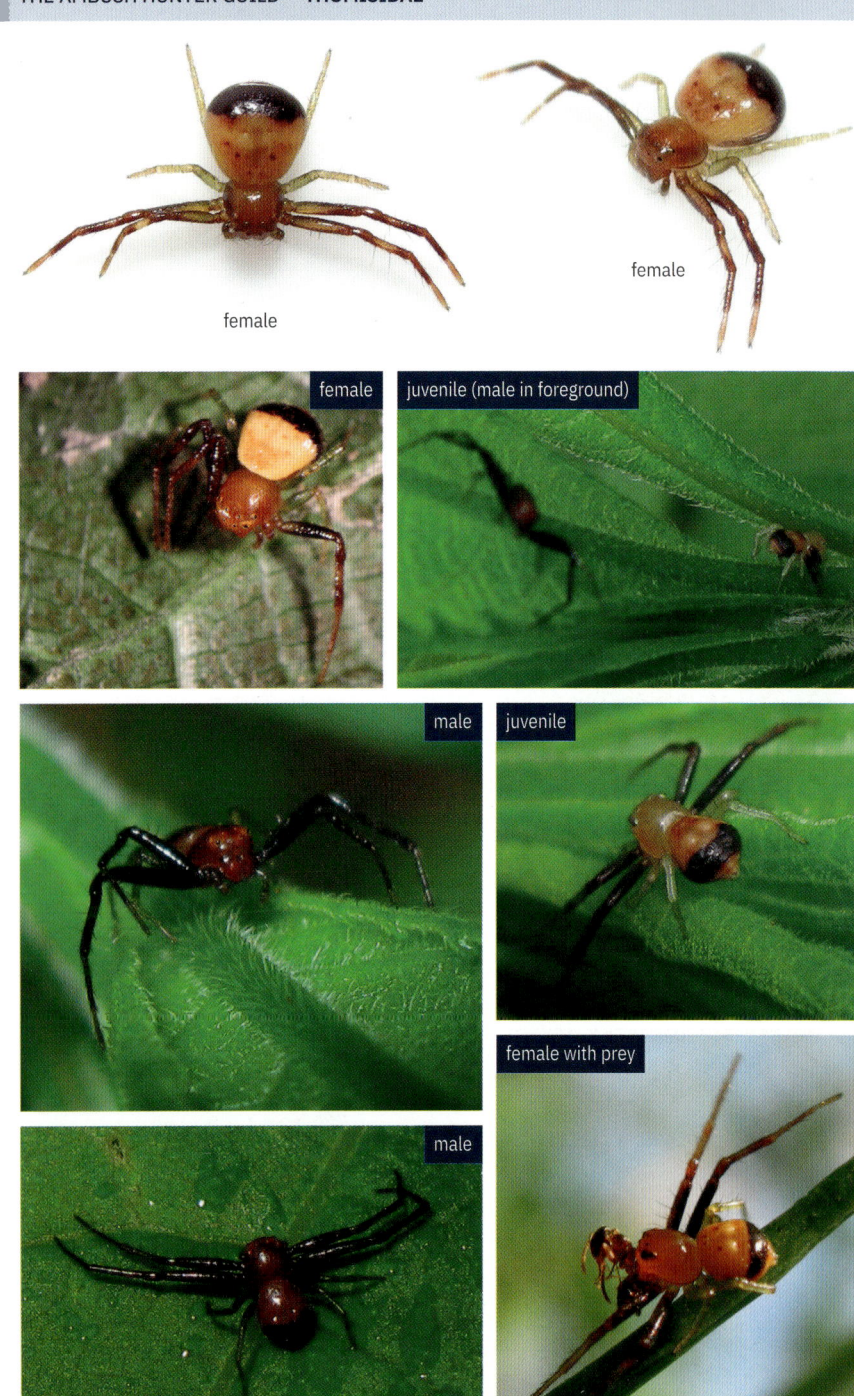

female

female

female

juvenile (male in foreground)

male

juvenile

male

female with prey

up the first two pairs of its legs in a way that resembled fly wings. It was not observed capturing the fly, but the behavior was interesting to watch. Looking at the spider from the posterior, the dark band does seem superficially to resemble fly eyes.

Known records from ON (Canada); AL, AR, CA, DC, FL, GA, IN, KS, LA, MD, MO, NC, NJ, NM, OH, OK, VA (United States)

Synema viridans (Banks, 1896)

Common Name This species does not have a common name.

Total Body Length Female 3.9–4.0 mm, Male 2.9–3.0 mm

Description The prosoma is vivid green to reddish green, with several black spines. A thin black line is around the edge; in some specimens, the line appears only on the anterior half. The lateral eyes are on white tubercles. The opisthosoma is cream white to pale green, with dark spines. The female has green legs. The male has legs that are green proximally and shift to a reddish orange distally. Based on the hairiness of this species, it could easily be confused with *Mecaphesa* species.

Habitat These spiders are often found on herbaceous vegetation.

Known records from AZ, FL, IA, TX

female

female

juvenile

male

Tmarus angulatus (Walckenaer, 1837)

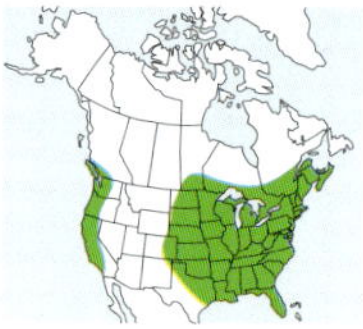

Common Name This species does not have a common name.

Other Colloquial Names Tuberculated crab spider

Total Body Length Female 4.5–7.0 mm, Male 3.0–5.0 mm

Description The prosoma is gray brown. There may be faint radial lines from the fovea. The eyes are set up on tubercles. The clypeus projects forward and, with the forward-oriented chelicerae, gives this genus a distinctive look. The opisthosoma is gray brown to golden brown. A large posteriorly pointing tubercle is on the end. The legs are tan to gray brown and speckled with fine spots. The spots are smaller than the markings seen in *Tmarus rubromaculatus*.

Habitat These spiders are often found on woody vegetation. When sitting on a tree branch or stick, this species is well camouflaged and difficult to see. The female lays her egg sac in a leaf that she then folds over and seals with silk.

Known records from BC, MB, NB, NS, ON, QC (Canada); AL, AZ, CA, CO, CT, DC, FL, IL, IN, KY, MA, ME, MN, MO, MS, MT, NC, NE, NH, NJ, NM, NY, OK, OR, PA, SD, TN, TX, UT, VT, WA (United States)

female with egg sac

female

female

male

male

Tmarus rubromaculatus Keyserling, 1880

Common Name This species does not have a common name.

Total Body Length Female 5.2–5.3 mm, Male 4.5 mm

Description The prosoma is gray brown. There may be faint radial lines from the fovea. The eyes are set up on tubercles. The clypeus projects forward and, with the forward-oriented chelicerae, gives this genus a distinctive look. The opisthosoma is gray brown to golden brown. A large posteriorly pointing tubercle is on the end. The legs are tan to gray brown and spotted. The markings tend to be larger than those seen in *Tmarus angulatus*. Otherwise, these species look very similar.

male

male

male

male

Habitat The spiders tend to be found on woody vegetation, where their coloration provides good camouflage.

Known records from AL, CT, DC, FL, GA, LA, MS, NC, NJ, TX

Xysticus auctificus Keyserling, 1880

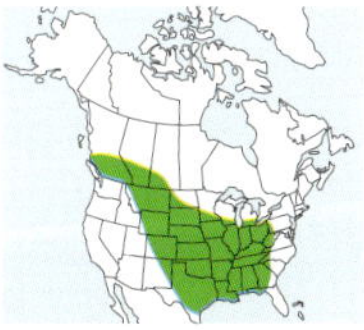

Common Name The genus *Xysticus* has the common name "ground crab spider."

Total Body Length Female 5.5 mm, Male 3.5 mm

Description The prosoma is yellow to light brown. A thin white V marking extends from the fovea to the posterior lateral eyes. Dusky bands are on either side of the midline. A dark spot is at the fovea, and a pair of dark spots is posterior to that. The opisthosoma is off-white to yellow. On either side of the midline is a pair of zigzag brown to dark brown stripes; within the stripes are three pairs of transverse dark brown to black dashes. The legs are yellow to light brown with black speckles.

Habitat These spiders are often found at the edges of fields.

Known records from AB, ON, SK (Canada);
AL, AR, CO, GA, IA, IL, IN, KS, LA, MS,
OH, OK, SD, TX, WY (United States)

female

female

female

male

male

Xysticus elegans Keyserling, 1880

Common Name Elegant crab spider

Other Colloquial Names The genus *Xysticus* has the
common name "ground crab spider."

Total Body Length Female 8.0–10.0 mm, Male 6.0–7.0 mm

Description The female prosoma is orange to brown. A pair of darker bands are on either side
of the midline. Some white to light brown coloration is usually immediately next to the eyes.
The opisthosoma is tan. A large brown marking covers the dorsal surface. A pale line splits the
brown marking down the middle. Two transverse pale lines cross the opisthosoma: one about
halfway down, the other three-quarters of the way down. The area just posterior to each of the
transverse lines is shaded in dark brown, almost black. The legs are tan to brown and have fine
speckling. The male prosoma is dark brown, often with a pair of pale lines that start joined at the

female
male

posterior end and fork out toward the posterior eye row. The marking of the opisthosoma is similar to that of the female, but the colors are more contrasting. The legs are tan to brown and so heavily speckled that, in some cases, they look to be a solid dark color.

Habitat These spiders are often found close to the ground, under logs, in leaf litter, or under debris, but they can also be found in low vegetation.

Known records from AB, BC, MB, NB, NL, NS, ON, QC, SK (Canada); AK, CO, CT, DC, GA, IA, IL, IN, KS, KY, MA, MD, ME, MI, MN, MO, MT, ND, NE, NH, NJ, NM, NY, OH, PA, TX, VT, WI, WV, WY (United States)

Xysticus ferox (Hentz, 1847)

Common Name The genus *Xysticus* has the common name "ground crab spider."

Other Colloquial Names Tan crab spider

Total Body Length Female 6.0–7.0 mm, Male 5.0–6.0 mm

Description The female prosoma is brown, with a tan midline that darkens toward the anterior and a tan submarginal band. The eye region is tan. The males are similar but usually darker. The opisthosoma is tan, and the dorsal surface is outlined in black or dark brown. The midline is slightly paler, and there are two to three transverse cream and black dashes on the lateral edges. The males tend to be darker and have more contrasting colors. The legs are tan

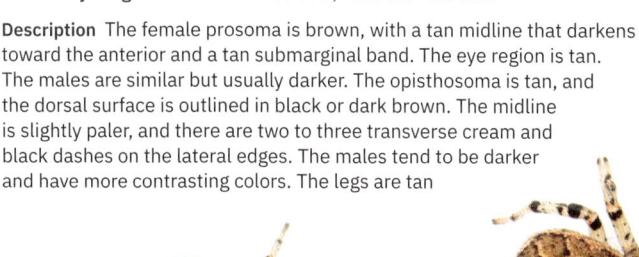

female

female

female

female

female with egg sac

spiderlings ballooning

male

male

and spotted, heavily in some areas. There are lines on the dorsal surface of legs I and II on the femurs, patellae, and tibiae; these may be obscured in darker specimens. In most males, legs I and II are deep brown, with pale metatarsi and tarsi, although some males look very much like the females.

Habitat These spiders are often found close to the ground, in leaf litter, under rocks and logs, or on low vegetation.

Known records from AB, BC, LB, MB, NB, NS, NT, ON, QC, SK, YT (Canada); AK, AL, AR, CO, CT, DC, FL, GA, IA, ID, IL, IN, KS, KY, LA, MA, MD, ME, MI, MN, MO, MS, MT, NC, NE, NH, NJ, NM, NY, OH, OK, PA, RI, SC, TN, TX, UT, VA, VT, WI, WV, WY (United States)

Xysticus fraternus Banks, 1895

Common Name The genus *Xysticus* has the common name "ground crab spider."

Other Colloquial Names Brotherly ground crab spider

Total Body Length Female 3.7–6.0 mm, Male 3.5–4.0 mm

Description The prosoma is brown, with a broad tan medial line. The prosoma is darker in the males. The opisthosoma is tan to brown. Three transverse lines are in the posterior half, with dark markings within them. The male opisthosoma is much darker and has similar light-colored transverse lines. Legs I and II are usually darker colored than the other legs. In the females, they are brown; in the males, they are dark brown to black. The females have pale tarsi; in the males,

female

female

male ventral

the tibiae, metatarsi, and tarsi are pale. The other legs are pale, with some spotting and occasional banding.

Habitat This species is found close to the ground, in leaf litter, under rocks, or in low vegetation.

Known records from NS, ON, QC (Canada); AL, AR, CO, CT, FL, GA, IL, IN, KY, LA, MA, MD, ME, MI, MN, MO, MS, MT, NC, NJ, NM, NY, OH, PA, SD, TN, TX, VA, WI, WV (United States)

male

male

Xysticus funestus Keyserling, 1880

Common Name The genus *Xysticus* has the common name "ground crab spider."

Other Colloquial Names Deadly ground crab spider (One has to wonder why this species is called "deadly," as it is not of any medical concern to humans.)

Total Body Length Female 6.0–7.0 mm, Male 4.0 mm

Description The prosoma is tan to orange. The sides are usually darker, and a thin black line and a yellow line are around the edge. The opisthosoma is tan, with a large brown marking on the dorsal surface. A pale midline may be within the brown marking. There is mottling and speckling throughout. The legs are tan, with extensive spotting and occasional bands.

Habitat These spiders are often found on low vegetation, fences, or the ground in leaf litter.

Known records from ON (Canada); AL, AR, CO, CT, DC, DE, FL, GA, IL, IN, KS, MA, MD, MI, MO, MS, MT, NC, NE, NJ, NM, NY, OH, OK, PA, RI, TN, TX, VA, WI, WV (United States)

female

female

female

Xysticus texanus
Banks, 1904

Common Name The genus *Xysticus* has the common name "ground crab spider."

Total Body Length
Female 5.4 mm, Male 4.3–4.4 mm

Description The prosoma is dark brown, with a pale midline band within which is a dark thin midline stripe. The opisthosoma is tan, with a series of increasingly pale rings. Three pairs of dark spots are on the posterior half. The legs are yellow to tan. Legs I and II have darkened areas from the distal end of the femur to the metatarsi.

Habitat These spiders are usually found close to the ground.

Known records from CO, FL, GA, KS, LA, NM, OH, TX

male

male

The spiders of the Ground Active Hunter Guild primarily run on the ground to pursue prey and do not rely on a web for prey capture. All the spiders in this guild are Araneomorphae. This guild includes the following families: Corinnidae, Dysderidae, Gnaphosidae, Homalonychidae, Liocranidae, Lycosidae, Myrmecicultoridae, Oonopidae, Phrurolithidae, Plectreuridae, Trachelidae, and Zodariidae.

KEY TO FAMILY

1a. Six eyes clustered, pair of tracheal spiracles near epigastric furrow ----------------------------- **2**

1b. Eight eyes, single tracheal spiracle near spinnerets -- **3**

Dysdera crocata

Agroeca sp.

2a. Elongated chelicerae, long fangs, specializes on isopods, prosoma usually red and opisthosoma pale tan/gray, 9.0–15.0 mm total body length ---- **Dysderidae** (page 339)

2b. Chelicerae and fangs not elongated, tends to be overall orange or yellow, 1.0–3.0 mm total body length --- **Oonopidae** (page 406)

Dysdera crocata

Dysdera crocata

Escaphiella hespera

3a. Posterior median eyes large and posterior lateral eyes placed back on carapace, making eyes appear to be in three rows, 2.2–35.0 mm total body length --- **Lycosidae** (page 362)

3b. Eyes not as above --- **4**

3a

Trochosa terricola

4a. Pentagonal opisthosoma, broad prosoma that narrows anteriorly,
6.5–12.8 mm total body length------------------------------------- **Homalonychidae** (page 358)
4b. Opisthosoma and prosoma not as above -- **5**

a

Homalonychus selenopoides

5a. Widely spaced, usually cylindrical spinnerets; posterior median eyes
usually modified (rectangular, triangular, etc.); 2.0–17.0 mm total
body length -- **Gnaphosidae** (page 341)
5b. Spinnerets not as above --- **6**

a

Drassyllus depressus

6a. Femur I curved and more robust, male with clasping
spur on tibia I, 4.2–16.5 mm total body length --- **Some Plectreuridae (*Plectreurys*)** (page 416)
6b. Femurs not as above --- **7**

7a. Femur I longer than prosoma, many dorsal spines,
6.7–19.0 mm total body length -------------------- **Some Plectreuridae (*Kibramoa*)** (page 415)
7b. Femur not as above --- **8**

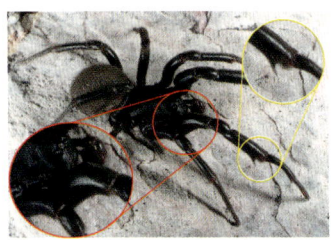

Plectreurys tristis

Kibramoa paiuta

8a. Anterior lateral spinnerets enlarged, other spinnerets reduced, 2.0–14.0 mm total body length -- **Zodariidae** (page 420)

8b. Anterior lateral spinnerets not enlarged -- **9**

Zodarion rubidum

9a. Precoxal triangles present -- **10**

9b. Precoxal triangles absent --- **13**

Trachelas tranquillus

Agroeca ornata

10a. Lacks ventral spines on anterior tibiae -- **11**

10b. Ventral spines present on anterior tibiae --- **12**

Trachelas tranquillus

Castianeira variata

11a. Cusps on anterior tibiae, tends to have a dark red prosoma and a pale
tan/gray opisthosoma, 3.0–10.0 mm total body length ------------------ **Trachelidae** (page 416)

11b. Paired rows of bristles on anterior tibiae
--------------------------------- **Some Liocrandiae (*Hesperocranum*), not covered in this guide**

Trachelas tranquillus

12a. Small spiders, 1.0–4.0 mm total body length, four pairs of ventral
spines on anterior tibiae -- **Phrurolithidae** (page 409)

12b. Medium-sized spiders, 4.0–15.0 mm total body length, several
pairs of ventral spines on anterior tibiae ---------------------------------- **Corinnidae** (page 331)

Scotinella fratrella   *Castianeira variata*

13a. Lives within harvester ant colonies, posterior median eyes set
at a 90° angle, 2.7–2.9 mm total body length -------------------- **Myrmecicultoridae** (page 405)

13b. Eyes otherwise, does not live within harvester ant colonies,
anterior tibiae with one to five pairs of ventral spines,
2.0–10.0 mm total body length --- **Liocranidae** (page 360)

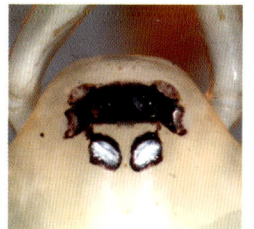
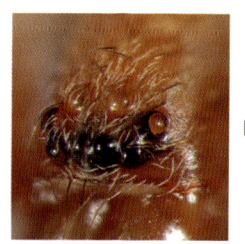

Myrmecicultor chihuahuensis   *Agroeca* sp.

CORINNIDAE > ANTLIKE RUNNERS

Other Colloquial Names Antmimic spiders, corinnid sac spiders

Genera in This Family *Castianeira* (26), *Creugas* (2), *Falconina* (1), *Mazax* (2), *Myrmecotypus* (1), *Septentrinna* (2), and *Xeropigo* (1)

Total Body Length 4.0–15.0 mm

Description These fast-running spiders are entelegyne and ecribellate. They have eight eyes in two rows. The anterior eye row is usually recurved, and the posterior eye row is procurved. The spiders have a pair of book lungs and a single tracheal spiracle near the spinnerets. They have precoxal triangles. Most corinnids have several pairs of ventral tibial spines on the anterior legs. They are somewhat ant-like, especially in the way they run.

eye configuration
(*Castianeira variata*)

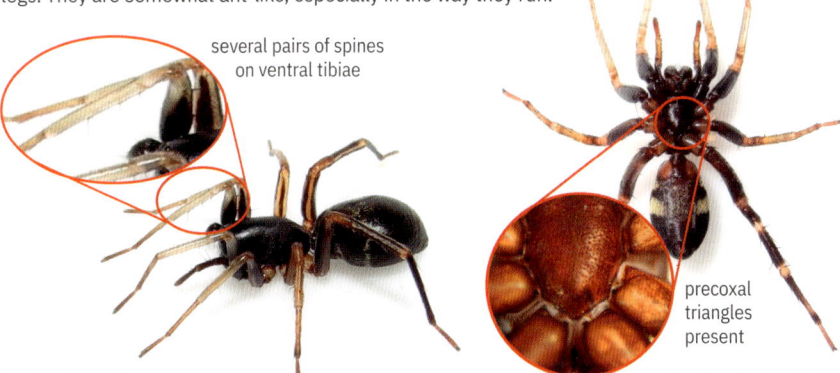

several pairs of spines
on ventral tibiae

precoxal
triangles
present

Castianeira variata

Castianeira longipalpa

Similar Families Clubionidae, Gnaphosidae, Liocranidae, Phrurolithidae, Salticidae, Trachelidae, Zodariidae

How to distinguish Corinnidae from other similar families:

Most corinnids lack the stiff bristles that are seen on the anterior of the opisthosoma of clubionids. Corinnids can be distinguished from gnaphosids and zodariids by their spinnerets. Most gnaphosids have widely spaced, cylindrical anterior lateral spinnerets, and zodariids have enlarged anterior lateral spinnerets while the others are reduced. The spinnerets in corinnids are clustered and conical, and only the posterior median spinnerets are enlarged. Some corinnids can look very similar to *Micaria* species in the family Gnaphosidae, whose spinnerets are not obviously cylindrical and widely spaced. The similar-looking corinnids do not have iridescent scales, the posterior eye row is procurved, and the posterior median eyes are round. These traits differ in *Micaria* species. Corinnids can be distinguished from some liocranids by the presence of precoxal triangles, which are absent in most liocranids. The liocranids that have precoxal triangles also have two rows of dense bristles on the anterior tibiae; corinnids have paired ventral spines on the anterior tibiae. Phrurolithids tend to be smaller (1.0–4.0 mm total body length) than most corinnids (4.0–15.0 mm total body length). Salticids have greatly enlarged anterior median eyes, which is unlike the eye configuration of corinnids. Corinnids can be distinguished from trachelids by the presence of several pairs of spines on the ventral tibiae, as trachelids have only short cusps. Also, some corinnids (especially males) have a scutum on the anterior dorsal opisthosoma, while clubionids, liocranids, trachelids, and zodariids do not have scuta.

Castianeira amoena (C. L. Koch, 1841)

Common Name Orange antmimic

Other Colloquial Names Orange antmimic sac spider, tiger antmimic, velvet ant mimic spider

Total Body Length Female 7.0–8.8 mm, Male 5.7–6.8 mm

Description The prosoma is bright orange. The opisthosoma is bright orange, with transverse black stripes; the orange fades to white on the anterior end. The first two pairs of legs have black femurs, and the distal segments are pale. Legs III and IV are banded in black, yellow, and orange. The coloration may be mimicking velvet ants, which have a very painful sting. The mimicry might reduce the risks of predation on these dramatically colored spiders.

female

Habitat These spiders are often found under rocks and on the ground in forested areas.

Known records from AL, AR, CO, FL, GA, IL, KS, LA, MO, MS, NC, OK, TN, TX, WY

Castianeira cingulata
(C. L. Koch, 1841)

Common Name Twobanded antmimic

Total Body Length Female 6.7–8.0 mm, Male 6.4–7.0 mm

Description The prosoma is jet-black. The opisthosoma is jet-black, with two transverse yellow to white bands on the anterior half. The legs are yellow to orange, with longitudinal black stripes on the femurs. Legs III and IV tend to be a bit darker than the anterior legs. The egg sac is slightly domed, white, and attached to the surface. Two bands are visible even on the opisthosoma of young that have just emerged from the egg sac. These spiders can look very similar to *Micaria* species in the family Gnaphosidae. *C. cingulata* does not have iridescent scales, the posterior eye row is procurved, and the posterior median eyes are round. These traits differ in *Micaria* species.

female

female

Habitat These spiders are often found in forested habitats and prairies in leaf litter, under logs, under rocks, or under other debris.

spiderlings in egg sac

spiderling just emerged from egg sac

Known records from MB, NB, NS, ON, QC (Canada); AL, AR, CT, FL, IA, IL, IN, KS, KY, MA, ME, MI, MN, MO, NH, NJ, NY, OH, PA, SC, SD, TN, VA, WI, WV (United States)

Castianeira descripta (Hentz, 1847)

Common Name Redspotted antmimic

Other Colloquial Names Red-spotted ant-mimic sac spider

Total Body Length Female 8.0–10.0 mm, Male 6.2–7.6 mm

Description The prosoma is dark red brown and covered with short hairs. The opisthosoma is black laterally and has a red to orange-red midline band that widens toward the posterior. In some cases, the band is limited to the posterior end. There is a small scutum on the anterior opisthosoma. The femurs on all legs are red brown. Otherwise, legs I and II are yellow, leg III is orange brown, and leg IV is red brown, with yellow tarsi. The female has longitudinal stripes on the patellae and tibiae of legs III and IV. Several species of *Castianeira* bear a red marking on the posterior opisthosoma (for example, *C. crocata*, *C. occidens*, and *C. floridana*). Note that *C. occidens* has a white median stripe on the prosoma that is never

female

seen in *C. descripta*, although the scales can easily be rubbed off. Note that *C. floridana* is found only in Florida. In *C. crocata*, the red marking on the opisthosoma stops anterior of the spinnerets, whereas in *C. descripta*, it continues all the way to the spinnerets. To confirm an identification, you would need to view the genitalia under a microscope.

Habitat These spiders are often found in leaf litter, under rocks, or on the ground in dry pastures and fields.

Known records from AB, MB, NB, NS, ON, QC, SK (Canada); AR, CO, CT, DC, FL, IA, IL, KY, MA, MD, ME, MI, MS, MT, NC, NH, NJ, NM, NY, OH, PA, TN, TX, UT, VA, VT, WI, WV (United States)

Castianeira longipalpa (Hentz, 1847)

Common Name Manybanded antmimic

Total Body Length Female 7.0–9.0 mm, Male 5.5–6.0 mm

Description The prosoma is dark brown to black and often covered with white to golden hairs. The opisthosoma is black, with a series of light-colored (white to yellow) bands. Four or more bands usually go completely across the opisthosoma, and many others are shorter. Some of the shorter ones may appear diamond shaped. The first two pairs of legs have black femurs, while the distal segments are lighter. The other legs are black and usually have yellow to white bands. The markings of this species can vary quite a bit, and the white scales that make up the bands can often be rubbed off, making identification difficult. This species looks very similar to *Castianeira variata*, but *C. longipalpa* is usually found in moister habitats, and it tends to mature in late summer and early fall, while *C. variata* tends to mature in early summer. To confirm an identification, one would need to look at the genitalia under a microscope.

Habitat These spiders are often found in moist leaf litter, under rocks, or in other damp habitats.

Known records from AB, BC, MB, NB, NS, ON, QC, SK (Canada); AR, CO, CT, DE, FL, GA, ID, IL, IN, KS, KY, LA, MA, MD, ME, MI, MN, MO, MT, NC, ND, NH, NJ, NM, NY, OH, OK, OR, PA, RI, TX, UT, VA, VT, WA, WI, WV (United States)

female

female

female

female
ventral

male

male

Castianeira occidens Reiskind, 1969

Common Name This species does not have a common name.

Other Colloquial Names Antmimic spider

Total Body Length Female 7.0–10.0 mm, Male 6.2–7.3 mm

Description The prosoma is reddish brown and covered with hairs, it has a broad white median stripe (these scales can be rubbed off easily in some specimens), and it sometimes is white around the edges. The opisthosoma is black, with a median stripe that starts as white or yellow and quickly shifts to bright orange and red. This stripe widens posteriorly. White bands on the lateral edges may continue for the length of the opisthosoma or stop halfway down. A scutum is on the anterior opisthosoma. The femurs are reddish brown, with black and white hairs. The other segments on legs I and II are yellow. Leg III is reddish brown. Leg IV is reddish brown, with yellow tarsi. This species can easily be distinguished from the other red-marked *Castianeira* species by the broad white median stripe on the prosoma.

female

male

male

Habitat These spiders are often found in dry grass habitats. They have also been found in mesquite. They have been collected at elevations up to 8,000 ft.

Known records from AZ, CA, CO, NM, NV, TX, UT

Castianeira variata
Gertsch, 1942

Common Name This species does not have a common name.

Other Colloquial Names Variegated ant-mimic sac spider

Total Body Length Female 7.2–10.6 mm, Male 5.7–7.2 mm

Description The prosoma is dark brown to black and often covered with white to golden hairs. The opisthosoma is black, with a series of light-colored (white to yellow) bands. More than eight bands usually go completely across the opisthosoma. The first two pairs of legs have black femurs, and the distal segments are lighter. The other legs are black and usually have yellow to white bands. The markings of this species can vary quite a bit, and the white scales that make up the bands can often be rubbed off, making identification difficult. This species looks very much like *Castianeira longipalpa*, but *C. variata*

female

female

female

female

female

female

female

female
ventral

male

juvenile

is usually found in drier habitats, and it tends to mature in early summer, while *C. longipalpa* tends to mature in late summer and early fall. To confirm an identification, one would need to look at the genitalia under a microscope.

Habitat These spiders are often found in dry woodland and open areas.

Known records from BC, ON (Canada); AR, CO, CT, IL, KS, LA, MA, MD, MO, PA, TX, UT, VA (United States)

Falconina gracilis
(Keyserling, 1891)

female

Common Name This species does not have a common name.

Total Body Length
Female 5.9–8.9 mm, Male 4.6–6.4 mm

Description The prosoma is brown to almost black. The opisthosoma is dark brown to magenta. There are eight pale splotches: one diamond-shaped marking is toward the anterior, three pairs of markings are on either side of the midline (sometimes the third pair is very faint), and there is one larger posterior marking. The legs are orange brown.

female | male

male

Habitat These spiders are often found in damp habitats under rocks, in water meter covers, or under logs.

Known records from AL, CA, FL, LA, MS, TX This is an introduced species.

Septentrinna bicalcarata (Simon, 1896)

Common Name This species does not have a common name.

Total Body Length Female 5.0–5.1 mm, Male 4.2–4.3 mm

Description The prosoma is a deep reddish brown and darker in the eye region. Faint radial lines may be seen. The opisthosoma is cream to yellow, with a slightly darker heartmark. The legs are reddish brown and unbanded.

Habitat These spiders are often found associated with ants.

Known records from AZ, CA, NM, TX

male

male

DYSDERIDAE ▸ WOODLOUSE SPIDERS

Other Colloquial Names Woodlouse hunters, sowbug-eating spiders, cell spiders

Genera in This Family *Dysdera* (1)

Total Body Length 9.0–15.0 mm

Description These spiders are haplogyne, ecribellate, and bicolor (red prosoma and tan to gray opisthosoma). They have six eyes that are clustered on the anterior prosoma. (The anterior median eyes are absent.) They have elongated chelicerae with long fangs. They have a pair of book lungs and a pair of tracheal spiracles near the epigastric furrow. These spiders specialize in preying upon isopods.

Similar Families Caponiidae, Segestriidae, Trachelidae, Trogloraptoridae

How to distinguish Dysderidae from other similar families:

Dysderids can be distinguished from caponiids by the presence of book lungs, which caponiids lack. Also, caponiids have either only two eyes or eight eyes in a cluster, whereas dysderids have six eyes in a cluster. Dysderids can be distinguished from segestriids and trogloraptorids by their eye configuration.

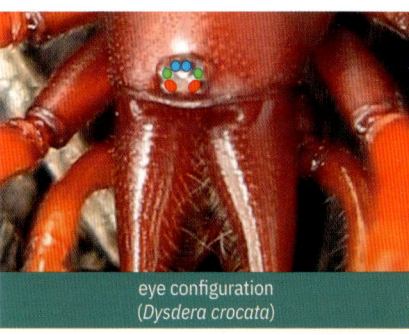

eye configuration
(*Dysdera crocata*)

pair of tracheal spiracles near book lungs, elongated chelicerae and fangs

Dysdera crocata

Segestriids and trogloraptorids have only six eyes (all are missing the anterior median eyes), while in dysderids, the eyes are clustered and occupy less than a third of the width of the prosoma. Segestriid eyes are in two rows that occupy over half the width of the prosoma. Trogloraptorids' lateral eyes are located adjacent to one another, and the posterior median eyes are between them. Additionally, segestriids rest with the first three pairs of legs directed anteriorly, whereas dysderids usually sit with only the first two pairs of legs directed anteriorly. The coloration of trachelids and dysderids is similar (red prosoma and pale opisthosoma), but there are several differences: Trachelids have eight eyes in two rows, while dysderids have six eyes in a cluster. Trachelids have a pair of book lungs and a single tracheal spiracle near the spinnerets, whereas dysderids have a pair of book lungs and a pair of tracheal spiracles located adjacent to the book lungs. Trachelids are entelegyne; dysderids are haplogyne. Trachelids do not have the elongated chelicerae and longs fangs that are seen in dysderids.

Dysdera crocata C. L. Koch, 1838

Common Name Woodlouse hunter

Other Colloquial Names Baked-bean spider, pill bug hunter, sow-bug killer, stiletto spider, woodlouse spider

Total Body Length Female 12.0–15.0 mm, Male 9.0–13.0 mm

Description The prosoma is orange to red. The chelicerae are elongated and extend forward. There are six eyes, arranged in a cluster that occupies less than a third of the width of the prosoma. The opisthosoma is tan to brown. A pair of tracheal slits are just posterior of the

female

female

female ventral

juvenile

female with isopod prey

male

book lungs. The legs are yellow to orange, without banding. This species is often confused with the spiders in the family Trachelidae. Trachelidae have eight eyes in two broad rows, whereas *D. crocata* has six eyes in a cluster. Trachelidae also lack the elongated chelicerae and have only a single spiracle located near the spinnerets.

Habitat These spiders are often found in moist habitats where isopods (their preferred prey) would be found. They are often associated with human structures. Contrary to popular belief,

male

male

the long fangs are not used to penetrate the hard armor of the isopids; rather, the spiders use their fangs and chelicerae to manipulate the prey item, so that a bite can be delivered to the softer underside (Řezáč, Pekár, and Lubin 2008).

Known records from BC, ON, QC (Canada); CA, CO, CT, IA, ID, IL, IN, KS, MA, NE, NM, OH, PA, TX (United States)
This is an introduced species.

GNAPHOSIDAE > STEALTHY GROUND SPIDERS

Other Colloquial Names Flat-bellied ground spiders, ground spiders

Genera in This Family *Callilepis* (7), *Camillina* (2), *Cesonia* (10), *Drassodes* (7), *Drassyllus* (47), *Eilica* (1), *Gertschosa* (1), *Gnaphosa* (20), *Haplodrassus* (9), *Herpyllus* (13), *Heser* (2), *Litopyllus* (2), *Marinarozelotes* (4), *Micaria* (41), *Neozimiris* (1), *Nodocion* (6), *Orodrassus* (3), *Parasyrisca* (1), *Prodidomus* (1), *Scopoides* (7), *Scotophaeus* (1), *Sergiolus* (16), *Sosticus* (3), *Synaphosus* (2), *Talanites* (5), *Urozelotes* (1), and *Zelotes* (45)

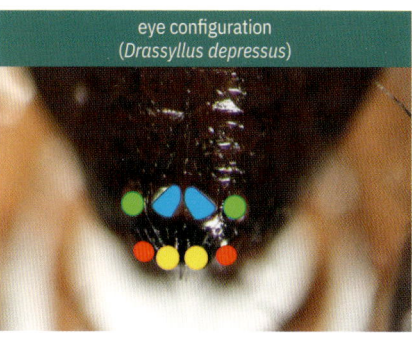

eye configuration
(*Drassyllus depressus*)

concave endites (*Gnaphosa muscorum*)

widely spaced cylindrical spinnerets
(*Drassyllus depressus*)

Total Body Length 2.0–17.0 mm

Description These extremely fast-running spiders are entelegyne and ecribellate. They have eight eyes in two rows, and the posterior median eyes are often modified (flattened, oval, or irregular in shape). They have concave endites. Most gnaphosids have enlarged, cylindrical, and widely spaced anterior lateral spinnerets. They have a pair of book lungs and a single tracheal spiracle near the spinnerets. Gnaphosids are one of the harder families to identify to genus or species without collecting a specimen. The characteristics that are most helpful to make a good identification are viewed more easily under a microscope. A few species are distinctive, but the majority of species seem to be plainly marked and difficult to identify.

Similar Families Liocranidae, Caponiidae, Cithaeronidae, Corinnidae, Hahniidae, Myrmecicultoridae, Phrurolithidae

How to distinguish Gnaphosidae from other similar families:

Enlarged anterior median spinnerets that are cylindrical and widely spaced distinguish gnaphosids from all other similar families. Liocranids have two rows of dense bristles or paired spines on the anterior tibiae; the bristles or spines are not seen in gnaphosids. Caponiids lack book lungs, which are present in gnaphosids. Additionally, the eye configuration of caponiids (only two eyes or eight eyes in a cluster) is unlike the eye configuration of gnaphosids. Cithaeronid tarsi have a pseudosegment, which gnaphosids lack. The spinnerets of many Hahniidae are in a transverse row, which is very different than the spinnerets in gnaphosids. Those species in the genus *Cicurina* (which had been moved from several different families before being assigned currently to Hahniidae) can be distinguished from gnaphosids by the lack of the enlarged cylindrical anterior lateral spinnerets that are characteristic of most gnaphosids. Those gnaphosids that lack the prominent spinnerets differ from *Cicurina* species, as they have iridescent scales, which *Cicurina* lack. Other than the differences in the spinnerets, phrurolithids differ from gnaphosids by lacking the modified posterior median eyes, and they also tend to be smaller (1.0–4.0 mm total body length) than most gnaphosids (2.0–17.0 mm total body length).

Callilepis pluto Banks, 1896

Common Name This species does not have a common name.

Total Body Length Female 4.8–6.2 mm, Male 3.0–4.2 mm

Description The prosoma is orange to orange red and sometimes covered with golden hairs. The opisthosoma is black and quite hairy; some specimens have transverse golden bands or stripes. The legs are orange brown. Legs I and II have darker patellae and tibiae, and the distal segments are light. Leg IV has some banding. There are visible spines on the legs.

Habitat These spiders are often found under rocks and other debris. There is some evidence that they may be ant specialists.

male

male

Known records from AB, BC, MB, NS, NT, ON, QC, SK (Canada); AL, CO, CT, GA, IA, ID, IL, IN, MA, ME, MI, MN, MT, NC, ND, NH, NJ, NY, OH, OR, PA, SD, TN, VA, VT, WA, WI, WV, WY (United States)

Cesonia bilineata (Hentz, 1847)

Common Name This species does not have a common name.

Other Colloquial Names Two-lined stealthy ground spider

Total Body Length Female 4.3–8.0 mm, Male 3.5–6.0 mm

Description The prosoma and opisthosoma have a central white stripe bordered with a black stripe and then another white stripe on the lateral edges. The legs are tan to yellow.

Habitat These spiders are usually found in the leaf litter of both coniferous and deciduous forests. There are also records from grasslands and sand dunes. The spiders can also be found in and around human structures.

Known records from MB, ON, QC (Canada); AL, AR, CT, FL, GA, IL, KY, LA, MA, MD, MI, MO, MS, NC, NJ, NM, NY, OH, OK, PA, SC, TN, TX, VA, WI, WV (United States)

Drassodes neglectus (Keyserling, 1887)

Common Name This species does not have a common name.

Other Colloquial Names Neglected ground hunter spider

Total Body Length Female 8.8–13.0 mm, Male 5.5–12.0 mm

Description The prosoma is reddish brown and covered with silvery-gray hairs. The opisthosoma is covered with similarly shaded hairs, and there is a dark heartmark. A series of chevrons is sometimes on the posterior portion of the opisthosoma.

Habitat These spiders are often found under rocks, under debris, in fields, or in and around human structures. The males and females will sometimes share a retreat. They overwinter as adult or subadults. They have been found at elevations up to 14,300 ft.

Known records from AB, BC, MB, NB, NL, NS, NT, ON, QC, SK, YT (Canada); AK, AZ, CA, CO, CT, ID, MA, ME, MI, MN, MT, ND, NH, NM, NY, OK, OR, SD, UT, VT, WA, WI, WV, WY (United States)

female

male

Drassyllus depressus
(Emerton, 1890)

Common Name This species does not have a common name.

Total Body Length
Female 4.7–6.0 mm, Male 5.0 mm

Description The prosoma is brown to nearly black and usually darker in the males. Faint radial lines and a V marking near the fovea that opens toward the posterior eyes may be visible in lighter specimens. The posterior median eyes are large and oval to rectangular in shape and sit in a V shape that opens toward the anterior. The opisthosoma is dark brown to black and lacks any markings. The male has a scutum on the anterior opisthosoma. The femurs, patellae, and tibiae are usually dark brown, while the metatarsi and tarsi are orange; this is most distinct in legs I and II.

modified posterior median eyes (female)

female

Habitat These spiders are found in a wide variety of habitats, including in and around human structures.

Known records from AB, BC, MB, NB, NS, ON, QC, SK (Canada); AR, AZ, CO, CT, IA, ID, IL, IN, KS, KY, MA, MD, ME, MI, MN, MO, ND, NE, NH, NJ, NM, NY, OH, OR, PA, SD, TX, UT, VA, VT, WA, WI (United States)

female ventral

juvenile

male

male

Gertschosa amphiloga (Chamberlin, 1936)

Common Name This species does not have a common name.

Total Body Length Female 5.7–5.8 mm

Description The prosoma is black. The opisthosoma is black, with two transverse white bands: one on the anterior and one approximately halfway down. All the femurs and patellae are dark to black, while the other segments are usually lighter. The specimen photographed here has been confirmed to be of the genus *Gertschosa*, but species-level identification is still pending. The males of *G. amphiloga* have yet to be described. This demonstrates how much is still unknown about spiders.

Habitat The specimen photographed here was found inside a home. Other specimens have been collected up in trees.

Known records from TX

female

female

Gnaphosa muscorum (L. Koch, 1866)

Common Name This species does not have a common name.

Total Body Length Female 12.0–15.0 mm, Male 10.0–12.0 mm

Description The prosoma is dark brown, and dark lines in the furrows radiate from the fovea. The opisthosoma is tan and heavily speckled with black spots, especially toward the lateral edges. The legs are dark brown and unbanded. They are quite hairy.

Habitat These spiders are often found under rocks and other debris.

Known records from AB, BC, LB, MB, NB, NL, NS, NT, NU, ON, QC, SK, YT (Canada); AK, AZ, CA, CO, CT, ID, MA, MD, ME, MI, MN, MT, ND, NH, NJ, NM, NV, NY, OR, PA, SD, UT, VT, WA, WI, WV, WY (United States)

female with egg sac

female with egg sac

male

male

Haplodrassus hiemalis (Emerton, 1909)

Common Name This species does not have a common name.

Total Body Length Female 5.8–7.1 mm, Male 5.6–6.3 mm

Description The prosoma is reddish brown, and faint radial lines are often visible. The posterior median eyes are somewhat triangular. The posterior lateral eyes are oval. The opisthosoma is gray to tan. The legs are pale brown.

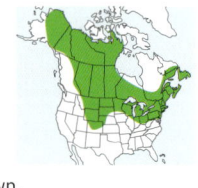

male

male

male

male

male
ventral

The retrolateral apophysis on the male pedipalp is dorsally oriented and tapers to a rounded tip.

Habitat These spiders tend to be found in leaf litter. Some have been collected from hardwood forested areas, grasslands, and wetlands. They have been found at elevations up to 15,250 ft.

Known records from AB, BC, MB, NB, NL, NS, NT, NU, ON, QC, SK, YT (Canada); AK, CO, CT, IL, MA, ME, MI, MN, ND, NH, NJ, NY, OH, PA, VT, WI, WY (United States)

Herpyllus ecclesiasticus
Hentz, 1832

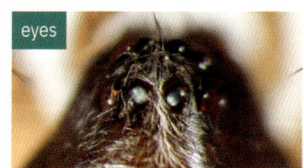

Common Name Eastern parson spider

Total Body Length Female 6.4–9.0 mm, Male 4.8–6.1 mm

Description The prosoma is black, with a broad white midline. The opisthosoma is black. There is a white heartmark that posteriorly resembles the shape of a parson's cravat. A white diamond is also just anterior to the spinnerets. The femurs are black and hairy, while the other leg segments are dark brown. This species looks very similar to the western species *Herpyllus propinquus* (western parson spider). Range can be helpful to distinguish the two. Otherwise, one would need to look at the genitalia under a microscope.

female

female

female

Habitat These spiders are often found under rocks and logs and in and around human structures. They can be found at elevations up to 8,000 ft.

Known records from AB, MB, NS, ON, QC, SK (Canada); AL, AR, AZ, CO, CT, DC, DE, FL, GA, IA, IL, IN, KS, KY, LA, MA, MD, ME, MI, MN, MO, MS, MT, NC, ND, NH, NJ, NY, OH, OK, PA, RI, SC, SD, TX, VA, VT, WI, WV, WY (United States)

female

male

Micaria gosiuta
Gertsch, 1942

Common Name This species does not have a common name.

Total Body Length Female 4.2–5.8 mm, Male 3.9–4.7 mm

Description The prosoma is golden to reddish brown and slightly darker in the eye region. The opisthosoma is golden to reddish brown. The markings are somewhat variable, but pairs

female

of white patches are often above the pedicel and about halfway down the opisthosoma. There may or may not be a visible constriction. Paired dark dashes or chevron markings are sometimes toward the posterior end. The legs are golden to reddish brown and slightly lighter at the tips.

Habitat These spiders can often be found running on the ground. They run with an ant-like gait.

Known records from AZ, CA, ID, NM, NV, OR, UT, WY

Micaria longipes Emerton, 1890

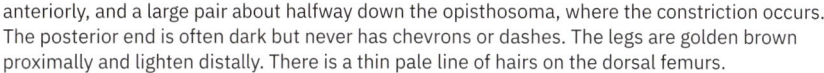

Common Name This species does not have a common name.

Total Body Length Female 4.4–5.5 mm, Male 3.7–4.8 mm

Description The prosoma is golden brown, with a radiating median mark from the fovea to the eye region. Iridescent hairs cover the prosoma. The opisthosoma is golden brown. There is a pair of transverse white dashes anteriorly, and a large pair about halfway down the opisthosoma, where the constriction occurs. The posterior end is often dark but never has chevrons or dashes. The legs are golden brown proximally and lighten distally. There is a thin pale line of hairs on the dorsal femurs.

Habitat These spiders are often found in agricultural fields, prairies, and oak litter and in and around human structures. They have been found at elevations up to 7,600 ft.

Known records from AB, BC, NB, ON, QC, SK (Canada); AL, AZ, CO, CT, IA, IL, KS, MA, MD, MI, MN, MO, MS, MT, NC, ND, NH, NJ, NM, NY, OH, OK, PA, SD, TN, TX, UT, VA, WI, WV (United States)

Micaria nye
Platnick &
Shadab, 1988

Common Name This
species does not have a
common name.

Total Body Length Female 3.1–4.2 mm,
Male 2.4–3.2 mm

female

female

female

Description The prosoma is reddish brown and
cloaked with white hairs. The opisthosoma is
orange brown. The anterior half is cloaked with
white hairs; the posterior half is cloaked with
iridescent hairs. The femurs are dark brown,
while the distal segments are lighter.

Habitat These spiders can be found running on the ground with an ant-like gait.

Known records from AZ, CA, CO, ID, NM, NV, TX, UT

Micaria pulicaria (Sundevall, 1831)

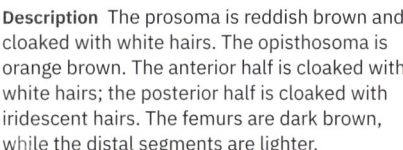

Common Name This species does not have a common name.

Other Colloquial Names Glossy ant spider

Total Body Length Female 3.4–4.0 mm, Male 2.4–4.0 mm

Description The prosoma and opisthosoma are dark brown to black. The
opisthosoma has an overall iridescent quality to it. One or two transverse white bands are on
the opisthosoma: one above the pedicle, and another about midway down. The femurs are dark
brown to black; the distal segments are lighter. These spiders look very similar to ants from the
family Formicidae, with which they seem to associate.

Habitat These spiders can often be found in leaf litter and under rocks. They run with an ant-like
gait. They have been found at elevations up to 12,000 ft.

male

male

male

Known records from AB, BC, MB, NB, NL, NS, NT, ON, QC, SK, YT (Canada); AK, AZ, CA, CO, CT, ID, IL, ME, MI, MN, MT, ND, NH, NM, NV, NY, OH, OR, SD, TX, UT, WA, WI, WY (United States)

Formicidae ant that might be the model that is being mimicked

Nodocion eclecticus
Chamberlin, 1924

Common Name This species does not have a common name.

Total Body Length
Female 7.1–9.8 mm, Male 4.9–6.0 mm

Description The prosoma is dark reddish brown and covered with pale hairs; it is slightly darker in the eye region. The posterior median eyes are irregularly ovoid. The opisthosoma is gray brown and covered with hairs. The males have a scutum on the anterior end; the scutum is highlighted well when the specimen is observed using ultraviolet light. The legs are golden brown.

Habitat These spiders tend to be found under debris, such as fallen trees, stones, etc.

Known records from BC (Canada); AZ, CA, CO, NM, TX, UT (United States)

male viewed with UV light

male

Scotophaeus blackwalli (Thorell, 1871)

Common Name European mouse spider

Other Colloquial Names Mouse spider

Total Body Length Female 7.0–12.0 mm, Male 4.7–9.3 mm

Description The prosoma is light brown to reddish brown. The opisthosoma is light brown and covered with a dense coating of silver hairs. The legs are light brown to reddish brown and often slightly darker at the distal ends. The shiny silver hairs on this species are quite distinctive and said to be reminiscent of the hair of some small rodents.

female

Habitat These spiders are often found in and around human structures. They can also be found in shrubs and chaparrals and under bark.

Known records from BC (Canada); CA, CO, OR, WA (United States)
This is an introduced species.

female

female ventral

female

Sergiolus capulatus (Walckenaer, 1837)

Common Name This species does not have a common name.

Other Colloquial Names Variegated ground spider

Total Body Length Female 6.0–10.0 mm, Male 5.5–7.0 mm

Description The prosoma is orange to red and usually darker in the eye region. The opisthosoma is black, with three white markings: one white band is just posterior of the pedicel; another white band that has a T on the anterior side is just over halfway down; and a curvy white band is just above the spinnerets. The legs are orange red. Legs I and II have black femurs and patellae. Legs III and IV have black banding at the distal end of each segment.

male

Habitat These spiders are often found in leaf litter, in meadows and lawns, and in and around human structures.

Known records from ON, QC (Canada); AR, CT, DC, FL, GA, IA, IL, IN, KS, MA, ME, MI, MN, MO, NC, NJ, NM, NY, OH, OK, PA, TN, TX, VA, WI, WV (United States)

Sergiolus montanus (Emerton, 1890)

Common Name This species does not have a common name.

Other Colloquial Names Common patterned ant-mimic ground spider

Total Body Length Female 7.3–8.6 mm, Male 4.4–5.8 mm

Description The prosoma is usually almost black, with a very broad midline of white hairs. In some specimens, almost the entire prosoma is white; in others, it may be reddish brown. The opisthosoma is black. Two anterior white patches are on either side of the opisthosoma and sometimes connected in the middle. Two roughly diamond-shaped white marks are on either side, about halfway down. A white marking is also on the posterior end. The legs are orange brown, with dark femurs; faint banding may be visible.

female

Habitat These spiders can be found beneath bark, under rocks, and in and around human structures. There are records of specimens being found at elevations up to 10,000 ft.

Known records from AB, BC, MB, NB, ON, QC, SK (Canada); AK, AR, AZ, CA, CO, ID, IL, MI, ME, MN, MO, MT, NC, ND, NH, NM, NV, NY, OH, OR, PA, SC, SD, TX, UT, WA, WI, WY (United States)

Sergiolus ocellatus (Walckenaer, 1837)

Common Name This species does not have a common name.

Total Body Length Female 4.6–6.8 mm, Male 4.1–5.2 mm

Description The prosoma is orange to red and slightly darker in the eye region. The opisthosoma is black, with three white markings: a white band is at the anterior end, another is about halfway down, and a third is just above the spinnerets. The middle band often has a curve that comes to a point that is directed anteriorly. Between the middle and anterior bands are two dots, one on each side of the midline. The legs are orange yellow. The femurs and patellae of legs I and II are black, and there is some banding on leg IV.

Habitat These spiders are often found in leaf litter, bogs, and prairies, and in and around human structures.

Known records from AB, MB, NB, NS, ON, SK (Canada); AL, AR, CA, FL, GA, IL, IN, LA, MA, MI, ME, MN, MO, MS, NC, ND, NJ, NY, OH, PA, TX, VA, WI, WV (United States)

male

male

Sergiolus tennesseensis
Chamberlin, 1922

Common Name This species does not have a common name.

Total Body Length
Female 5.0–6.5 mm,
Male 3.5–5.5 mm

male

Description The prosoma is black but covered with dense white hairs. The opisthosoma is black, with three broad white bands: one anteriorly, one about halfway down, and one posteriorly. The femurs are black, and the distal segments are orange to pink.

Habitat These spiders can be found in leaf litter, under rocks, and beneath bark.

male

male

Known records from AR, CO, IL, IN, MI, MO, ND, TN, TX, VA

Sosticus insularis (Banks, 1895)

Common Name This species does not have a common name.

Total Body Length Female 5.5–7.7 mm, Male 4.5–6.0 mm

Description The prosoma is brown and darker in the eye region. The opisthosoma is dark gray. An anterior scutum is present in the males. The legs are orange brown to brown. Legs I and II have darker femurs, patellae, and tibiae.

Habitat These spiders can be found in leaf litter, beneath bark, and in and around human structures.

Known records from ON, QC (Canada); CO, CT, FL, GA, IA, IL, IN, MI, MN, MO, NY, OH, OK, PA, TX, WI, WV (United States)

female

female

female ventral

female

egg sacs

Zelotes fratris
Chamberlin, 1920

Common Name
This species does not
have a common name.

Total Body Length
Female 6.3–7.7 mm, Male 5.5–6.6 mm

Description The prosoma and opisthosoma are
jet-black and fairly shiny. The posterior median
eyes are somewhat triangular. The males have
a scutum on the anterior of the opisthosoma.
The legs are jet-black, with slightly lighter
distal segments. There are no markings.
There are more than forty species of *Zelotes*,
and they are difficult to identify without a
specimen, as many tend to be uniformly black.

female

female

female

female ventral

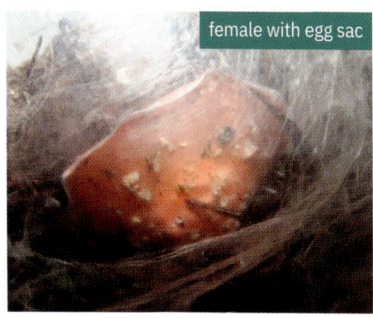

female with egg sac

male

Habitat These spiders can be found in shrubs, agricultural fields, and leaf litter, and under logs and rocks. They have been found at elevations up to 11,500 ft.

Known records from AB, BC, LB, MB, NB, NL, NS, NT, ON, PE, QC, SK, YT (Canada); AK, AZ, CA, CO, CT, IA, ID, IL, IN, MA, ME, MI, MN, MT, NC, ND, NE, NH, NJ, NM, NV, NY, OH, OR, PA, SD, UT, VT, WA, WI, WY (United States)

Zelotes monachus Chamberlin, 1924

Common Name This species does not have a common name.

Total Body Length Female 4.9–6.7 mm, Male 4.5–6.0 mm

Description The prosoma and opisthosoma are jet-black and fairly shiny. The posterior median eyes are somewhat triangular. The males have a scutum on the anterior of the opisthosoma. The legs are jet-black, with slightly lighter distal segments. There are no markings. There are more than forty species of *Zelotes*, and they are difficult to identify without a specimen, as many tend to be uniformly black.

subadult male with termite prey

Habitat These spiders are often found in duff, under stones, and under human debris. They have been observed feeding on termites, which may be their preferred prey. They have been found at elevations up to 9,000 ft.

Known records from AZ, CA, CO, NM, NV, TX, UT

Zelotes tenuis
(L. Koch, 1866)

Common Name
This species does not have a common name.

Total Body Length
Female 4.6–6.6 mm, Male 3.3–5.0 mm

Description The prosoma is tawny brown, often with faint markings. The posterior median eyes are somewhat rectangular and obliquely

female prosoma

oriented. The opisthosoma is dark gray. The males have a small scutum on the anterior of the opisthosoma. The legs are mostly tan to brown. Dusky bands are sometimes on the distal femurs. Legs I and II of the males are often slightly darker than the other legs.

Habitat These spiders are often found in leaf litter or under debris. They are sometimes found in and around human structures.

female ventral

female

female

Known records from CA, UT

This is an introduced species, previously documented only from CA. The specimen photographed here was found in UT.

HOMALONYCHIDAE ▶ DUSTY DESERT SPIDERS

Genera in This Family *Homalonychus* (2)

Total Body Length 6.5–12.8 mm

Description These cryptic spiders are entelegyne and ecribellate. Their prosoma is broad and as long as it is wide, and it narrows anteriorly. They have eight eyes. The posterior eye row is strongly recurved. The spiders have strongly convergent endites. Their opisthosoma is pentagonal. Stiff hairs on the prosoma, opisthosoma, and legs trap small pieces of dirt and sand, providing these spiders with good camouflage. Freshly molted individuals have been observed digging a shallow pit and then rolling over, so they are upside down, and then rocking back and forth, gently coating themselves with sand and dirt. They are usually found under rocks and plant debris.

eye configuration
(*Homalonychus selenopoides*)

Similar Families Pisauridae, Sparassidae, Zodariidae

How to distinguish Homalonychidae from other similar families:

Homalonychids can be distinguished from pisaurids, sparassids, and zodariids by the shape of the prosoma. Homalonychid prosomas narrow anteriorly to a greater extent than is seen in other similar families. The art of using debris stuck in their stiff hairs to camouflage themselves is also unique to this family.

prosoma narrows anteriorly; opisthosoma pentagonal shaped (*Homalonychus selenopoides*)

Homalonychus selenopoides Marx, 1891

Common Name This species does not have a common name.

Other Colloquial Names Dusty spider, encrusted spider

Total Body Length Female 8.0–9.0 mm, Male 7.0–7.7 mm

Description The prosoma is gray brown to reddish, with black spots; it narrows in the eye region. The opisthosoma is pentagonal and gray brown to reddish, with black markings. The spiders are covered with modified hairs that often collect sand and debris, giving them a dusty appearance and often obscuring the markings. The legs are a similar color: gray brown to reddish.

Habitat These spiders tend to live in desert or chaparral habitats and can be found under rocks and other debris. They will sometimes bury themselves partially in the sand or soil, so only their outstretched legs are visible. The females adorn their egg sacs with dirt and debris.

freshly molted before dusting

freshly molted after dusting

Known records from AZ, CA, NV

LIOCRANIDAE > SPINYLEGGED SAC SPIDERS

Genera in This Family *Agroeca* (6), *Apostenus* (2), *Hesperocranum* (1), *Liocranoeca* (1), and *Neoanagraphis* (2)

Total Body Length 2.0–10.0 mm

Description These spiders are entelegyne and ecribellate. They have eight eyes in two straight to slightly curved rows. They do not have precoxal triangles. They have a pair of book lungs and a single tracheal spiracle located near the spinnerets. In some species, the anterior tibiae and metatarsi have two rows of dense bristles on the ventral surface. Other species have one to five pairs of spines. They tend to live under rocks and in leaf litter.

Similar Families Anyphaenidae, Clubionidae, Corinnidae, Gnaphosidae, Phrurolithidae, Sparassidae, Trachelidae

How to distinguish Liocranidae from other similar families:

Liocranids can be distinguished from anyphaenids, clubionids, corinnids, phrurolithids, and trachelids by their lack of precoxal triangles and the dense rows of bristles or paired spines on the anterior tibiae

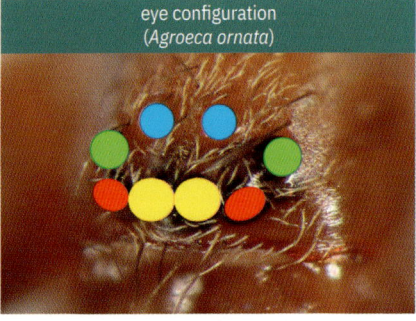

eye configuration (*Agroeca ornata*)

lacks precoxal triangles (*Agroeca ornata*)

and metatarsi. Anyphaenids, clubionids, corinnids, and phrurolithids all have precoxal triangles, and these families have paired spines on the anterior tibiae. Liocranids can be distinguished from anyphaenids by the placement of their tracheal spiracle. In liocranids, the spiracle is located near the spinnerets, whereas in anyphaenids, the tracheal spiracle is located more anteriorly. The widely spaced, cylindrical spinnerets that are seen in gnaphosids distinguish them from liocranids. They can be distinguished from sparassids by their prograde legs; the legs of sparassids are laterigrade.

Agroeca ornata Banks, 1892

Common Name This species does not have a common name.

Total Body Length Female 5.0–6.6 mm, Male 4.4 mm

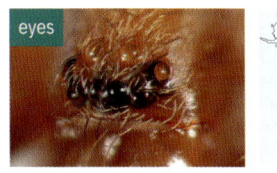
eyes

female

male

Description The prosoma is orange brown to brown. A faint pattern of markings radiates from the fovea. The opisthosoma is brown, and two rows of indistinct darker markings run down the length of the opisthosoma. The legs are orange brown, with visible spines.

Habitat These spiders are often found in leaf litter. The egg sacs are spherical and covered with dirt and debris and suspended from rocks, plants, or other structures near the ground.

Known records from AB, BC, MB, NB, NL, NS, NT, ON, PE, QC, SK, YT (Canada); AK, CA, CO, CT, IA, ID, IL, MA, ME, MI, MN, NH, NJ, NY, PA, UT, VT, WI, WV (United States)

Neoanagraphis chamberlini Gertsch & Mulaik, 1936

Common Name This species does not have a common name.

Total Body Length Female 5.5–9.7 mm, Male 3.8–9.1 mm

Description The prosoma is orange brown and covered with hairs. The opisthosoma is gray brown and covered with hairs. A faint heartmark

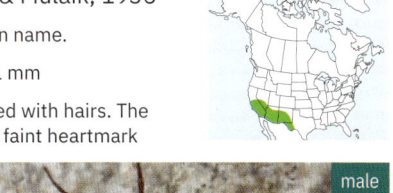

is usually visible. The legs are orange brown, with many spines. The tarsal claws on legs III and IV are extremely long.

Habitat Not much is known about the habitat preferences of this species, but some of these spiders have been found under debris and in sand dune habitats.

Known records from AZ, CA, NM, NV, TX

male

LYCOSIDAE ▶ WOLF SPIDERS

Genera in This Family *Acantholycosa* (1), *Allocosa* (17), *Alopecosa* (8), *Arctosa* (11), *Camptocosa* (2), *Geolycosa* (19), *Gladicosa* (5), *Hesperocosa* (1), *Hogna* (19), Lycosa (9), *Melocosa* (1), *Paratrochosina* (1), *Pardosa* (80), *Pirata* (24), *Piratula* (5), *Rabidosa* (5), *Schizocosa* (27), *Sosippus* (6), *Tigrosa* (5), *Trabeops* (1), *Trebacosa* (1), *Trochosa* (4), and *Varacosa* (5)

Total Body Length 2.2–35.0 mm

Description These spiders are entelegyne and ecribellate. They have eight eyes that appear to be in three rows. They usually have enlarged posterior median eyes. The posterior eyes make a quadrangle that is usually narrower anteriorly than it is posteriorly. They have a pair of book lungs and a single tracheal spiracle near the spinnerets. The females carry the egg sacs by attaching them to the spinnerets. When the young emerge, they ride on the female's opisthosoma for a period of time before dispersing. Lycosids in general can be difficult to identify without looking at the genitalia.

Similar Families Agelenidae, Cheiracanthiidae, Ctenidae, Miturgidae, Pisauridae, Salticidae, Trechaleidae, Zoropsidae

How to distinguish Lycosidae from other similar families:

Lycosids have a distinctive eye configuration. The four anterior eyes are usually small and arranged in a row across the front of the prosoma. The posterior median eyes are usually quite large. The posterior lateral eyes are placed back on the prosoma, giving the appearance of three eye rows. This is very different from the eye configuration seen in agelenids, cheiracanthiids, miturgids, salticids,

eye configuration
(*Hogna baltimoriana*)

lines drawn through posterior lateral and posterior median eyes cross in front of prosoma (*Hogna baltimoriana*)

babies riding on female's abdomen (*Schizocosa* sp.)

and zoropsids. Ctenids and pisaurids can have a somewhat similar-looking eye configuration. If you were to draw a line through the posterior lateral and median eyes on each side of the prosoma and extend those lines forward, they would cross beyond the front of the spider in lycosids, whereas those lines would cross on the prosoma in pisaurids. Ctenids have small anterior lateral eyes, and the anterior eye row is recurved, making the anterior lateral eyes appear to be in a row with the posterior median eyes. Both lycosids and trechaleids attach their egg sacs to their spinnerets, but in lycosids, when the young emerge, they ride on their mother's opisthosoma. In contrast, in trechaleids, the young ride on the egg sac. Also note that trechaleids are represented in North America (north of Mexico) by only one species, and it is found only in Arizona.

Allocosa funerea (Hentz, 1844)

Common Name This species does not have a common name.

Total Body Length Female 5.0–6.0 mm, Male 3.5–4.5 mm

Description The prosoma is shiny black. The opisthosoma is dusky yellow, with purplish-brown markings and a heartmark that is either dark outlined or completely dark and occasionally is plain black. The femurs are shiny black. The distal segments are yellow orange, with banding.

female
ventral

female
(parasitic mite
on prosoma)

female

female with
egg sac

Habitat These spiders can be found in lawns, fields, meadows, and forested habitats, and in and around human structures.

Known records from AL, AR, CT, DC, DE, FL, GA, IL, IN, KS, KY, LA, MA, MD, MI, MO, MS, NC, NJ, NY, OH, OK, PA, SC, TN, TX, VA, WV

Alopecosa kochi (Keyserling, 1877)

Common Name This species does not have a common name.

Other Colloquial Names Koch's wolf spider

Total Body Length Female 9.0–16.0 mm, Male 6.6–11.0 mm

Description The prosoma is brown, with a broad tan median line. The median line constricts slightly at the fovea and widens again before

female

female with egg sac

male

male

male

constricting again at the eye region. A few darker lines radiate from the fovea. There may be a faint submarginal band. The opisthosoma is darkest on the lateral edges. There may or may not be a visible heartmark. A pale median line often runs the length of the opisthosoma. The legs are reddish brown, and the femurs have indistinct dark banding. They can be quite variable in their markings and coloration.

Habitat These spiders are often found under rocks and in leaf litter.

Known records from AB, BC, MB, ON, QC, SK (Canada); AR, AZ, CA, CO, FL, ID, IN, KS, MA, ME, MI, MN, MT, ND, NE, NH, NJ, NM, NV, NY, OK, OR, PA, SD, TX, UT, WA, WY (United States)

Arctosa littoralis (Hentz, 1844)

Common Name Shoreline wolf spider

Total Body Length Female 11.2–14.7 mm, Male 9.6–12.8 mm

Description The prosoma can range from very pale gray to brown to almost orange, with mottled markings throughout. The eye region is usually darker. The coloration of the opisthosoma is similar to that of the prosoma. A slighter darker heartmark may be visible. The legs are tan and brown and banded.

Habitat This species is known to live on sandy beaches of both freshwater and saltwater bodies, where its coloration hides it perfectly. Often, you will not see these spiders until they decide to

female

female

female

move. They have been found sheltering within driftwood but are also known to burrow.

Known records from BC, MB, NB, NS, ON, PE, QC, SK (Canada); AK, AR, AZ, CA, CO, CT, DE, FL, GA, IA, ID, IL, IN, KS, MA, MD, MI, MN, MO, MS, MT, NC, NH, NJ, NM, NV, NY, OH, OK, OR, PA, RI, SC, SD, TX, UT, VA, WA, WI, WY (United States)

Geolycosa missouriensis (Banks, 1895)

Common Name The genus *Geolycosa* has the common name "burrowing wolf spider."

Other Colloquial Names Missouri burrowing wolf spider

Total Body Length Female 15.8–21.0 mm, Male 15.0–16.3 mm

female

female

Description The prosoma is tan to brown or light gray and highly domed between the fovea and the posterior eye row. The chelicerae are covered with thick yellow to dark golden hairs. The opisthosoma is tan to brown, sometimes gray. The legs are tan to brown and unbanded. The ventral surface of the patellae, tibiae, metatarsi, and tarsi of legs I and II is cloaked with black hairs, and the ventral femurs are yellow.

Habitat These spiders like to dig their burrows in sandy soil and may or may not have a turret at the entrance. They will readily rear up when disturbed.

Known records from AB, MB, ON, QC, SK (Canada); AZ, CO, FL, IA, IL, IN, KS, MI, MN, MO, MT, NC, NE, NJ, NM, NY, OH, OK, SD, TX, UT, WI (United States)

female ventral

female in burrow

female (defensive pose)

Geolycosa turricola (Treat, 1880)

Common Name The genus *Geolycosa* has the common name "burrowing wolf spider."

Total Body Length Female 22.0 mm, Male 20.0 mm

Description The female prosoma is dark brown gray, with a lighter median band. The chelicerae may have yellow hairs, especially toward the ends. The opisthosoma is tan gray, with a darker median band. The legs are dark gray and unbanded. The ventral legs I and II are darker on the patellae, tibiae, metatarsi, and tarsi. The ventral femurs are light in color. The sternum and coxa are light brown. The males are light tan gray, with a brown paramedian marking. The chelicerae are often cloaked in yellow hairs. The heartmark is dark. The legs are tan, with dark ventral patellae, tibiae, metatarsi, and tarsi. The ventral femurs are light in color. The sternum and coxae are tan.

female

female

subadult male

subadult male ventral

burrow entrance

Habitat These spiders typically build their burrows in open fields and grasslands. The burrows can be more than 8 in deep, and a turret is usually at the burrow entrance.

Known records from CT, FL, GA, MA, MD, MI, MN, NC, NH, NJ, NY, OH, PA, TN, VA

Gladicosa pulchra (Keyserling, 1877)

Common Name This species does not have a common name.

Total Body Length Female 15.3–16.5 mm, Male 12.0–12.7 mm

Description The prosoma is brown to gray. A pale median band that is triangular, posterior to the fovea (the tip pointing to the pedicel), followed by a round area posterior of the eyes, and then encompassing the eye region. The edges are pale, and the darkest area is near the pedicel. The opisthosoma is gray to brown. A pair of black markings are on each side of the opisthosoma at the anterior end. The dorsal surface is pale gray to tan, and some light spots are present. The ventral opisthosoma

female

female

is dark brown to black posterior of the epigastric furrow and yellowish anteriorly. The edges are darker. The legs are gray to brown and banded in tan. This species looks very similar to *Gladicosa gulosa*, but the ventral opisthosoma of *G. gulosa* is lighter (yellow to light brown posterior of the epigastric furrow), and *G. pulchra* tends to be slightly larger than *G. gulosa* (females are 11.0–14.0 mm, and males are 10.0–11.0 mm).

Habitat These spiders are sometimes found on tree trunks but can also be found under stones or logs.

Known records from AL, AR, CO, FL, GA, IL, KS, KY, LA, MO, MS, NC, NJ, NY, OH, OK, TN, TX, VA

female

female
ventral

juvenile

juvenile

male

male

Hesperocosa unica (Gertsch & Wallace, 1935)

Common Name This species does not have a common name.

Total Body Length Female 5.9 mm, Male 4.0 mm

Description The prosoma is dark brown, almost black. There is a pale gray median band that remains the same width for the length of the prosoma. The opisthosoma is dark brown, almost black, with a pale gray median band that covers most of the dorsal surface. The legs are uniformly black, except for the tarsi, which are yellow to orange.

Habitat Not much is known about the habitat preferences of this species.

Known records from AZ, NM, TX

Hogna antelucana
(Montgomery, 1904)
and **Hogna labrea**
(Chamberlin & Ivie, 1942)

Common Name These species do not have common names.

Total Body Length
Female 13.5–19.0 mm, Male 13.0–18.0 mm

Description The prosoma is gray brown, with a tan median stripe and edges. Dark lines radiate from the fovea. Two dark dashes may be in the median stripe near the posterior eyes. The anterior eye row sits almost on the edge of the clypeus. The markings on the opisthosoma can vary. It can be tan to brown, with faint black spotting and no heartmark, or it can have a dark heartmark, dark bands on the anterior-lateral edges, and a series of connected spots. The males tend to be more dramatically marked than the females. The legs are tan with faint banding. The ventral opisthosoma, sternum, and coxae are jet-black. When *H. labrea* was described,

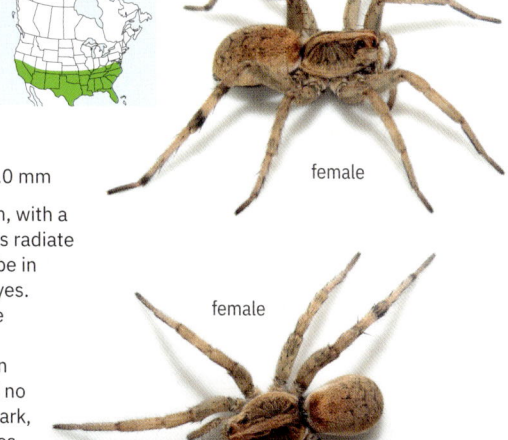

female

female

369

female

female
ventral

it was noted that its outward appearance was the same as that of *H. antelucana*. The species can be distinguished only by looking at the genitalia.

Habitat These spiders are often found under stones or other debris.

Known records of *H. antelucana* from AR, AZ, CA, CO, FL, KY, MS, NM, OK, SC, TX, UT; known records of *H. labrea* from AZ, CA

Hogna baltimoriana (Keyserling, 1877)

Common Name This species does not have a common name.

Other Colloquial Names Unbanded wolf spider

Total Body Length Female 15.0–18.0 mm, Male 14.0–17.5 mm

Description The prosoma is dark brown, with a thin tan stripe from the fovea to the eye region. Tan lines radiate from the fovea, and the edges are also tan. The chelicerae are cloaked with thick yellow hairs. The opisthosoma is tan, with a dark heartmark bordered by six white spots. There is brown mottling on the dorsal surface. The ventral opisthosoma has a dark marking posterior of the epigastric furrow, and the sternum is black. The legs are tan, with two pairs of longitudinal dashes on the dorsal femurs. The undersides of the patellae are black.

Habitat These spiders are often found in grassy areas but can also be found in and around human structures.

female

female ventral

Known records from ON (Canada); AL, AR, AZ, CO, CT, IA, IL, IN, LA, MA, MD, MI, MN, MO, MS, MT, NC, NJ, NM, OH, PA, RI, SD, TX, VA, WI (United States)

Hogna carolinensis (Walckenaer, 1805)

Common Name Carolina wolf spider

Total Body Length Female 22.0–35.0 mm, Male 18.0–20.0 mm

Description The prosoma can vary in color from gray to dark reddish brown. A wide pale median band ends at the posterior eyes. The edges are also pale. There may be faint radial lines, but they usually are not as distinct as those seen in other species. The chelicerae are cloaked in yellow hairs. The opisthosoma is tan, with a dark heartmark. Faint spots and chevrons may be on the posterior end. The ventral opisthosoma is black, as are the sternum and coxae. The females have

female ventral | male ventral

male | male

black markings on the ventral-distal femurs, while the males have black markings distally on the ventral femurs, patellae, tibiae, and the metatarsi. The tarsi are black ventrally. The undersides of the patellae are black.

Habitat These spiders are often found in pastures and fields. This is a burrowing species. Its burrow usually has a turret adorned with grass and debris. The spiders tend to hunt at night but can occasionally be found during the day.

Known records from ON (Canada); AL, AR, AZ, CA, CO, CT, DE, FL, GA, IL, IN, KS, KY, LA, MA, MD, ME, MI, MO, MS, MT, NC, NE, NH, NJ, NM, NY, OH, OK, OR, PA, RI, SC, TN, TX, UT, VA, VT, WI, WY (United States)

Hogna coloradensis (Banks, 1894)

Common Name This species does not have a common name.

Other Colloquial Names Colorado wolf spider

Total Body Length Female 10.5–20.1 mm, Male 9.5–13.6 mm

Description The opisthosoma is dark brown, with a tan to yellow-orange median band that widens slightly at the eye region and wide tan edges. The chelicerae are cloaked with yellow hairs. The opisthosoma is brown, with a dark brown heartmark. There is a wide yellow-tan midline stripe with paired white spots that are faintly connected. The ventral opisthosoma is tan, with black anterior of the epigastric furrow and a black mark near the spinnerets. There may be some black spots between the epigastric furrow and the spinnerets, but they are usually scattered. The sternum and coxae are black. The legs are gray to tan without banding. The ventral distal femurs and tibiae have black markings.

subadult male
ventral

subadult male

subadult
male

subadult male

Habitat These spiders are often found in sandy habitats, where they build a shallow burrow.

Known records from AZ, CA, CO, KS, NE, NM, TX, WA, WY

Hogna frondicola
(Emerton, 1885)

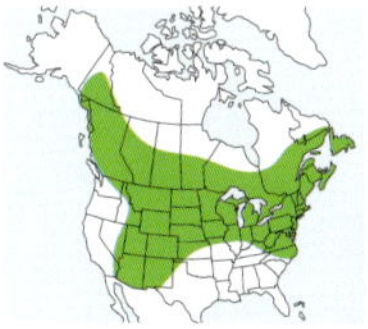

Common Name
This species does not have a common name.

Other Colloquial Names Forest wolf spider

Total Body Length
Female 11.0–14.0 mm, Male 9.0–12.0 mm

Description The prosoma is dark reddish brown, with a wide pale median stripe that runs parallel until the posterior lateral eyes, where it narrows angularly between the eyes. The edges are tan. The opisthosoma is gray brown, with black on the anterior margins and sometimes two to four black spots. A few dark transverse marks may be on the posterior. The heartmark is absent or indistinct. The ventral opisthosoma is suffused with black. The sternum and coxae are reddish brown suffused with black.

female

female at burrow entrance

female at burrow entrance

The legs are reddish brown, with indistinct pale banding.

Habitat These spiders are often found in leaf litter. The females build a shallow burrow when they are guarding their egg sacs.

Known records from AB, BC, MB, NB, NL, NS, ON, QC, SK, YT (Canada); AR, AZ, CO, CT, FL, IA, IL, IN, KS, MA, MD, ME, MI, MS, MT, NC, NE, NH, NJ, NM, NY, OH, PA, RI, TX, UT, VT, WA, WI, WY (United States)

Pardosa bucklei Kronestedt, 1975

Common Name The genus *Pardosa* has the common name "thinlegged wolf spiders."

Total Body Length Female 6.8–9.0 mm, Male 6.0–7.6 mm

Description The prosoma is dark brown. There is a faint pale gray median band, although it is not contiguous. There is a faint, broken, pale gray submarginal band. The opisthosoma is dark brown, almost black. A series of irregular pale gray markings is on the dorsal surface. The legs are dark brown, with light brown banding on the distal segments. Numerous long spines are seen on the legs, which is common in this genus.

female

Habitat These spiders are often found in prairies or grassy habitats. They have been found at elevations up to 12,000 ft.

Known records from AB, BC, SK (Canada); AZ, CA, CO, ID, MT, NE, NM, OR, UT, WY (United States)

male

male

male

male

Pardosa milvina (Hentz, 1844)

Common Name The genus *Pardosa* has the common name "thinlegged wolf spiders."

Other Colloquial Names Shore spider

Total Body Length
Female 5.2–6.5 mm, Male 4.0–4.9 mm

Description The prosoma is brown, with a yellow median stripe and borders. The median stripe is lobed and ends at the posterior eyes. The opisthosoma is brown, with two pairs of indistinct dark spots and overall brown and tan mottling. The legs are yellow, with dusky banding. Numerous long spines are seen on the legs, which is common in this genus. The males have black pedipalps, and white hairs are scattered on the pedipalp femurs and patellae.

female

375

Habitat These spiders can be found in habitats ranging from dry meadows to pond edges, forests to agricultural fields. They can also be found in and around human structures.

Known records from ON, QC (Canada); AL, AR, CA, CT, DE, FL, GA, IA, IL, IN, KS, KY, LA, MA, MD, ME, MI, MN, MO, MS, NC, NE, NH, NJ, NY, OH, OK, PA, RI, SC, TN, TX, VA, VT, WI, WV (United States)

female

female

female
ventral

female with
egg sac

male

female with
young

male

male

Pardosa moesta Banks, 1892

Common Name The genus *Pardosa* has the common name "thinlegged wolf spiders."

Total Body Length Female 5.1–6.1 mm, Male 4.4–5.3 mm

Description The prosoma is dark brown to almost black and shiny. A few light-colored hairs may be around the edge. The opisthosoma is dark brown, with light-colored hairs that create faint dashes across the posterior half. The legs are dark brown proximally and lighten somewhat distally. A few indistinct bands may be seen.

Habitat These spiders are often found in open habitats, such as fields, lawns, beaches, and meadows. They are also sometimes found in forested habitats.

Known records from AB, BC, LB, MB, NB, NL, NS, NT, ON, PE, QC, SK, YT (Canada); AK, AR, CO, CT, DE, IA, IL, IN, MA, MD, ME, MI, MN, MT, ND, NE, NH, NJ, NY, OH, OR, PA, RI, SD, TN, UT, VA, VT, WA, WI, WV, WY (United States)

female

female ventral

female

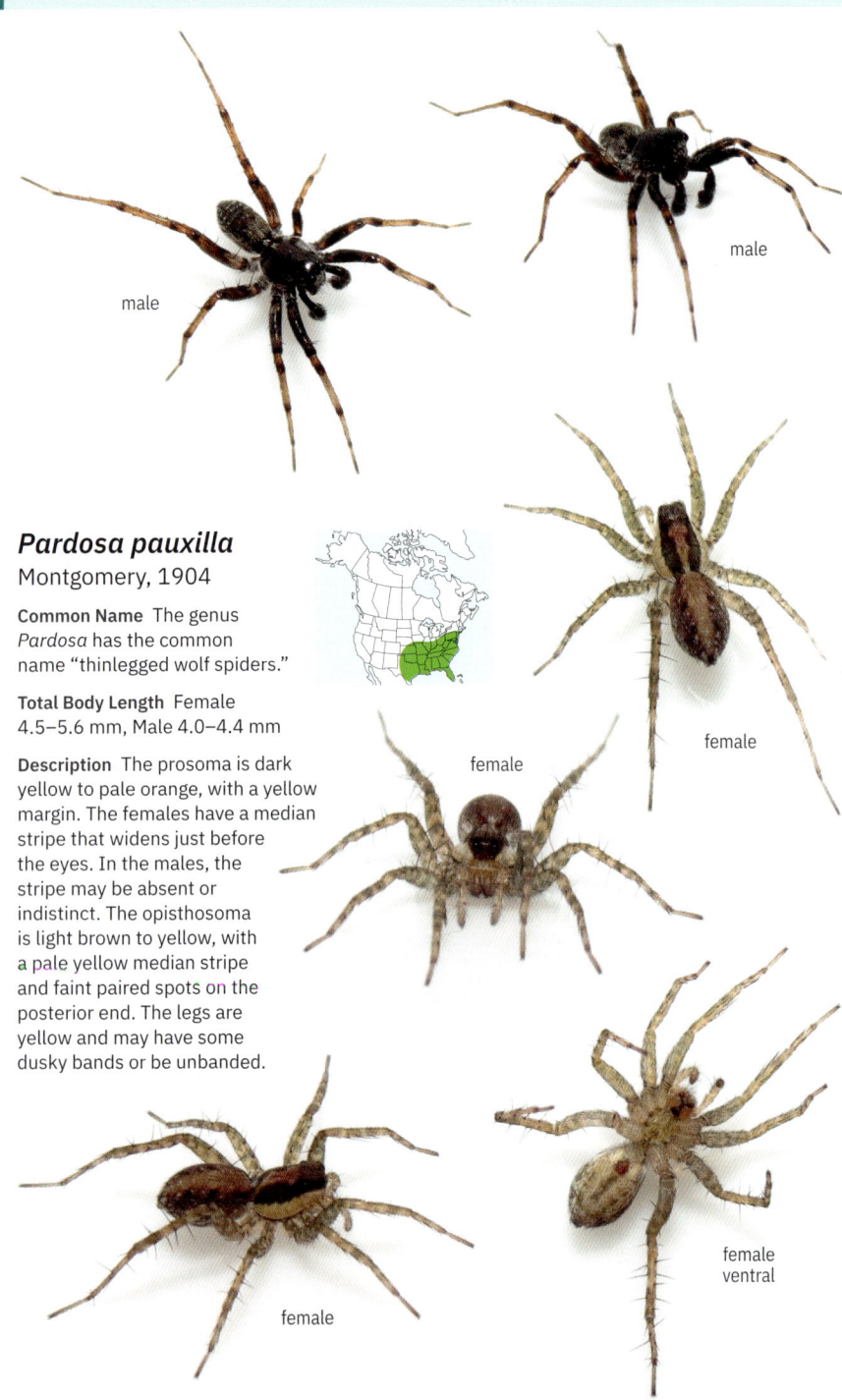

male

male

Pardosa pauxilla
Montgomery, 1904

Common Name The genus *Pardosa* has the common name "thinlegged wolf spiders."

Total Body Length Female 4.5–5.6 mm, Male 4.0–4.4 mm

Description The prosoma is dark yellow to pale orange, with a yellow margin. The females have a median stripe that widens just before the eyes. In the males, the stripe may be absent or indistinct. The opisthosoma is light brown to yellow, with a pale yellow median stripe and faint paired spots on the posterior end. The legs are yellow and may have some dusky bands or be unbanded.

female

female

female

female
ventral

female

male

male

male

Habitat These spiders can be found in agricultural fields, forested habitats, and grassy areas, and sometimes in and around human structures.

Known records from AR, FL, GA, KS, LA, MD, MS, NC, NE, NJ, NM, OH, OK, PA, SC, TX, VA

Pirata aspirans Chamberlin, 1904

Common Name The genus *Pirata* has the common name "pirate wolf spiders."

Total Body Length Female 4.0–4.6 mm, Male 3.9 mm

Description The prosoma is brown, with dusky edges and an irregular yellow median band that forks anterior of the fovea, creating a tuning-fork pattern that touches the posterior lateral eyes. The opisthosoma is dark brown, with a paler heartmark. There are four pairs of white spots on the posterior half. The lateral edges are highlighted with white hairs. The legs are yellow, with dusky bands. The tibiae are usually darker than the other segments; this characteristic is most obvious on leg I.

Habitat These spiders tend to be found in moist habitats, such as wet meadows, bogs, and swamps.

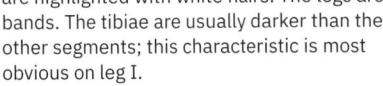

male

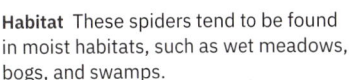

male

male

male

Known records from MB, NS, ON, QC (Canada); CO, CT, GA, IA, IL, IN, KS, MA, ME, MI, MN, MO, NJ, NY, OH, VA, WI (United States)

Pirata montanus Emerton, 1885

Common Name The genus *Pirata* has the common name "pirate wolf spiders."

Other Colloquial Names Dark-legged pirate wolf spider

Total Body Length Female 4.7–5.3 mm, Male 4.3–5.3 mm

Description The female prosoma is dark brown, with a tan median stripe. Within the median stripe may be an indistinct brown V that opens toward the eye region from the fovea. The edge of the prosoma has a thin line of white hairs. The opisthosoma is brick red to reddish brown. The heartmark is gray to dark brown. There is a series of dark chevrons posterior to the heartmark. At the edge on either side of each chevron is a white to yellow spot. The legs are yellow to light brown. A dusky band is at the distal end of the femurs, tibiae, and metatarsi. The male prosoma is dark brown to black. A white median stripe starts out narrow at the posterior edge and gradually widens as it approaches the eye region. The edge of the prosoma has a thin line of white hairs. The opisthosoma is dark brown to black, with a white spot at the anterior end just before the dark heartmark, which is bordered in white. There is a series of white chevrons with white spots on each edge. The lateral edges are spotted with white. The legs are yellow to light brown. The femurs of legs I and II are dark, almost black. There are dusky bands at the femurs of legs III and IV. The legs become paler distally.

Habitat These spiders tend to be found in areas with moist conditions, including both coniferous and deciduous forests, bogs, marshes, and wet grassy areas.

Known records from MB, NB, NL, NS, ON, QC (Canada); AR, CT, IL, IN, KY, MA, MD, ME, MI, MN, MS, NC, NH, NJ, NY, OH, PA, TN, VA, VT, WI, WV (United States)

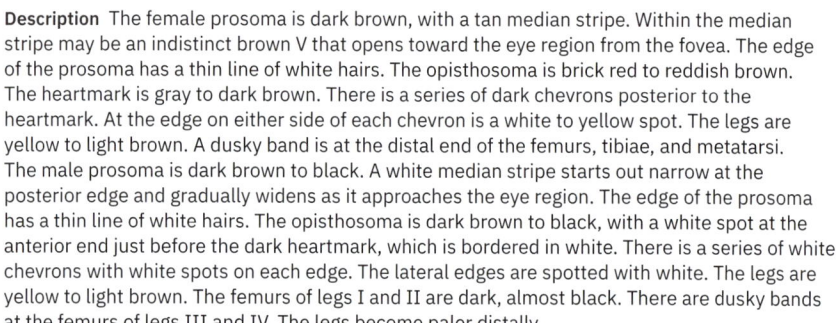

female

male

male

male

Pirata sedentarius Montgomery, 1904

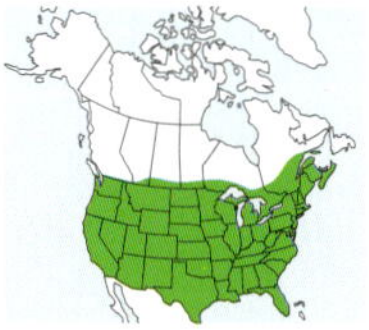

Common Name The genus *Pirata* has the common name "pirate wolf spiders."

Total Body Length Female 4.4–5.0 mm, Male 4.5 mm

Description The prosoma is brown, with pale edges and a pale median line that forks abruptly at the fovea, creating a marking shaped like a tuning fork that leads to the posterior lateral eyes. The opisthosoma is brown. There is an orange heartmark. A series of triangular yellow to orange shapes point toward the heartmark and have white dots at each base corner. The lateral edges are variegated in brown and yellow. The legs are yellow, with dusky banding.

Habitat These spiders tend to be found in forested habitats, marshes, and lake shores.

Known records from AB, MB, NB, NS, ON, QC, SK (Canada); AL, AR, AZ, CA, CO, CT, FL, GA, IA, ID, IL, IN, KS, KY, MA, MD, ME, MI, MO, MS, MT, NC, ND, NE, NH, NJ, NM, NV, NY, OH, OK, OR, PA, RI, SD, TN, TX, UT, VA, VT, WA, WI, WV (United States)

female

female

female
ventral

female

female with
egg sac

Piratula canadensis (Dondale & Redner, 1981)

Common Name This species does not have a common name.

Total Body Length Female 3.7–3.8 mm, Male 3.0–3.5 mm

Description The prosoma is brown, with a dark brown median stripe that makes a tuning-fork shape starting at the fovea. The points go to the posterior lateral eyes. There is a dark submarginal band. There may or may not be a dark edge to the prosoma. The opisthosoma is brown, with a slightly lighter heartmark (that may not be visible). There may be some indistinct markings, but there are no prominent markings on the opisthosoma. The ventral opisthosoma is light brown, with two coalescing longitudinal stripes. The legs are tan to brown, and there may be some faint dusky banding. The distal segments are usually darker. The tarsi of leg I in the males have a distinct curve to them.

male

male

male

Habitat These spiders tend to be found in moist habitats within wetlands and forests.

Known records from AB, BC, MB, NB, NL, NS, ON, QC, SK (Canada); AK, IL, OH (United States)

Piratula minuta (Emerton, 1885)

Common Name This species does not have a common name.

Total Body Length Female 2.8–3.7 mm, Male 2.5–3.1 mm

Description The prosoma is dark brown to black. A very dark yellow median stripe contains the tuning-fork mark seen on most *Pirata* and *Piratula* species. The very edge has a thin line of white hairs and a thin dark yellow line. Some specimens are so dark that the prosoma looks almost completely black unless the lighting is just right. The opisthosoma is dark brown to black, and five to six pairs of white to yellow spots are on either side of the midline on the dorsal surface. White spots may be scattered on the lateral surfaces. The legs are dark yellow, with broad dark brown to black bands; sometimes these are so broad that the legs look almost completely dark. In the males, the first pair of legs often looks iridescent. This is one of the smaller wolf spiders, and it is often overlooked due to its small size.

Habitat These spiders are often found in forested habitats, wetlands, and lawns. The egg sacs are often brilliant white.

Known records from MB, NB, NL, NS, ON, PE, QC, SK (Canada); AR, CT, DE, FL, IA, ID, IL, IN, MA, MD, ME, MI, MN, MS, NC, NE, NH, NJ, NY, OH, PA, SC, TN, UT, VA, VT, WI, WV (United States)

female

female with egg sac

female with egg sac

female
ventral

male

male

Rabidosa punctulata (Hentz, 1844)

Common Name Dotted wolf spider

Total Body Length Female 11.0–17.0 mm, Male 13.0–15.0 mm

Description The prosoma is dark brown, with a tan median stripe that narrows at the eye region and continues to the clypeus. The chelicerae are dark brown, sometimes with pale hairs on the lateral edges. Broad tan bands are on either side of the prosoma edges, and the very edge of the prosoma is also tan. The opisthosoma has a broad dark midline stripe that does not have chevrons or other markings in it. The stripe is bordered on each side by tan stripes. The lateral edges are more mottled. The ventral opisthosoma is tan, with scattered to dense dark brown to almost black dots. The legs

female

Rabidosa carrana ventral markings

female
ventral

juvenile

female

are tan, with faint, dusky, paired dorsal longitudinal stripes. This species is sometimes confused with *Rabidosa carrana*, which is found in Florida, Georgia, and North Carolina. Note that in *R. carrana*, three sets of paired white dots are within a large black marking on the ventral opisthosoma.

Habitat These spiders are often found in grassy habitats.

Known records from AL, AR, CT, FL, GA, IL, IN, KS, LA, MA, MD, MI, MO, MS, NC, NE, NJ, NY, OH, OK, PA, RI, SC, TN, TX, VA

female

Rabidosa rabida (Walckenaer, 1837)

Common Name This species does not have a common name.

Other Colloquial Names Rabid wolf spider

Total Body Length Female 16.0–21.0 mm, Male 11.0–12.0 mm

Description The prosoma is dark brown, and a tan median stripe narrows at the eye region and continues to the clypeus. Broad tan bands are on

male

male

male

male

male

juvenile

juvenile ventral

juvenile

either side of the prosoma edges, and the very edge of the prosoma is also tan. The chelicerae are pale, and dark stripes continue the pattern from the prosoma. The opisthosoma has a broad dark midline stripe, on the edges of which paired light-colored dots appear as you approach the posterior end, giving it the appearance of chevron shapes. The stripe is bordered by tan stripes on each side. The lateral edges are more mottled. The ventral opisthosoma is tan and has no markings. The legs are tan, and faint, dusky, paired, dorsal longitudinal stripes are on the femurs. In the mature males, the first pair of legs is completely dark brown.

Habitat These spiders are often found in forested habitats, areas of tall grass, and agricultural fields.

Known records from ON (Canada); AL, AR, AZ, CT, DE, FL, GA, IA, IL, IN, KS, KY, LA, MA, MD, MI, MN, MO, MS, NC, NE, NJ, NY, OH, OK, PA, SC, TN, TX, VA, WV (United States)

Schizocosa avida (Walckenaer, 1837)

Common Name This species does not have a common name.

Other Colloquial Names Lance wolf spider

Total Body Length Female 6.6–14.7 mm, Male 6.3–9.8 mm

Description The prosoma is dark brown, with a tan median line that is completely parallel its entire length. The edges of the prosoma are also tan. Within the brown, shaded radial lines are visible. The opisthosoma is gray brown. The heartmark is dark brown. A tan midline stripe tapers at the posterior end. Paired dark spots or dashes are on either side

female

female

female ventral

male

male

male ventral

of the tan midline starting about halfway down. The ventral opisthosoma of the females often has a bright yellow central marking that is surrounded by black. At times, this may look like a central yellow stripe bordered in black. The males tend to have a pale yellow ventral opisthosoma. The legs are tan, and faint dusky bands are sometimes visible.

Habitat These spiders are often found in pastures, meadows, and lawns.

Known records from MB, NS, ON, QC (Canada); AL, AR, AZ, CO, CT, DE, FL, GA, IA, IL, IN, KS, KY, LA, MA, MD, ME, MI, MO, MS, MT, NC, NE, NH, NJ, NY, OH, OK, PA, RI, SC, SD, TN, TX, UT, VA, WI, WV, WY (United States)

Schizocosa crassipes (Walckenaer, 1837)

Common Name This species does not have a common name.

Total Body Length Female 6.2–8.7 mm, Male 5.0–6.5 mm

Description The female prosoma is dark brown, with a broad cream median band. The very edge of the prosoma is tan. The female opisthosoma is variegated with browns, blacks, and tans. There is the hint of a slightly paler median stripe, but it is not well defined. The female legs are brown, striped with tan. The metatarsi and tarsi are yellow. The male prosoma is dark black brown, with a gray-tan median stripe. The very edges of the prosoma are tan, and there is a tan submarginal band. The opisthosoma is dark black brown, with a broad gray-tan median stripe that becomes

male

less defined posteriorly. The first pair of legs are dark, except for the distal metatarsi and tarsi, which are yellow. The tibiae of leg I are covered in dense hairs, creating a brush. The other legs are brown, with faint banding on the proximal segments and paler metatarsi and tarsi. The first pair of legs is used in courtship displays. This species looks very similar to *Schizocosa ocreata*. Note that in *S. crassipes*, the tibial brushes on the male leg I are limited to just the tibia, while in *S. ocreata*, they are somewhat longer and dark hairs are often seen on the metatarsi.

Habitat These spiders are often found in leaf litter away from water.

Known records from AL, AR, CT, DE, FL, GA, IA, IL, IN, KS, LA, MD, MI, MO, MS, NC, NE, OH, PA, SC, TN, TX, WI

Schizocosa mccooki (Montgomery, 1904)

Common Name This species does not have a common name.

Total Body Length Female 9.6–22.7 mm, Male 9.1–15.5 mm

Description The prosoma is dark to reddish brown, with a wide pale median band and pale submarginal bands. Dark radial lines are often visible. The opisthosoma is gray brown, and a dark heartmark is usually present. The heartmark often has a pale border. A tan median stripe with paired white spots is on either side. The lateral edges are tan. The legs are tan to brown and unbanded.

Habitat These spiders can be found in a variety of habitats, including lake edges, grasslands, woodlands, and sagebrush.

Known records from AB, BC, MB, ON, SK (Canada); AZ, CA, CO, IA, ID, IL, KS, MI, MN, MT, ND, NE, NM, NV, OH, OK, OR, SD, TX, UT, WA, WI, WY (United States)

male

male

male

male

juvenile   juvenile

Schizocosa ocreata (Hentz, 1844)

Common Name Brushlegged wolf spider

Other Colloquial Names Brush-legged split wolf spider

Total Body Length Female 7.3–10.4 mm, Male 5.7–10.0 mm

Description The female prosoma is dark brown, with a broad cream median band. The very edge of the prosoma is tan. The female opisthosoma is variegated with browns, blacks, and tans.

female

female

female with egg sac

female with young

juvenile

male

male

male

There is the hint of a slightly paler median stripe, but it is not well defined. The female legs are brown, striped with tan. The metatarsi and tarsi are yellow. The male prosoma is dark black brown, with a gray-tan median stripe. The very edges of the prosoma are tan, and there are faint hints of a tan submarginal band. The opisthosoma is dark black brown, with a broad gray-tan median stripe that becomes less defined posteriorly. The first pair of legs are dark, except for the distal metatarsi and tarsi, which are almost white. The tibiae of leg I are covered in dense hairs, creating a brush, and long black hairs are also seen on the metatarsi to some extent. The other legs are brown, with dusky banding on the proximal segments and paler metatarsi and tarsi. The first pair of legs is used in courtship displays. This species looks very similar to *Schizocosa crassipes*, but note that the tibial brushes on the male leg I are somewhat longer, and that there are dark hairs on the metatarsi. In *S. crassipes*, the brushes are somewhat shorter and do not extend to the metatarsi. The genitalia of *S. ocreata* are indistinguishable from those of *Schizocosa rovneri* and *Schizocosa stridulans*. Females can be conclusively identified only if collected with a male. The males are distinguished by the pigmentation and presence of tibial brushes on leg I.

Habitat These spiders are commonly found among the leaf litter of forested habitats or in open fields near wooded areas.

Known records from ON (Canada); AL, AR, CO, CT, DE, FL, GA, IA, IL, IN, KS, KY, LA, MA, MD, MI, MN, MO, MS, NC, NE, NJ, NY, OH, OK, PA, RI, SC, SD, TN, TX, VA, WI, WV (United States)

Schizocosa perplexa Bryant, 1936

Common Name This species does not have a common name.

Total Body Length Female 12.0–13.0 mm, Male 9.0–12.0 mm

Description The prosoma is dark black brown, with a tan-brown median stripe that is more contrasting in the males. The very edge of the prosoma is tan, and there are tan hairs around the edge. The opisthosoma is dark black brown. There are black markings in the anterior-lateral area.

female

female

female
ventral

female with
egg sac

male

female
with young

male

male

male
ventral

subadult male

The males have a tan-brown median stripe, with paired dark dots in the posterior half of the median stripe at the edges. The ventral opisthosoma is black, with a golden midline from the epigastric furrow to the spinnerets. The first pair of legs in the males have dark bristles on the tibiae, but the bristles are far sparser than those seen in other *Schizocosa* species. The other legs are tan brown, mottled, and unbanded. This species looks very similar to *Gladicosa bellamyi*. Direct comparisons to *G. bellamyi* have not been completed at this time. In the literature, Brady (1987) noted that some *G. bellamyi* specimens examined had a yellow median line on the ventral opisthosoma (it is speculated that this may be the distinguishing characteristic between the two species), and as there seemed to be subtle differences in specimens from different locales, there are two sets of figures for the genitalia. It is possible that some of those specimens were in fact *S. perplexa*, and this issue should be resolved in the future. There is also some debate as to whether *S. perplexa* belongs in the genus *Schizocosa*, and the species may be moved when further research is completed. Research is ongoing.

Habitat Not much is recorded in the literature about the habitat preferences of this species. These spiders have been collected from field edges near forested areas.

Known records from OH, OK, TX

Schizocosa rovneri Uetz & Dondale, 1979

Common Name This species does not have a common name.

Total Body Length Female 6.0–8.0 mm, Male 6.5–8.1 mm

male

Description The female prosoma is dark brown, with a broad cream median band. The very edge of the prosoma is tan. The female opisthosoma is variegated with browns, blacks, and tans. There is the hint of a slightly paler median stripe, but it is not well defined. The female legs are brown, striped with tan. The metatarsi and tarsi are yellow. The male prosoma is dark black brown, with a gray-tan median stripe. The very edges of the prosoma are tan, and there are faint hints of a tan submarginal band. The opisthosoma is dark black brown, with a broad gray-tan median stripe that becomes less defined posteriorly. The tibiae of leg I are not dark and do not have thick brushes. The lack of dark pigmentation distinguishes *S. rovneri* from *Schizocosa stridulans*, and the lack of brushes distinguishes it from *Schizocosa ocreata*. The metatarsi and tarsi of leg I are almost white. The other legs are brown, with dusky banding on the proximal segments and paler metatarsi and tarsi. The genitalia of this species are indistinguishable from those of *S. ocreata* and

S. stridulans, but the males are easily distinguished by their lack of brushes and lack of pigmentation on tibiae I. The females can be conclusively identified only if collected with a male.

Habitat These spiders are often found in bottomlands and floodplains.

Known records from AL, IL, KY, LA, MO, MS, OH, TX

Schizocosa saltatrix (Hentz, 1844)

Common Name This species does not have a common name.

Total Body Length Female 7.0–11.8 mm, Male 6.5–8.2 mm

Description The prosoma is dark brown, with a pale tan median stripe that widens slightly toward the eye region. There is a tan submarginal stripe. The opisthosoma is mottled with brown, with a darker heartmark and some indistinct chevrons. The coloration of the males is slightly more contrasting than that of the females. The legs are tan to brown and banded.

male

male

Habitat These spiders are often found in hardwood forests.

Known records from NS, ON, QC (Canada); AL, AR, CO, CT, DE, FL, GA, IA, IL, IN, KS, KY, LA, MA, MD, ME, MI, MN, MO, MS, NC, NE, NH, NJ, NM, NY, OH, OK, PA, RI, SC, SD, TN, TX, UT, VA, VT, WI, WV (United States)

Schizocosa stridulans Stratton, 1984

Common Name This species does not have a common name.

Total Body Length Female 6.6–11.4 mm, Male 5.0–6.8 mm

Description The female prosoma is dark brown, with a broad cream median band. The very edge of the prosoma is tan. The female opisthosoma is variegated with browns, blacks, and tans. There is the hint of a slightly paler median stripe, but it is not well defined. The female legs are brown, striped with tan. The metatarsi and tarsi are yellow. The male prosoma is dark black brown, with a gray-tan median stripe. The very edges of the prosoma are tan. The opisthosoma is dark gray brown, with a broad gray-tan median stripe that becomes less defined posteriorly. The tibiae, patella, and distal ends of the femur of leg I are dark, but they lack the thick brushes seen in *Schizocosa ocreata*. The metatarsi and tarsi of leg I are almost white. The other legs are yellow brown, with dusky banding on the proximal segments and paler metatarsi and tarsi. The genitalia of this species are indistinguishable from those of *S. ocreata* and *S. rovneri*, but the males are easily

male

male

distinguished by the pigmented leg I and lack
of brushes. The females can be confidently identified
only if collected with a male.

Habitat These spiders tend to be found among the leaf litter in upland forests, typically in oak
or oak hickory forests.

Known records from AL, AR, IL, KY, MO, MS, OH, TN, TX

Sosippus californicus Simon, 1898

Common Name This species does not have a common name.

Other Colloquial Names The genus *Sosippus* is often referred to as "funnel
web wolf spiders."

Total Body Length Female 13.7–18.7 mm, Male 12.8–15.0 mm

Description The prosoma is pale gray to light brown. The edges are off-white, with a very thin
median stripe and radiating lines in white with black shading. The opisthosoma is pale gray to
light brown, with a slightly darker gray to brown median stripe. Paired white dots are on either
side of the median stripe, some with connecting white bands. The legs are gray to brown and
unbanded. Unlike most other wolf spiders, the spiders of the genus *Sosippus* build a large sheet
web with a funnel retreat similar to Agelenidae.

Habitat These spiders are often found in and around vegetation associated with wet habitats
(floodplains and drainage areas).

Known records from AZ, CA

female

female

female

female

Tigrosa annexa (Chamberlin & Ivie, 1944)

Common Name This species does not have a common name.

Total Body Length Female 10.8–15.6 mm, Male 10.5–17.4 mm

Description The prosoma is brown, with a tan median strip flanked by two small dashes just behind the posterior lateral eyes. There are light and dark shaded radiating lines. The edge is tan. The opisthosoma is brown, with a tan midline stripe in which the dark heartmark lies. The lateral edges are mottled with more tan than brown. Paired light dots may be on either side of the midline toward the posterior. The legs are tan to brown, and faint banding may be seen on the proximal segments. The ventral opisthosoma is tan, occasionally with a few scattered dark spots and never as suffused with dark spots as the closely related *Tigrosa helluo*.

female

female

female

female
ventral

Habitat These spiders are often found in grassy habitats and moist environments.

Known records from AR, FL, GA, IL, LA, MS, NC, OH, TX

Tigrosa aspersa (Hentz, 1844)

Common Name This species does not have a common name.

Other Colloquial Names Woodland giant wolf spider

Total Body Length Female 16.6–25.0 mm, Male 10.6–18.0 mm

Description The prosoma is light brown to almost black. A pale yellow to orange median line is limited to just the eye region. There may be faint radial lines, which often are not visible on darker specimens. The opisthosoma is light brown to almost black. Sometimes there is some mottling, but there are few to no distinct markings. The male is often lighter than the female. The ventral opisthosoma is pale and has some irregular black spotting. The legs are brown to almost black. The femurs have yellow to orange banding.

Habitat These spiders tend to live in burrows, hunt at night, and live in forested habitats.

Known records from ON (Canada); AL, AR, CO, CT, DC, DE, GA, IL, IN, KS, MA, MD, MI, MO, MS, NC, NJ, NY, OH, PA, RI, TN, TX, VT, WI, WV, WY (United States)

female in burrow

female with egg sac in burrow

female in burrow

female in burrow

female ventral in burrow (defensive pose)

female with young

Tigrosa georgicola (Walckenaer, 1837)

Common Name This species does not have a common name.

Total Body Length Female 18.0–24.0 mm, Male 18.0–22.0 mm

Description The prosoma is light to dark brown. There are dark radiating lines. A yellow median stripe runs from the anterior median eyes to the end of the prosoma. There are pale submarginal bands. The opisthosoma is light to dark brown, with a dark brown heartmark. Some faint chevron patterning may be seen. The ventral opisthosoma has three dark stripes made up of spots from the epigastric furrow to the spinnerets. The legs are yellow to light brown, with darker bands.

female

Habitat These spiders are often found in deciduous forests or at the edge of forested habitats. They do not travel far from their burrow when hunting.

Known records from AL, AR, FL, GA, IA, IL, KS, LA, MA, MD, MS, NC, NY, OH, SC, TX, VA, WV

female ventral

female with young

Tigrosa helluo (Walckenaer, 1837)

Common Name This species does not have a common name.

Other Colloquial Names Wetland giant wolf spider

Total Body Length Female 14.9–21.5 mm, Male 9.6–12.9 mm

Description The prosoma is dark brown. There is a pale median stripe. The edges are tan; this is more pronounced in the males. There are some dark radiating lines. The opisthosoma is dark brown, with a darker heartmark. The males have a pale median stripe that is broken into chevron shapes toward the posterior. The ventral opisthosoma is suffused with black in the females, and tan with scattered black spots in the males. In the females, the sternum and coxae are suffused with black; they are tan in the males.

Habitat These spiders are often found in lawns, agricultural fields, forested areas, and marshes. They can also sometimes be found in and around human structures.

Known records from ON, QC (Canada); AL, AR, CO, CT, DE, FL, GA, IA, IL, IN, KS, LA, MA, MD, ME, MI, MO, MS, NC, NE, NH, NJ, NM, NY, OH, OK, PA, RI, SC, TN, TX, UT, VA, VT, WI (United States)

female

female ventral

female

female

young riding on female's opisthosoma

juvenile

juvenile

male

male ventral

male

male

Trabeops aurantiacus (Emerton, 1885)

Common Name This species does not have a common name.

Total Body Length Female 3.0–3.5 mm, Male 2.5–3.0 mm

Description The prosoma is brown to reddish brown, with a faint pale median stripe, and there are pale submarginal bands, which may be broken. The opisthosoma is yellow to reddish orange and in some specimens is completely unmarked. There may be a yellow heartmark, and a series of yellow markings that look like broken transverse bands may also cross the opisthosoma. The legs are dark yellow, but the color fades toward the distal segments. In the females, legs I and II are darker, with pale, almost white, metatarsi and tarsi. The males have very dark femurs on leg I, while the other segments are pale, almost white. This is one of the smaller wolf spiders, and it is often overlooked due to its small size.

Habitat These spiders can often be found in forested habitats or moist environments, such as marshes and near water bodies.

Known records from NS, ON (Canada); AR, AZ, CT, DE, FL, IL, GA, MA, MD, ME, MI, MO, MS, NC, NE, NJ, NY, OH, SC, VA, WI (United States)

female

female with egg sac on penny for scale

female on quarter for scale

female with young

male

male

Trochosa ruricola (De Geer, 1778)

Common Name This species does not have a common name.

Other Colloquial Names Rustic wolf spider

Total Body Length Female 9.0–14.0 mm, Male 7.0–9.0 mm

Description The prosoma is dark brown, with a pale median stripe that
widens just before the eye region and contains a dark dash on each side of the midline.
The submarginal band is tan and contiguous. There are pale radial lines. The opisthosoma is
tan to brown. The heartmark is tan to off-white and bordered in dark brown. A series of paired
black dots may be on the posterior end. The legs are tan to dusky light brown. Some dusky bands
may be visible on the proximal segments. The males have dark tibiae, metatarsi, and tarsi on the
first pair of legs. The mature males can easily be distinguished
from *Trochosa terricola* by the dark legs I. In both the males
and females of *T. ruricola*, the submarginal band on the

female

female

female
with young

male

male ventral

male

prosoma is contiguous, whereas in *T. terricola*, the submarginal band is broken, although this can be difficult to see on live specimens.

Habitat These spiders are often found in fields, grassy habitats, and lawns.

Known records from BC, NB, NS, ON, PE, QC (Canada); CO, CT, IL, IN, OH, FL, MA, ME, PA, RI, WI (United States)
This is an introduced species.

Trochosa sepulchralis (Montgomery, 1902)

Common Name This species does not have a common name.

Other Colloquial Names Rustic wolf spider

Total Body Length Female 10.8–11.2 mm, Male 7.9–8.5 mm

Description The prosoma is golden brown, with a pale median stripe.
Unlike most *Trochosa* species, this one does not have the two parallel dashes within the median band. There are faint radial lines and a pale tan submarginal band. The opisthosoma is mottled with brown. There may be a visible heartmark. A series of chevrons is sometimes on the posterior end. The ventral opisthosoma is uniformly black, as are the sternum and coxae. The legs are golden brown, with faint banding visible.

Habitat These spiders seem to prefer edge habitats and are often found on the boundary of wooded areas.

Known records from AR, FL, GA, IN, KS, LA, MD, MS, OH, OK, PA, TX

female

female

male

female
ventral

male

male
ventral

Trochosa terricola Thorell, 1856

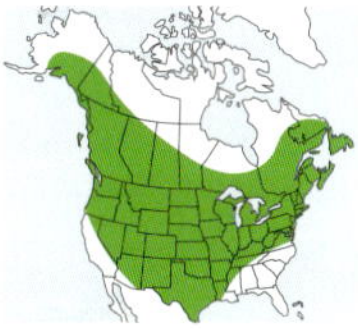

Common Name This species does not have a common name.

Other Colloquial Names Ground wolf spider

Total Body Length Female 8.4–14.0 mm, Male 7.5–12.0 mm

Description The prosoma is dark brown, with a pale median stripe that widens just before the eye region and contains a dark dash on each side of the midline. The submarginal band is tan and dashed or broken. There are pale radial lines. The opisthosoma is tan to brown. The heartmark is tan to off-white and bordered in dark brown. A series of paired black dots may be on the posterior end. The legs are tan to dusky light brown. Some dusky bands may be visible

female

female

female with young emerging from egg sac

female

on the proximal segments. The mature males can easily be distinguished from *Trochosa ruricola* by the lack of dark legs I. In both the males and the females of *T. terricola*, the submarginal band on the prosoma is broken, whereas in *T. ruricola*, the submarginal band is contiguous or unbroken, although this can be difficult to see at times on a live specimen.

Habitat These spiders are often found in fields, grassy habitats, forest edges, and lawns.

Known records from AB, BC, LB, MB, NB, NL, NS, NT, ON, PE, QC, SK (Canada); AK, AR, AZ, CA, CO, CT, FL, IA, ID, IL, IN, KS, MA, ME, MI, MN, MS, MT, ND, NE, NH, NJ, NM, NY, OH, OK, OR, PA, RI, SC, SD, TX, UT, VT, WA, WI, WY (United States)

Varacosa avara
(Keyserling, 1877)

Common Name
This species does not
have a common name.

Total Body Length
Female 7.9–13.0 mm,
Male 6.0–9.0 mm

Description The prosoma is yellow brown, with a light median band that narrows in the eye region. A pair of dark dashes is sometimes on either side of the midline within the median band. There may be a faint submarginal band. The opisthosoma is gray brown, with an indistinct heartmark and a mottled overall appearance. The legs are reddish brown, with faint dusky bands.

Habitat These spiders are often found under rocks in prairies, grassy areas, and forest edges.

Known records from ON, QC (Canada); AL, AR, AZ, CO, CT, DE, FL, GA, IL, IN, KS, LA, MA, MD, ME, MI, MO, MS, NC, NE, NH, NJ, NY, OH, OK, PA, RI, SC, TN, TX, VA, WI, WV (United States)

MYRMECICULTORIDAE > ANTCOLONY SPIDERS

Genera in This Family *Myrmecicultor* (1)

Total Body Length 2.7–2.9 mm

Description These small spiders are
entelegyne and ecribellate. Their posterior
median eyes are nearly rectangular and
oriented at a 90° angle. Both eye rows are
procurved, the posterior eye row strongly so
(giving the appearance of three eye rows).
They have serrula on endites. They have a pair
of book lungs and a wide tracheal spiracle that
is located somewhat anterior of the spinnerets.
They lack precoxal triangles. They are overall
pale. They have been found living in
association with harvester ant colonies
(*Pogonomyrmex rugosus*, *Novomessor
albisetosis*, and *Novomessor cockerelli*). This
family has been described only recently.

Similar Families Gnaphosidae, Zodariidae

How to distinguish Myrmecicultoridae from
other similar families:

Myrmecicultorids can be distinguished from
zodariids, as they have serrula on their endites,
which zodariids lack. They can be distinguished
from most gnaphosids by the lack of widely
spaced, cylindrical anterior lateral spinnerets.

eye configuration
(*Myrmecicultor chihuahuensis*)

posterior median eyes rectangular and set
at almost 90° (*Myrmecicultor chihuahuensis*)

Myrmecicultorids do not have a scutum, which is sometimes seen in gnaphosids. Additionally,
gnaphosids have concave endites, which is not a trait seen in myrmecicultorids.

Myrmecicultor chihuahuensis
Ramírez, Grismado & Ubick, 2019

Common Name This species does not have a common name.

Total Body Length Female 2.7 mm, Male 2.9 mm

Description The prosoma is yellow to orange,
and the eyes are surrounded in black. The eyes
are in two procurved rows. The posterior eye
row is strongly procurved (almost giving the
appearance of three eye rows). The posterior
median eyes are rectangular and obliquely
oriented. The opisthosoma is dark yellow to
orange. The legs are golden to orange. This
species and family have been described only
recently.

Habitat This species is associated with
harvester ants. The spiders live in the ant
colony but do not seem to feed on the ants.

males and females
near ant colony

males and females near ant colony

males and females near ant colony

They also do not mimic the ants. The ants seem to tolerate them. The nature of the relationship between the ants and the spiders is not known at this time, but research is ongoing. These spiders have been found in areas with loose sand mixed with stones, as well as in areas with desert pavement.

Known records from TX

OONOPIDAE ▷ GOBLIN SPIDERS

Other Colloquial Names Dwarf armored spiders, dwarf hunting spiders, dwarf sixeyed spiders, taco spiders

Genera in This Family *Brignolia* (3), *Cinetomorpha* (4), *Escaphiella* (3), *Heteroonops* (1), *Ischnothyreus* (2), *Noonops* (9), *Oonopoides* (3), *Oonops* (1), *Opopaea* (5), *Orchestina* (8), *Pelicinus* (1), *Scaphiella* (1), *Scaphioides* (1), *Stenoonops* (1), *Tapinesthis* (1), *Triaeris* (1), and *Xestaspis* (1)

Total Body Length 1.0–3.0 mm

Description These small spiders are haplogyne and ecribellate. They have six eyes in a cluster. They have a pair of book lungs and paired tracheal spiracles near the epigastric furrow. They tend to be overall yellowish to orange.

Similar Families Caponiidae, Linyphiidae, Ochyroceratidae, Telemidae, Trogloraptoridae

How to distinguish Oonopidae from other similar families:

Oonopids can be distinguished from caponiids by the presence of book lungs (which

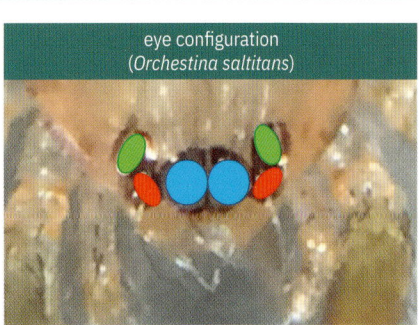

eye configuration (*Orchestina saltitans*)

overall orange in color (*Escaphiella hespera*)

caponiids lack). Oonopids can be distinguished from caponiids, linyphiids, ochyroceratids, telemids, and trogloraptorids by their eye configuration. Oonopids have six eyes in a cluster. In most caponiids, only the anterior median eyes are present, or, if the spiders have eight eyes, they are clustered together. None of the caponiids has six eyes. Linyphiids usually have eight eyes in two rows, although some males have highly modified carapace and eye arrangements, but these would look unlike any oonopids. Ochyroceratids and telemids have six eyes in three diads. Trogloraptorids have six eyes. (The anterior median eyes are absent.) The lateral eyes are located adjacent to one another, and the posterior median eyes are between them.

Cinetomorpha bandolera Ott & Harvey, 2019

Common Name This species does not have a common name.

Total Body Length Female 1.9 mm, Male 1.8–1.9 mm

Description The prosoma is orange brown and somewhat darker at the edges. A reddish-brown scutum with light-colored hairs covers the entire dorsal surface of the opisthosoma. The legs are reddish orange and unbanded.

Habitat These spiders are often found under rocks and can sometimes be found in and around human structures.

Known records from AZ, NM, TX

Escaphiella hespera (Chamberlin, 1924)

Common Name This species does not have a common name.

Other Colloquial Names Taco spiders

Total Body Length Female 1.5–1.6 mm, Male 1.5–1.6 mm

Description The prosoma is reddish orange and slightly darker around the border. The spiders have short wide chelicerae that lack hairs. The opisthosoma is pale orange to white, with two orange lateral scuta. These spiders are sometimes referred to as "taco spiders," as the lateral scuta look somewhat like a taco shell. Dark hairs are visible on the opisthosoma. The legs are orange to yellow orange and unbanded.

Habitat These spiders are often found in leaf litter and under rocks and can sometimes be found in and around human structures.

Known records from AZ, CA, NM, NV, TX, UT

female

female

Orchestina saltitans Banks, 1894

Common Name This species does not have a common name.

Total Body Length Female 1.4 mm, Male 0.9 mm

Description The prosoma is pale yellow to pale orange, with a netlike pattern in pale purple or brown. The opisthosoma is pale orange to gray. The legs are pale yellow and unbanded. The femurs of leg IV are robust, enabling the spider to move in short hops.

Habitat These spiders are usually found in leaf litter but are often overlooked due to their small size.

Known records from AL, CT, DC, GA, MA, MI, MO, NC, NJ, NM, NY, OH, PA

female

female

PHRUROLITHIDAE > DWARF ANTRUNNER SPIDERS

Other Colloquial Names
Phrurolithid spiders, guardstone spiders

Genera in This Family *Drassinella* (6), *Phrurolithus* (26), *Phruronellus* (5), *Phrurotimpus* (22), *Piabuna* (5), and *Scotinella* (16)

Total Body Length 1.0–4.0 mm

Description These small spiders are entelegyne and ecribellate. They have eight eyes, which are usually in two straight rows. They have precoxal triangles. The anterior tibiae have four pairs of ventral spines. They have a pair of book lungs and a tracheal spiracle near the spinnerets. They are commonly found under rocks or in leaf litter and are sometimes associated with ants.

Similar Families Clubionidae, Corinnidae, Gnaphosidae, Liocranidae, Trachelidae

How to distinguish Phrurolithidae from other similar families:

Phrurolithids can be distinguished from clubionids by the ventral spines on the anterior tibiae: phrurolithids have four (or more) pairs of ventral spines, whereas clubionids have fewer than four pairs of ventral spines. Phrurolithids are usually smaller (1.0–4.0 mm total body length) than most corinnids (4.0–15.0 mm total body length). Phrurolithids lack the widely spaced, cylindrical anterior median spinnerets that are found in gnaphosids. Phrurolithids can be distinguished from liocranids by the presence of precoxal triangles

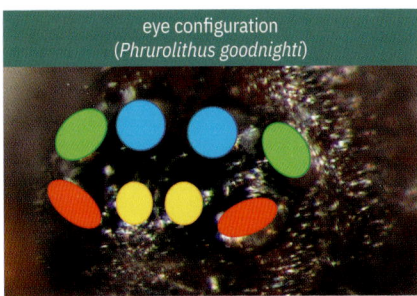
eye configuration
(*Phrurolithus goodnighti*)

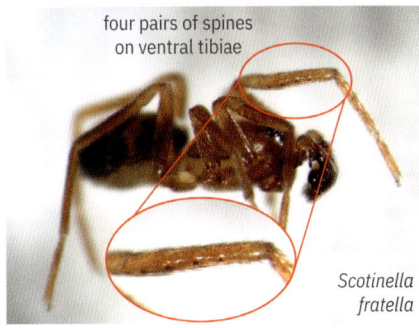
four pairs of spines
on ventral tibiae

*Scotinella
fratella*

precoxal triangles present

*Scotinella
fratella*

(which liocranids lack) and the lack of two rows of dense bristles on the anterior tibiae and metatarsi; the bristles are characteristic of some liocranids. The paired spines on the ventral anterior tibiae distinguish phrurolithids from trachelids, as trachelids lack spines on the ventral surfaces of the anterior tibiae; rather, they have short cusps.

Phrurolithus goodnighti
Muma, 1945

Common Name This species does not have a common name.

Total Body Length
Female 2.1 mm, Male 1.6 mm

Description The prosoma is dark brown, even darker around the edges, and usually quite shiny. There are some dark radiating lines, and a trident-shaped marking is often anterior of the fovea, with the points directed toward the eyes. The opisthosoma is dark brown, with two transverse yellow bands: the first anteriorly, the second about halfway down the opisthosoma. The second band is curved toward the middle and often touches the first band. The male opisthosoma is covered with a large scutum. The legs are yellow orange, with a dark band around the distal femur. These small spiders can be difficult to identify, and often a specimen is needed for microscopic examination for confirmed species-level identification.

Habitat These spiders are often found in leaf litter or under rocks.

Known records from IL, MD, OH

female

female ventral

male

male ventral

Phruronellus formica (Banks, 1895)

Common Name This species does not have a common name.

Total Body Length Female 2.1–2.8 mm, Male 2.3–2.6 mm

Description The prosoma is dark brown, with dark radiating lines. The edges are slightly darker, as is the eye region. The opisthosoma of both the male and female is covered with a dark shiny scutum and will look black and shiny. The legs are orange brown, with light bands near the joints.

female

female

female ventral

Habitat This species is often found living close to *Crematogaster lineolata* ants.

Known records from CT, FL, GA, IL, LA, MA, NC, NY, OH, OK, TX

Phrurotimpus borealis (Emerton, 1911)

Common Name This species does not have a common name.

Other Colloquial Names Greater ant-mimic Corinne spider

Total Body Length Female 3.5 mm, Male 2.5–2.6 mm

Description The prosoma is orange brown and often shiny and iridescent, with a netlike network of dark lines. A thin cream to white line is around the edge. The opisthosoma is dark gray, with a series of chevrons. The male opisthosoma is dorsally covered with a scutum. The opisthosoma often has an iridescent sheen to it. The legs are yellow, with a white band at the distal end of each tibia. The first pair of legs has a dark marking from the mid-femur to the white band at the distal end of the tibia. The males have brushes of black setae on the tibia of leg I.

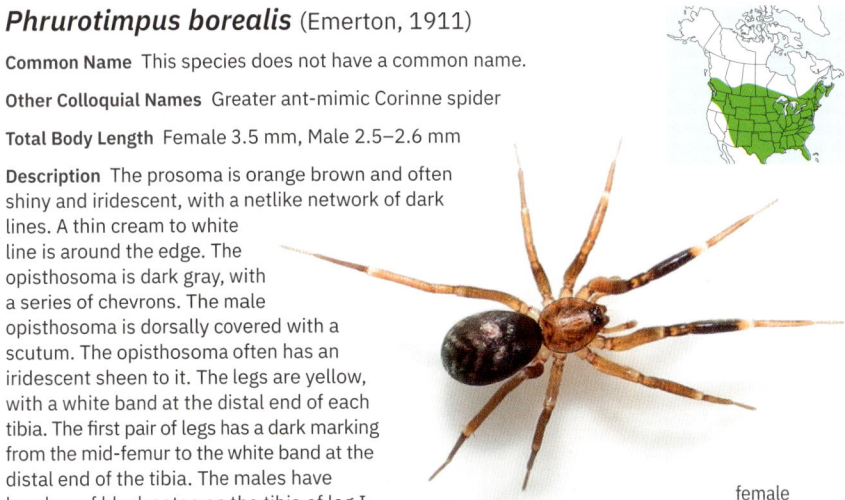

female

female

female with egg sac

male

male

male
ventral

female and male mating

Habitat These spiders are often found in leaf litter, under rocky debris, and in and around human structures.

Known records from AB, BC, MB, NB, NS, ON, PE, QC, SK (Canada); AL, AR, AZ, CO, CT, FL, IL, IN, KY, LA, MA, MI, ME, MT, NC, NH, NY, OH, PA, RI, SC, SD, TX, UT, VA, VT, WI, WV, WY (United States)

Phrurotimpus palustris (Banks, 1892)

Common Name This species does not have a common name.

Total Body Length Female 2.3–2.9 mm, Male 1.7–2.2 mm

Description The prosoma is orange brown, with a brown submarginal band. The opisthosoma is pale orange to pale brown. A series of chevrons down the midline corresponds with the lateral transverse bands. The male

female

female

female

egg sac

opisthosoma has a large scutum that covers nearly all the dorsal surface. The legs are yellow to orange, with dusky banding. The proximal half of tibia I is often darker. This species was previously misidentified as *Phrurotimpus alarius* in much of the scientific literature.

Habitat These spiders are often found in leaf litter of deciduous forests, meadows, and grasslands, under logs and other debris, and near human structures.

Known records from NS, ON, QC (Canada); CT, DC, IA, IL, IN, KS, KY, MA, MD, ME, MI, MN, MO, NC, NE, NH, NJ, NY, OH, PA, RI, TN, VA, VT, WI, WV (United States)

Scotinella fratrella (Gertsch, 1935)

Common Name
This species does not have a common name.

Total Body Length
Female 2.0 mm, Male 1.7–1.8 mm

female

Description The prosoma is yellow, with a pair of dusky markings on either side of the midline from the fovea to the posterior lateral eyes and a dark line around the edge. The opisthosoma is mottled with dark brown, and two light-colored chevron-shaped bands curve at the midline to create points toward the anterior. The legs are yellow and unbanded. This is one of the few species in which the female has asymmetrical genitalia (which can usually be seen only with a microscope).

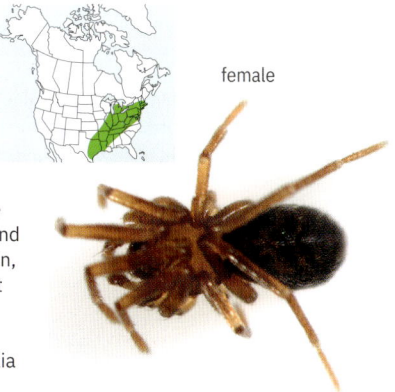

female
ventral

male

male

male

Habitat These spiders are found in leaf litter and under rocks and may be associated with ants. They are often found in young forested areas, fields, and grassy habitats.

Known records from ON (Canada); AR, IL, MA, NE, OH, TX (United States)

PLECTREURIDAE > SPURLIPPED SPIDERS

Genera in This Family *Kibramoa* (7) and *Plectreurys* (9)

Total Body Length 5.0–12.0 mm

Description These haplogyne spiders are ecribellate. They have eight eyes in two rows. The chelicerae are fused at the base. They have a pair of book lungs and a small vestigial spiracle near the spinnerets. They are found mainly in the Southwest in arid habitats.

Similar Families Caponiidae, Diguetidae, Filistatidae, Pholcidae

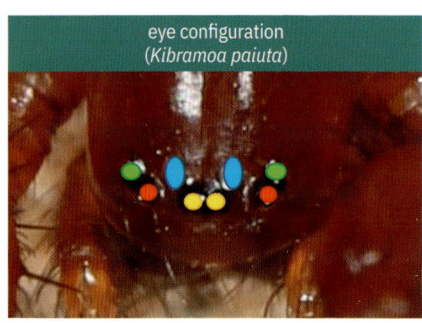

eye configuration
(*Kibramoa paiuta*)

How to distinguish Plectreuridae from other similar families:

Plectreurids can be distinguished from caponiids by the presence of book lungs, as caponiids have two pairs of tracheal spiracles and lack book lungs. Plectreurids are ground-hunting spiders and can be distinguished from diguetids, which build an extensive space web for prey capture. Plectreurids are ecribellate, which distinguishes them from filistatids, which are cribellate. They can be distinguished from pholcids by their shorter stockier legs; pholcids tend to have very long slender legs. Also, pholcids build a web for prey capture. Additionally, they can be distinguished

femur longer than prosoma
(*Kibramoa paiuta*)

robust curved femur I, male with clasping spur
(*Plectreurys tristis*)

from caponiids, diguetids, filistatids, and pholcids by their eye configuration: plectreurids have their eyes in two rows that occupy more than half the width of the prosoma, whereas caponiids have their eyes clustered (and most have only two eyes), diguetids have six eyes arranged in three diads, filistatids have their eyes in a cluster on a central mound, and the eyes of pholcids are in two triads and usually elevated.

Kibramoa paiuta Gertsch, 1958

Common Name This species does not have a common name.

Total Body Length Female unknown, Male 8.5 mm

Description The prosoma is dark brown and darker in the eye region. The opisthosoma is dark brown, with a dusky heartmark. Long black hairs are scattered on the opisthosoma. Femur I is longer than the prosoma. Eighteen to nineteen prolateral spines are on femur I, and about fifteen ventral spines are on tibia I. The femurs of all the legs are dusky brown, and the distal segments are a more golden brown. Faint

male

dusky bands occur distally on the tibiae. Only the males of this species have been described.

Habitat These spiders have been found in caves and under debris.

Known records from AZ, NV

male

juvenile

Plectreurys tristis Simon, 1893

Common Name This species does not have a common name.

Total Body Length Female 12.0–17.0 mm, Male 11.5–17.0 mm

Description The prosoma is dark reddish brown to black. The opisthosoma is dark gray to black and clothed in dark hairs. The legs are black and unbanded. The males have a distinctive clasping spur on the first tibia. The femurs of leg I are robust and visibly thicker than those of the other legs and are shorter than the prosoma.

Habitat These spiders are often found in arid habitats and have been found at elevations up to 6,500 ft.

Known records from AZ, CA, ID, NV, TX, UT

female

female

male

TRACHELIDAE ▶ CUSPULED SPIDERS

Other Colloquial Names Ground sac spiders, bullheaded sac spiders, broadheaded sac spiders

Genera in This Family *Meriola* (3) and *Trachelas* (8)

Total Body Length 3.0–10.0 mm

Description These bicolored spiders (usually with a red prosoma and yellow to gray opisthosoma) are entelegyne and ecribellate. They have eight eyes in two rows. They have a pair of book lungs and a single tracheal spiracle located near the spinnerets. They have precoxal triangles. They lack leg spines and have short cusps in their place. They can sometimes be found in human structures.

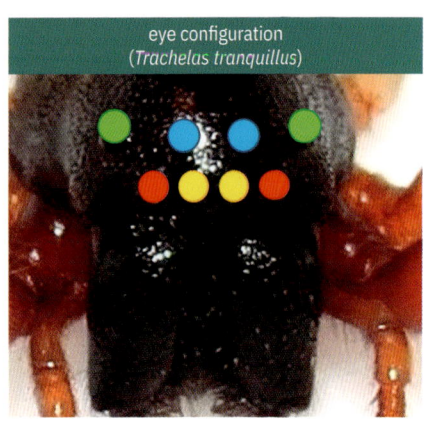
eye configuration
(*Trachelas tranquillus*)

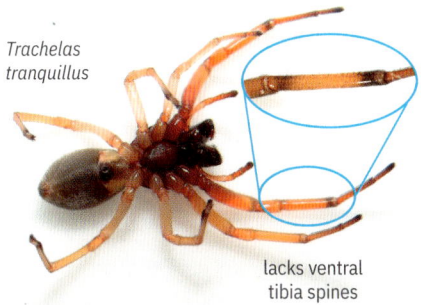

Trachelas tranquillus

lacks ventral tibia spines

precoxal triangles present
(*Trachelas tranquillus*)

Similar Families Corinnidae, Clubionidae, Dysderidae, Liocranidae, Phrurolithidae, Salticidae

How to distinguish Trachelidae from other similar families:

often found in and around human structures, this one was on a kitchen counter
(*Trachelas tranquillus*)

The lack of leg spines (with short cusps in their place) distinguishes trachelids from corinnids, clubionids, liocranids, and phrurolithids. The presence of precoxal triangles in trachelids distinguishes them from most liocranids, and those liocranids with precoxal triangles have paired bristles on the ventral surface of the anterior tibiae; the bristles are not seen in trachelids. Salticids have greatly enlarged anterior median eyes, unlike the eye configuration of trachelids. The coloration (red prosoma and pale opisthosoma) of trachelids and dysderids is similar, but there are several differences between the two. Trachelids have eight eyes in two rows; dysderids have six eyes in a cluster. Trachelids have a pair of book lungs and a single tracheal spiracle near the spinnerets, while dysderids have a pair of book lungs and a pair of tracheal spiracles located adjacent to the book lungs. Trachelids are entelegyne; dysderids are haplogyne. Also, dysderids have elongated chelicerae and long fangs; trachelids do not exhibit this trait.

Meriola decepta Banks, 1895

Common Name This species does not have a common name.

Total Body Length
Female 3.4–4.2 mm, Male 3.1–4.1 mm

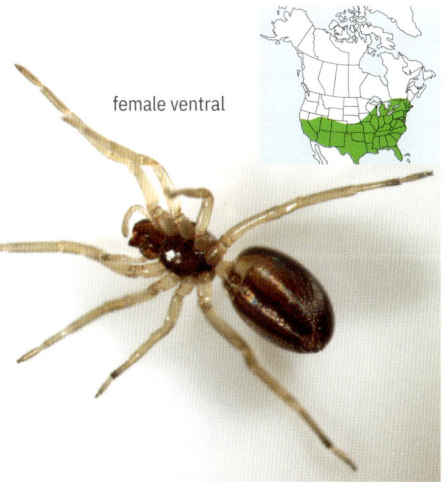

female ventral

Description The prosoma is dark orange to deep red. The posterior eyes are in a straight row. The opisthosoma is pale yellow to cream, with a dark heartmark and a dusky pattern, creating the appearance of pale chevrons posterior of the heartmark. The legs are orange to red. This species looks very similar to *Trachelas* species, but note that the posterior eye row is not recurved and the opisthosoma usually has markings in addition to a heartmark.

female

juvenile

Habitat These spiders are often found in leaf litter and agricultural fields.

Known records from BC, ON (Canada); AR, AZ, CA, CO, DC, FL, GA, IL, KS, KY, LA, MA, MD, MI, MO, MS, NC, NJ, NM, NV, NY, OH, OK, PA, SC, TN, TX, UT, VA (United States)

Trachelas pacificus Chamberlin & Ivie, 1935

Common Name This species does not have a common name.

Other Colloquial Names Bull-headed sac spider

Total Body Length Female 6.0–7.7 mm, Male 5.3–6.6 mm

Description The prosoma is dark red to maroon and shiny. It gradually elevates to the eye region. The posterior eye row is strongly recurved. (The lateral eyes are further back on the prosoma than the median eyes.) The chelicerae are robust but not elongated. The opisthosoma can range from tan to dark gray. Legs I are the darkest, red orange; legs II are slightly lighter; and legs III and IV are pale yellow. The coloration is similar to that of *Dysdera crocata*, for which *T. pacificus* seems to be confused often. *T. pacificus* can easily be distinguished from *D. crocata* by the lack of elongation of the chelicerae. *T. pacificus* also has eight eyes (*D. crocata* has only six), and it does not have the paired tracheal spiracles adjacent to the

male

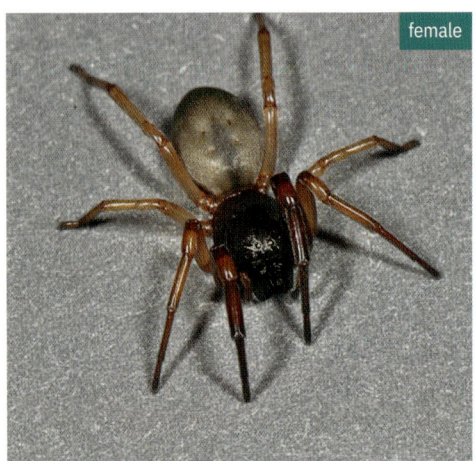

female

male

book lungs that are found in *D crocata*. *T. pacificus* looks like the other species in the genus *Trachelas*. To confirm an identification, you would need to examine the genitalia under a microscope.

Habitat These spiders are often found wandering at night hunting. During the day, they hide in a retreat. They can be found in and around human structures. They have also been found in agricultural fields and woodlands.

Known records from AZ, CA, CO, NV

Trachelas tranquillus (Hentz, 1847)

Common Name Bullheaded sac spider

Total Body Length Female 6.0–7.7 mm, Male 5.3–6.5 mm

Description The prosoma is dark red to maroon and shiny. It gradually elevates to the eye region. The posterior eye row is strongly recurved. (The lateral eyes are further back on the prosoma than the median eyes.) The chelicerae are robust but not elongated. The opisthosoma can range from tan to dark gray. Legs I are the darkest, red orange; legs II are slightly lighter; and legs III and IV are pale yellow. The coloration is similar to that of *Dysdera crocata*, for which *T. tranquillus* seems to be confused often. *T. tranquillus* can easily be distinguished from *D. crocata* by the lack of elongation of the chelicerae. *T. tranquillus* also has eight eyes (*D. crocata* has only six), and it does not have the paired tracheal spiracles adjacent to the book lungs that are found in *D. crocata*. *T. tranquillus* looks like the other species in the genus *Trachelas*. To confirm an identification, you would need to examine the genitalia under a microscope.

female

female

female
ventral

egg sac

Habitat These spiders are often found wandering at night hunting. During the day, they hide in a retreat. They can be found in and around human structures and in forested habitats. The egg sacs are a white disc attached to the surface of a rock or log and often adorned with debris. Rays of silk often extend from the sac.

Known records from NS, ON (Canada); AL, AR, AZ, CA, CT, GA, IA, IL, IN, KS, MA, MD, ME, MI, MN, MO, NC, NH, NJ, NY, OH, OK, PA, TN, TX, UT, VA, WI, WV (United States)

ZODARIIDAE ▸ IGLOO SPIDERS

Other Colloquial Names Ant spiders, ant-eating spiders, burrow-making spiders, sand-loving spiders, spotted ground spiders

Genera in This Family *Lutica* (4) and *Zodarion* (2)

Total Body Length 2.0–14.0 mm

Description These small to medium-sized spiders are entelegyne and ecribellate. They have eight eyes in two rows. They have a pair of book lungs and a single tracheal spiracle near the spinnerets. They have enlarged anterior lateral spinnerets (usually cylindrical and contiguous), and the other spinnerets are

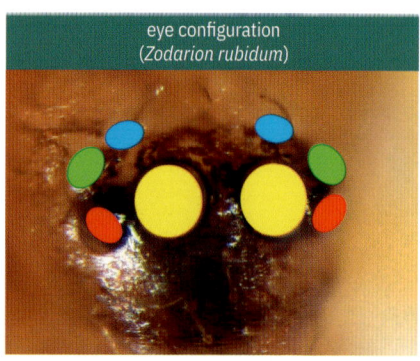

eye configuration
(*Zodarion rubidum*)

reduced. At least some members of this family are known to be ant predators. Some species in this family have been documented building igloo-shaped retreats.

Similar Families Amaurobiidae, Corinnidae, Homalonychidae, Myrmecicultoridae

How to distinguish Zodariidae from other similar families:

Zodariids have enlarged anterior lateral spinnerets, and the other spinnerets are reduced; this trait distinguishes them from other similar families. Zodariids are ecribellate, whereas amaurobiids are cribellate. Homalonychids have a prosoma that narrows anteriorly to a greater extent than is seen in the other zodariids. Myrmecicultorids have serulla on their endites, which zodariids lack.

Zodarion rubidum

enlarged anterior lateral spinnerets

Lutica nicolasia Gertsch, 1961

Common Name This species does not have a common name.

Total Body Length Female 14.0 mm, Male unknown

Description The prosoma is yellow, and dusky markings radiate from the fovea. The opisthosoma is pale yellow, with dark markings. The heartmark is slightly darker. There are two dark (almost black) markings on the anterior edge, followed by triangular dots on the lateral edge toward the posterior. A series of broken chevron shapes run down the midline of the dorsal opisthosoma. They may look like paired dots with a dash between them. The legs are pale yellow and unbanded. All species of *Lutica* look outwardly similar. To confirm a species-level identification, you would need to examine the genitalia. The male of this species has not been described.

female

Habitat These spiders are found in coastal dune areas, where they construct a silk-lined burrow. They can capture prey by biting through the walls of their burrow.

Known records from CA (San Nicolas Island)

Zodarion rubidum Simon, 1914

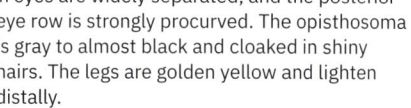

Common Name This species does not have a common name.

Other Colloquial Names European ant-eating spider

Total Body Length Female 2.8–4.4 mm, Male 2.6–3.1 mm

Description The prosoma is golden orange and gradually darkens toward the eye region, where it is reddish orange. The anterior median eyes are the largest; they are twice as large as any of the others. The posterior median eyes are widely separated, and the posterior eye row is strongly procurved. The opisthosoma is gray to almost black and cloaked in shiny hairs. The legs are golden yellow and lighten distally.

female

male
ventral

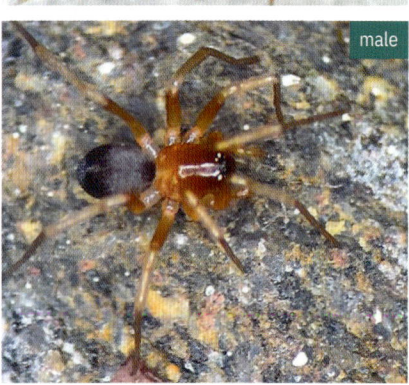

male

Habitat These spiders are often found under rocks. They are small and fast runners and are known to prey on ants.

Known records from BC, ON, QC (Canada); CO, IL, IN, MT, OH, PA (United States) This is an introduced species and likely present in more locations than have been recorded.

The spiders of the Other Active Hunter Guild do not use a web for prey capture. They hunt on various surfaces (plants, walls, etc.) and chase or pounce on their prey. All the spiders in this guild are Araneomorphae. This guild includes the following familes: Anyphaenidae, Cheiracanthiidae, Clubionidae, Ctenidae, Miturgidae, Oxyopidae, Philodromidae, Pisauridae, Salticidae, Scytodidae, Sparassidae, Trechaleidae, and Zoropsidae.

KEY TO FAMILY

1a. Six eyes in three diads, prosoma highly domed, 3.5–10.0 mm total body length --- **Scytodidae** (page 572)

1b. Eight eyes, prosoma not highly domed -- **2**

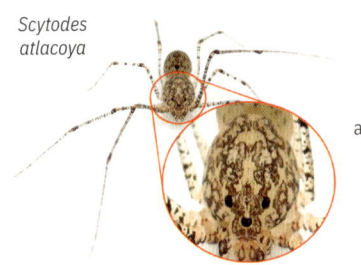

Scytodes atlacoya

a

2a. Anterior median eyes much larger than others, eyes in three rows with posterior eyes on lateral edges of carapace, 1.0–22.0 mm total body length -- **Salticidae** (page 481)

2b. Eyes otherwise --- **3**

 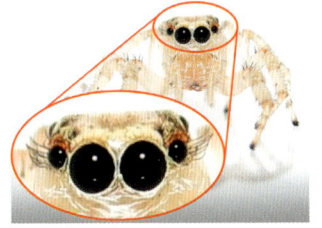

a

Colonus sylvanus   *Colonus sylvanus*

3a. Eyes arranged in a hexagon with anterior median eyes below, legs with prominent spines, 8.5–21.5 mm total body length ------------------ **Oxyopidae** (page 449)

3b. Eyes not arranged in a hexagon, legs with or without spines ------------------------------------ **4**

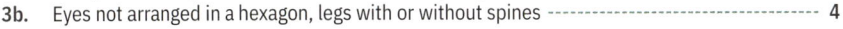

a

Oxyopes scalaris

4a. Anterior lateral eyes small and close to posterior lateral eyes, making
eyes appear to be in three rows, 6.0–30.0 mm total body length ---------- **Ctenidae** (page 444)

4b. Eyes in two rows --- **5**

*Anahita
punctulata*

a

5a. Tracheal spiracle located anteriorly from spinnerets, usually pale,
3.0–8.5 mm total body length --- **Anyphaenidae** (page 427)

5b. Tracheal spiracle located near spinnerets -- **6**

Anyphaena pectorosa

a

6a. Tarsi long and flexible; found in Gila River drainage area in AZ; egg
sac attached to spinnerets, and young ride on egg sac when they
emerge; 13.0–20.0 mm total body length ----------------------------- **Trechaleidae** (page 579)

6b. Tarsi not long and flexible -- **7**

a

Trechalea gertschi

7a. Legs laterigrade, may appear somewhat dorsoventrally flattened ------------------------------- **8**

7b. Legs prograde or, at most, slightly laterigrade, not dorsoventrally flattened ------------------ **10**

Elaver excepta

a

b

Lauiricius hooki

8a. Tends to be small (3.0–8.2 mm total body length), widespread,
leg II longest (sometimes extremely so) ------------------------------ **Philodromidae** (page 458)

8b. Tends to be larger (usually greater than 10 mm total body length) ------------------------------ **9**

Philodromus infuscatus

a

9a. Claw tufts present, southern distribution from CA to FL, 7.0–50.0 mm
total body length --- **Sparassidae** (page 575)

9b. Lacks claw tufts, found in Southwest, 10.0–18.0 mm total
body length --- **Some Zoropsidae (*Lauricius*)** (page 581)

a

b

*Olios
giganteus*

Lauiricius hooki

10a. Ecribellate, egg sac carried in chelicerae (but also attached to spinnerets), builds a nursery web for young, anterior eye row variable, posterior eye row recurved, posterior eyes larger than anterior eyes, 5.0–37.0 mm total body length ---------------------------- **Pisauridae** (page 470)

10b. Cribelate or ecribellate, egg sac not carried in chelicerae, does not build nursery web, eyes variable -- **11**

Dolomedes sp.    Dolomedes albineus   Dolomedes triton

11a. Precoxal triangles present -- **12**

11b. Precoxal triangles absent --- **13**

Elaver excepta   Zora pumila

12a. Stiff bristles on anterior of opisthosoma, tends to create retreat in curled leaf, 3.0–10.0 mm total body length ----------------------------- **Clubionidae** (page 439)

12b. Lacks stiff bristles on anterior of opisthosoma, 4.0–10.0 mm total body length, some species often found in homes ----------------- **Cheiracanthiidae** (page 435)

Clubiona sp.

Elaver excepta

Cheiracanthium inclusum

13a. Opisthosoma ovoid, usually with stripes or multiple markings, 3.0–17.0 mm total body length -- **Miturgidae** (page 447)

13b. Opisthosoma ovoid or round, usually with minimal markings and covered with short hairs, 2.0–14.0 mm total body length --------- **Some Zoropsidae** (page 581)

a

b

Zora pumila

Titiotus hansii

ANYPHAENIDAE > GHOST SPIDERS

Other Colloquial Names
Anyphaenid sac spiders

Genera in This Family *Anyphaena* (21), *Arachosia* (1), *Hibana* (7), *Lupettiana* (1), *Pippuhana* (1), and *Wulfila* (7)

Total Body Length 3.0–8.5 mm

Description These spiders are often pale, entelegyne, and ecribellate. They have eight eyes in two rows. They have a pair of book lungs and a single tracheal spiracle located anterior of the spinnerets (closer to the epigastric furrow than the spinnerets). They have precoxal triangles. They are often found on vegetation or under loose bark.

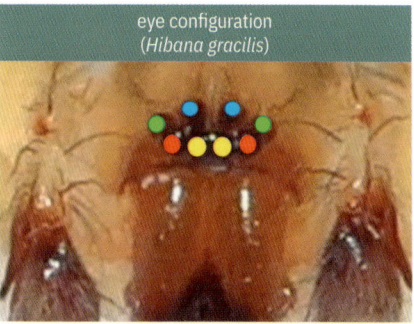

eye configuration
(*Hibana gracilis*)

usually found on vegetation
(*Anyphaeana*)

precoxal triangles present

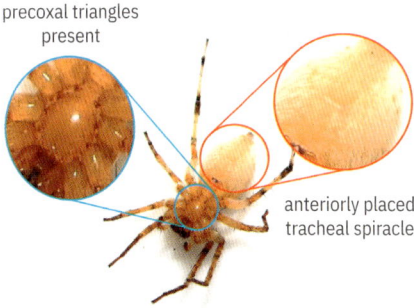

anteriorly placed tracheal spiracle

Anyphaena pectorosa

Similar Families Cheiracanthiidae, Cithaeronidae, Clubionidae, Liocranidae, Miturgidae, Sparassidae

How to distinguish Anyphaenidae from other similar families:

Anyphaenids can be distinguished from all similar families by the anteriorly placed tracheal spiracle, as in all listed similar families, the tracheal spiracle is located near the spinnerets.

Anyphaena catalina Platnick, 1974

Common Name This species does not have a common name.

Total Body Length Female 4.5–4.6 mm, Male 3.5 mm

Description The prosoma is dark orange, with light-colored hairs in the median band. There is a pair of dark paramedian bands, with radiating lines within them. The opisthosoma is cream white on the dorsal surface. The heartmark is reddish brown. There are the following reddish-brown markings: a thin U shape on the anterior, a pair of triangles on either side of the heartmark at its posterior end, and four to five markings down the median posterior of the heartmark and ending at the spinnerets. The lateral edges of the opisthosoma are golden. The tracheal spiracle is about midway between the spinnerets and the epigastric furrow. The legs are reddish brown and have indistinct banding.

female

Habitat These spiders have been found in coniferous forests and at elevations up to 10,500 ft.

Known records from AZ

Anyphaena fraterna
(Banks, 1896)

Common Name
This species does not have a common name.

Total Body Length
Female 5.0–5.7 mm, Male 4.3–5.0 mm

Description The prosoma is yellow to pale brown. A dusky marking radiates on either side of the midline. The opisthosoma is pale yellow to tan. The heartmark is golden yellow. Two short dark longitudinal lines are on the anterior on either side of the heartmark; the lines may appear broken. There are scattered dark spots, and at the posterior end, there are faint dusky chevrons. The tracheal spiracle is located about halfway between the spinnerets and the epigastric furrow. The males have two processes on coxae III (one blunt, the other short and pointed) and a pointed process

female

female

female
ventral

female

juvenile

on coxae IV. The legs are yellow to tan, with scattered spotting.

Habitat These spiders tend to be found on shrubs and other low vegetation.

Known records from AL, AR, FL, GA, IL, IN, KS, KY, MD, MO, NC, NJ, NY, OK, PA, TN, TX, WV

Anyphaena pectorosa L. Koch, 1866

Common Name This species does not have a common name.

Total Body Length Female 5.0–5.5 mm, Male 4.7–5.0 mm

Description The prosoma is pale yellow. There is a pair of dusky paramedian bands. Pale yellow hairs are in the median band. Two small, parallel, dusky dashes are just posterior of the posterior median eyes. The opisthosoma is pale yellow. The heartmark is golden yellow. Two short dark longitudinal lines are on the anterior on either side of the heartmark; the lines may appear broken. There are scattered dark spots, and at the posterior end, there are faint dusky chevrons. The tracheal spiracle is located about halfway between the spinnerets and the epigastric furrow. The males have two processes on coxae III (one blunt, the other pointed) and a longer pointed process on coxae IV. The legs are yellow, with dusky banding.

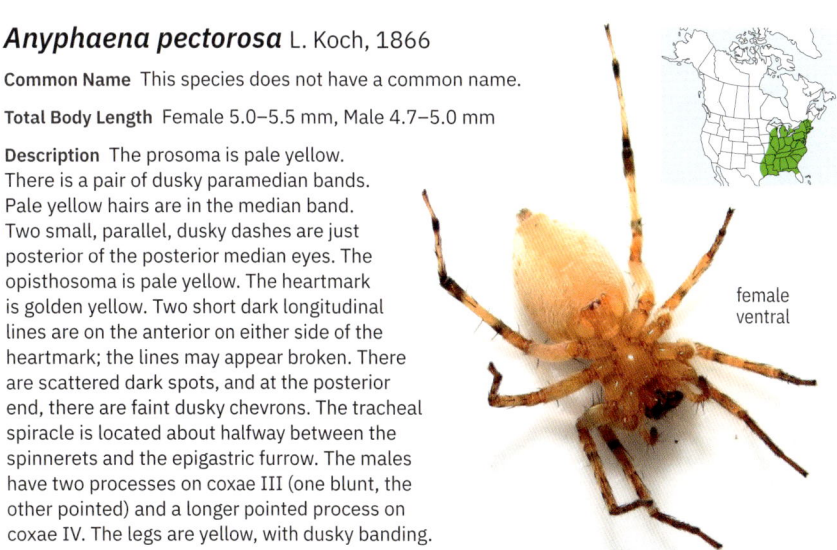

female
ventral

429

female

male ventral

male

male

Habitat These spiders are often found on trees and other vegetation. The females lay their egg sac within foliage leaves. They roll or fold the leaves over the egg sac and seal it with silk. Then they sit on the leaf, guarding the egg sac.

Known records from ON (Canada); AL, CT, FL, GA, IL, IN, KY, LA, MA, MD, MI, MO, MS, NC, NJ, NY, OH, PA, RI, SC, TN, TX, VA, WI, WV (United States)

Arachosia cubana (Banks, 1909)

Common Name This species does not have a common name.

Total Body Length Female 5.9 mm, Male 5.2 mm

Description The prosoma is tan. A light brown midline forks anterior of the fovea and runs through the posterior eye row. A series of three dark dots or dashes is on each side of the prosoma. The opisthosoma is tan, with a light brown midline stripe that often has a pale center. A light brown line on each lateral edge connects with the midline stripe just above the spinnerets. The tracheal spiracle is about halfway between the spinnerets and the epigastric furrow. The legs are cream to tan, with small spots. This species could easily be mistaken for a *Tibellus* species (Philodromidae). Note the dots around the prosoma, which are not seen in *Tibellus* species.

Habitat These spiders are often found in habitats containing tall grass.

Known records from ON (Canada); AR, AZ, CT, FL, IL, MA, MI, NC, NJ, OH (United States)

female

female

Hibana gracilis (Hentz, 1847)

Common Name Garden ghost spider

Total Body Length Female 6.4–7.0 mm, Male 5.7–6.5 mm

Description The prosoma is yellow to orange. There are thin dark orange paramedian bands. The eye region is slightly darker than the rest of the prosoma. The chelicerae are dark brown. The opisthosoma is tan to golden. Small reddish-brown longitudinal dashes sometimes give the appearance of two longitudinal bands. The tracheal spiracle is distinctly closer to the epigastric furrow than the spinnerets. On mature spiders, the legs are tan to light brown and unbanded; on immature spiders, the legs can have dusky bands.

female

female

female

female
ventral

juvenile

male

male

male ventral

male

Habitat These spiders are often found in shrubs and herbaceous vegetation and can be found in and around human structures.

Known records from ON (Canada); AL, AR, AZ, CT, DE, FL, GA, IA, IL, IN, KS, LA, MA, MD, MI, MO, MS, NC, NJ, NY, OH, OK, PA, RI, SC, TX, VA, WI (United States)

Lupettiana mordax
(O. Pickard-Cambridge, 1896)

Common Name This species does not have a common name.

male

Total Body Length
Female 3.0–5.5 mm,
Male 3.7–5.0 mm

Description The prosoma is dark brown, and some dark radial lines may be faintly visible. The chelicerae are dark brown, elongated, and porrect, which is most pronounced in the males and unlike any other anyphaenid in North America. The opisthosoma is yellow to light brown, with several dark transverse bands at regular intervals. The tracheal spiracle is about halfway between the spinnerets and the epigastric furrow. The legs are yellow and sometimes have dusky bands.

Habitat These spiders have been found in webworm nests, loblolly pine habitats, and low shrubs and trees.

Known records from AL, AR, CA, DC, FL, GA, LA, MD, MO, MS, NC, SC, TX

male

Wulfila albens (Hentz, 1847)

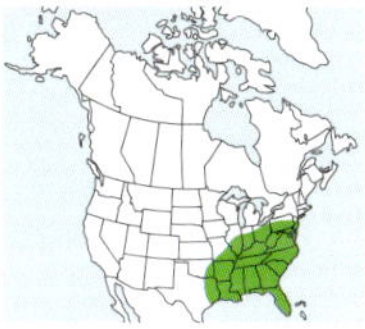

Common Name This species does not have a common name.

Total Body Length Female 4.0 mm, Male 3.6–3.7 mm

Description The prosoma is very pale, almost white, and lacks markings. The opisthosoma is very pale yellow, almost white, and also lacks markings. The color of the opisthosoma can be shifted temporarily when colorful prey is consumed. The ventral opisthosoma has a broad median stripe, the color of which can vary from dusky gray to deep green. The tracheal spiracle is about halfway between the spinnerets and the epigastric furrow. The sternum is uniformly colored and ranges from dusky gray to deep green. The legs are very pale, long, and slender. Leg I is especially long.

Habitat These spiders are often found on trees and shrubs.

Known records from AL, FL, GA, IL, KY, LA, MD, MS, NC, OH, TN, TX, VA

male

male ventral

male

Wulfila saltabundus (Hentz, 1847)

Common Name This species does not have a common name.

Total Body Length Female 3.7–4.2 mm, Male 2.9–3.5 mm

Description The prosoma is very pale, almost white. There are small black paramedian radial dashes. The opisthosoma is pale yellow. The color of the opisthosoma can be shifted temporarily when colorful prey is consumed. Paired dots run down the length of the opisthosoma, and other dots are scattered throughout. The ventral opisthosoma has a dusky median line from the pedicle to the epigastric furrow. Two dots are located on the midline between the epigastric furrow and the spinnerets. The second dot is usually elongated and may even be triangular. A pair of dusky dots are on either side of the midline about halfway down. A small dusky dot is just below the spinnerets. The tracheal

juvenile

female

female

female
ventral

male

male

spiracle is about halfway between the spinnerets and the epigastric furrow. The sternum is pale, with a wavy dusky border. The legs are very pale, long, and slender. Leg I is especially long. The male pedipalp retrolateral tibial apophysis is uniquely shaped.

Habitat These spiders are often found on shrubs and trees.

Known records from NS, ON, QC (Canada); AL, AR, CT, FL, IL, IN, LA, MA, MD, ME, MI, MN, MO, MS, NC, NE, NH, NJ, NY, OH, PA, RI, SC, TN, TX, VA, WI, WV (United States)

male ventral

male

CHEIRACANTHIIDAE ▸ YELLOW SAC SPIDERS

Other Colloquial Names Eutichurid spiders

Genera in This Family *Cheiracanthium* (2) and *Strotarchus* (3)

Total Body Length 4.0–10.0 mm

Description These spiders are often pale, entelegyne, and ecribellate. They have eight eyes in two rows; the posterior eye row is usually straight. They have a pair of book lungs and a single tracheal spiracle located near the spinnerets. The legs are usually long and slender; they have precoxal triangles. These spiders were once considered of medical concern, but studies have shown that they are not medically significant to humans.

Similar Families Anyphaenidae, Clubionidae, Lycosidae, Miturgidae, Sparassidae

How to distinguish Cheiracanthiidae from other similar families:

Cheiracanthiids can be distinguished from anyphaenids by the placement of the tracheal spiracle. In cheiracanthiids, it is located near the spinnerets, while in anyphaenids, it is placed anteriorly, usually halfway between the spinnerets and the epigastric furrow or closer

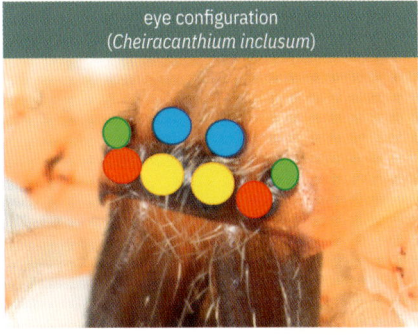

eye configuration
(*Cheiracanthium inclusum*)

precoxal triangles present

Cheiracanthium mildei

to the epigastric furrow. Cheiracanthiids can be distinguished from clubionids by the lack of the stout bristles on the anterior opisthosoma that are present in clubionids. They can be distinguished from lycosids by their eye configuration: cheiracanthiids have their eyes in two rows and all the eyes are of similar size, whereas lycosids have enlarged posterior median eyes and the posterior lateral eyes are set back, making it look like their eyes are in three rows. Cheiracanthiids can be distinguished from miturgids by the presence of precoxal triangles, which miturgids lack. They can be distinguished from sparassids by their prograde legs. Sparassids have laterigrade legs.

Cheiracanthium inclusum (Hentz, 1847)

Common Name Agrarian sac spider

Other Colloquial Names Yellow sac spider, black-toed yellow sac spider

Total Body Length Female 4.9–9.7 mm, Male 4.0–7.7 mm

Description The prosoma is pale yellow to light orange and darker in the eye region. The opisthosoma is pale yellow to light orange and has a darker heartmark. The legs are pale yellow to light orange, and the tarsi have dark tips, leading some people to call these spiders "black-toed yellow sac spiders." Absent are the stiff hairs on the anterior of the opisthosoma that one would find in the similar-looking clubionid species. This species is difficult to distinguish from the introduced *Cheiracanthium mildei* without looking at genitalia. A general rule of thumb is that if it is found away from human structures, it is likely *C. inclusum*, while if it is found near human structures, it is likely *C. mildei*.

Habitat These spiders are active during the night and can usually be found in a silken retreat during the day. They are usually found in the foliage of low-growing plants and can be common in agricultural fields. They have been documented feeding on plant nectar.

Known records from AB, BC, ON (Canada); AL, AR, AZ, CA, CO, CT, DC, FL, GA, IL, IN, LA, MA, MD, MI, MS, NC, NE, NJ, NM, NV, NY, OH, OK, OR, PA, SC, TN, TX, UT, WA (United States)

Cheiracanthium mildei L. Koch, 1864

Common Name The genus *Cheiracanthium* has the common name "longlegged sac spiders."

Other Colloquial Names Yellow sac spider, black-toed yellow sac spider

Total Body Length Female 6.0–10.7 mm, Male 5.8–8.5 mm

Description The prosoma is pale yellow to light orange and darker in the eye region. The opisthosoma is pale yellow to light orange and has a darker heartmark. The legs are pale yellow to light orange, and the tarsi have dark tips, leading some people to call these spiders "black-toed yellow sac spiders." Absent are the stiff hairs on the anterior of the opisthosoma that one would find in the similar-looking clubionid species. This species is difficult to distinguish from *Cheiracanthium inclusum* without looking at genitalia. A general rule of thumb is that if it is found away from human structures, it is likely *C. inclusum*, while if it is found near human structures, it is likely *C. mildei*.

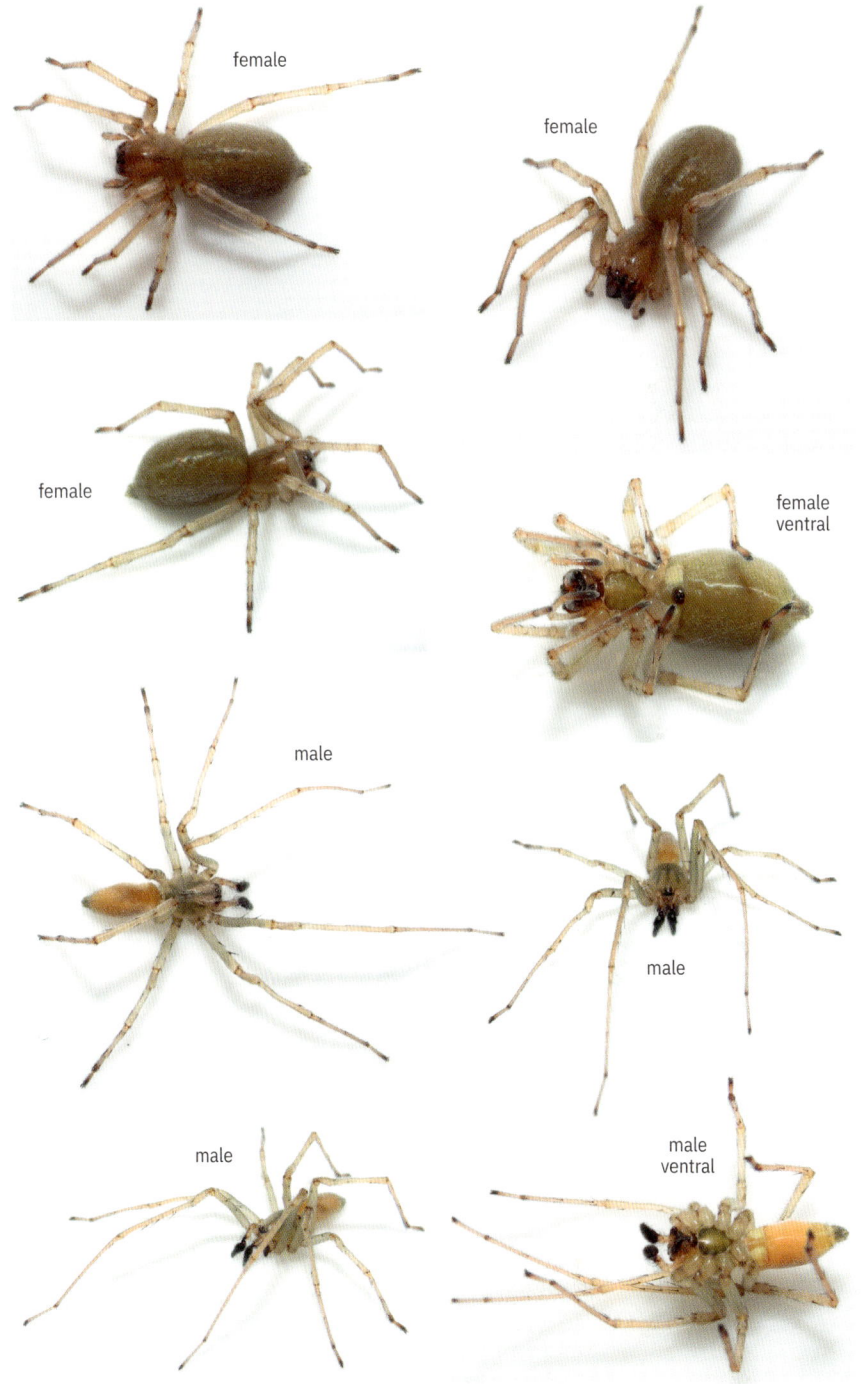

female

female

female

female
ventral

male

male

male

male
ventral

Habitat In North America, this species seems to be synanthropic and is usually found in and around human structures. These spiders are active during the night and can usually be found in a silken retreat during the day. They have been documented feeding on plant nectar. In buildings, they seem to like to build their silken retreats in the corner where walls meet the ceiling.

Known records from BC, ON, QC (Canada); AL, CA, CO, CT, IL, IN, MA, ME, NH, NJ, NY, OH, OR, PA, UT, WA (United States)
This is an introduced species.

Strotarchus beepbeep
Bonaldo, Saturnino, Ramírez & Brescovit, 2012

Common Name This species does not have a common name.

Total Body Length Female 9.8 mm, Male 6.0–8.4 mm

Description The prosoma is dark yellow to pale orange and darker in the eye region and paler at the sides. The opisthosoma is dark yellow to pale orange and covered with hairs. A heartmark is sometimes visible, although at times it is indistinct. The spinnerets are quite long, compared to other species in this family. Absent are the stiff bristles on the anterior opisthosoma that are seen in clubionids. The legs are dark yellow to pale orange and unbanded. The male pedipalps have an extended cymbium, but to a lesser degree than seen in *Strotarchus piscatorius*.

female

Habitat These spiders are often found near the ground.

Known records from AZ

Strotarchus piscatorius (Hentz, 1847)

Common Name This species does not have a common name.

Total Body Length Female 7.6–9.2 mm, Male 7.2–8.7 mm

Description The prosoma is dark yellow to pale orange and darker in the eye region. Faint dark lines radiate from the fovea. The opisthosoma is dark yellow to pale orange and covered with hairs. A heartmark is

male

male

sometimes visible, although at times it is indistinct. The spinnerets are quite long, compared to other species in this family. Absent are the stiff bristles on the anterior opisthosoma that are seen in clubionids. The legs are dark yellow to pale orange and unbanded. The male pedipalps have an extraordinary extended cymbium.

Habitat These spiders are often found near the ground. When disturbed, they pretend to be dead instead of fleeing.

Known records from NS, ON (Canada); AL, CT, FL, GA, IN, LA, MA, MD, ME, NJ, OH, TX, WV (United States)

CLUBIONIDAE ▶ SAC SPIDERS

Genera in This Family *Clubiona* (51) and *Elaver* (8)

Total Body Length 3.0–10.0 mm

Description These spiders are often pale, entelegyne, and ecribellate. They have eight eyes in two rows that occupy the width of the prosoma. They have precoxal triangles. They have a pair of book lungs and a single tracheal spiracle near the spinnerets. They often build a retreat by rolling a leaf and securing it with silk.

Similar Families Anyphaenidae, Cheiracanthiidae, Cithaeronidae, Corinnidae, Liocranidae, Miturgidae, Phrurolithidae, Sparassidae, Trachelidae

How to distinguish Clubionidae from other similar families:

Clubionids can be distinguished from anyphaenids by the placement of the tracheal spiracle: it is placed anteriorly in anyphaenids (halfway between the epigastric furrow and the spinnerets or closer to the epigastric furrow), whereas it is located near the spinnerets in clubionids. Clubionids can be distinguished from cheiracanthiids, corinnids, and trachelids by the stiff bristles that are found on the

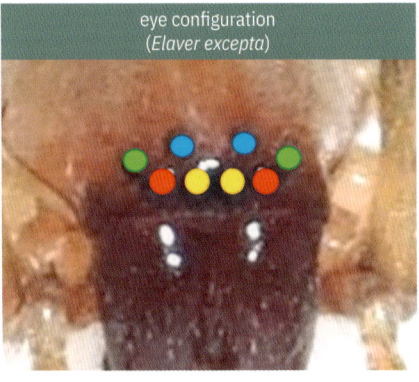

eye configuration
(*Elaver excepta*)

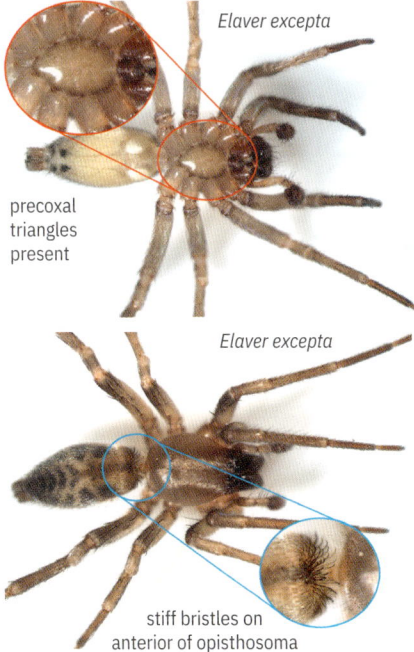

Elaver excepta

precoxal triangles present

Elaver excepta

stiff bristles on anterior of opisthosoma

creates retreat in rolled leaf (*Clubiona*)

anterior opisthosoma, as most cheiracanthiids, corinnids, and the trachelids lack this trait. Cithaeronids have pseudosegmented tarsi, a trait not seen in clubionids. Many male corinnids have a scutum, which is never seen in clubionids. Miturgids and most liocranids lack the precoxal triangles that are found in clubionids. The liocranids that have precoxal triangles have paired rows of bristles on the anterior tibiae; the bristles are not seen in clubionids. They can be distinguished from phrurolithids by the ventral spines on the anterior tibiae: phrurolithids have four (or more) pairs of ventral spines, whereas clubionids have fewer than four pairs of ventral spines. Clubionids can be distinguished from sparassids, as they have prograde legs and sparassids have laterigrade legs. They can be distinguished from trachelids by their coloration: trachelids tend to be bicolor, with a red or dark brown prosoma and a tan opisthosoma, while clubionids tend to be pale, and the prosoma and opisthosoma are similar in color.

Clubiona abboti L. Koch, 1866

Common Name The genus *Clubiona* has the common name "leafcurling sac spiders."

Total Body Length Female 4.0–5.4 mm, Male 3.7–4.4 mm

Description The prosoma is yellow to golden yellow and darker in the eye region. The opisthosoma is yellow to golden yellow and covered with silky-looking hairs. A slightly darker heartmark is often visible. Stiff dark bristles are on the anterior. Some indistinct shading is toward the posterior end. The legs are yellow to golden yellow and unbanded. There is one prolateral spine on femur I. *Clubiona* species are difficult to identify without looking at the genitalia.

Habitat These spiders are often found in forested habitats, leaf litter, and low herbaceous vegetation.

Known records from AB, MB, NB, NL, NS, ON, PE, QC, SK (Canada); AL, CO, CT, DC, FL, GA, IL, IN, KY, LA, MA, MD, ME, MI, MN, MS, MT, NC, NE, NH, NJ, NM, NY, OH, OK, OR, PA, TN, TX, UT, VA, WI (United States)

female

female

female
ventral

male

male

male
ventral

Clubiona obesa Hentz, 1847

Common Name The genus *Clubiona* has the common name "leafcurling sac spider."

Total Body Length Female 6.7–11.4 mm, Male 6.7–8.0 mm

male

male

male

male
ventral

Description The prosoma is yellow to pale orange, and the eye region is darker. The male chelicerae are somewhat elongated. The opisthosoma is dusky yellow to orange and covered with silver-gray hairs. Stout dark bristles are on the anterior. The heartmark is indistinct. The legs are pale yellow to pale orange and unbanded. There is one prolateral spine on femur I. *Clubiona* species are difficult to identify without looking at the genitalia.

Habitat These spiders are often found in deciduous woods on leaves and branches of low vegetation.

Known records from MB, NL, NS, ON, PE, QC (Canada); AL, CT, DC, IA, IL, IN, MA, ME, MD, MI, MN, MS, NC, NE, NH, NJ, NY, OH, PA, RI, VA, VT, WI, WV (United States)

Clubiona riparia L. Koch, 1866

Common Name The genus *Clubiona* has the common name "leafcurling sac spider."

Total Body Length Female 5.4–8.7 mm, Male 4.4–7.4 mm

Description The prosoma is dark yellow to golden brown and darker in the eye region. The opisthosoma is tan to grayish and, unlike in most *Clubiona* species, well marked. There is a dark heartmark. Posterior of the heartmark is a series of dots that lead to the spinnerets. The heartmark and these dots are surrounded by a paler area. The lateral edges of the opisthosoma are shaded darker, sometimes in a series of slanted lines. Stout bristles are on the anterior. The legs are yellow to golden brown, and the metatarsi and tarsi are darker than the other segments. There is one prolateral spine on femur I.

female

female ventral

egg sac in leaf retreat

Habitat These spiders are often found in vegetation near bodies of water (streams, wetlands, lakes, etc.). The female lays her egg sac on a leaf of herbaceous vegetation and then folds the leaf twice to create a three-sided chamber. The female stays within the leaf structure to guard the egg sac.

Known records from AB, BC, MB, NB, NL, NS, NT, ON, PE, QC, SK, QC, YT (Canada); AK, CO, CT, IL, MA, ME, MI, MN, MT, NH, NM, NV, NY, OH, PA, SD, UT, WA, WI, WY (United States)

Elaver excepta (L. Koch, 1866)

Common Name This species does not have a common name.

Other Colloquial Names White sac spider

Total Body Length Female 6.7–7.4 mm, Male 4.5–6.5 mm

Description The prosoma is yellow to orange and darker in the eye region. The opisthosoma is tan to grayish tan, with a series of markings that distinguish most *Elaver* species from *Clubiona* species. A dusky heartmark is followed by a series of chevron markings. Lateral to these are trapezoidal dusky-gray markings. Stout bristles are on the anterior. The legs are yellow to orange, and the metatarsi and tarsi are darker than the other segments. There are two prolateral spines on femur I.

Habitat These spiders are often found in deciduous forests and in leaf litter, under rocks, or under bark.

Known records from NS, ON, QC (Canada); AL, AR, AZ, CT, DC, DE, FL, GA, IL, IN, KS, LA, MA, MD, ME, MI, MN, MO, MS, NC, NE, NH, NJ, NY, OH, OK, PA, RI, TN, TX, VA, WI (United States)

female

female

female ventral

female with egg sac

male ventral

male

male

Elaver texana (Gertsch, 1933)

Common Name This species does not have a common name.

Total Body Length Female 9.2–11.7 mm, Male 7.6–10.5 mm

Description The prosoma is orange and slightly darker in the eye region. The opisthosoma is orange to brown, with dusky bands at the posterior end. These can be hard to see in dark specimens. Stout bristles are on the anterior. The legs are yellow to orange, and the metatarsi and tarsi I and II are darker. There are two prolateral spines on femur I.

Habitat These spiders are often found on low vegetation.

Known records from TX

male

male
ventral

male

male

CTENIDAE ▸ WANDERING SPIDERS

Genera in This Family *Acanthoctenus* (2), *Anahita* (1), *Ctenus* (5), and *Leptoctenus* (1)

Total Body Length 6.0–30.0 mm

Description These spiders are entelegyne. This family contains both cribellate and ecribellate genera. The spiders have eight eyes. The anterior lateral eyes are small and close to the posterior lateral eyes. The eye rows are strongly recurved, giving the appearance of three eye rows. They have a pair of book lungs and a single tracheal spiracle located near the spinnerets. They tend to be nocturnal hunters.

Similar Families Lycosidae, Miturgidae, Pisauridae, Zoropsidae

How to distinguish Ctenidae from other similar families:

Ctenids can be distinguished from other similar families by their eye configuration. The anterior lateral eyes are smaller than the other eyes, and they are located close to the posterior lateral eyes. The eye rows are strongly recurved, giving the appearance of three eye rows. Lycosid eyes are also placed so they appear to be in three rows, but lycosids have large posterior median eyes, and the anterior

eye configuration
(*Anahita punctulata*)

eye row is straight or nearly so. Miturgids have a straight or slightly recurved anterior eye row, and the anterior lateral eyes are not reduced in size. Pisaurid eyes can be unequal in size, but not to the extent of the reduced anterior lateral eye size seen in ctenids. Spiders in the genera *Zoropsis* and *Zorocrates* in the family Zoropsidae are superficially similar to spiders in *Acanthoctenus* (Ctenidae), but *Zoropsis* and *Zorocrates* spiders have a straight anterior eye row, and the anterior lateral eyes are not significantly reduced in size.

Anahita punctulata
(Hentz, 1844)

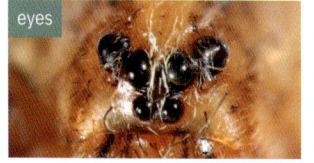

eyes

Common Name This species does not have a common name.

Other Colloquial Names Southeastern wandering spider

Total Body Length Female 7.0–9.9 mm, Male 6.0–8.0 mm

Description The prosoma is orange. A pale median stripe is bordered on each side by a shaded dark line. The paramedian bands are golden brown, with radial lines visible. There is then a pale submarginal band, and dusky shading is at the edges. The very edge of the prosoma has a thin dark line. The opisthosoma is brown. A pale median stripe is bordered on each side by a broken

female

female

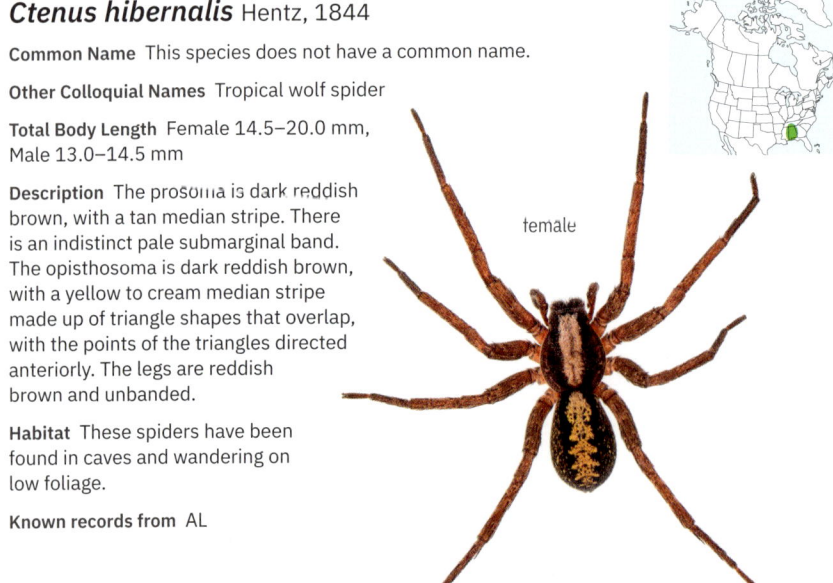

female

female
ventral

juvenile

white line. There is dark shading and dark spots throughout. The legs are golden brown, with spots and some dusky bands.

Habitat These spiders are often found in moist habitats and are known to hunt sometimes on the ground.

Known records from AL, AR, FL, GA, IN, MS, NC, OH, SC, TN, TX

Ctenus hibernalis Hentz, 1844

Common Name This species does not have a common name.

Other Colloquial Names Tropical wolf spider

Total Body Length Female 14.5–20.0 mm, Male 13.0–14.5 mm

Description The prosoma is dark reddish brown, with a tan median stripe. There is an indistinct pale submarginal band. The opisthosoma is dark reddish brown, with a yellow to cream median stripe made up of triangle shapes that overlap, with the points of the triangles directed anteriorly. The legs are reddish brown and unbanded.

Habitat These spiders have been found in caves and wandering on low foliage.

Known records from AL

female

Leptoctenus byrrhus Simon, 1888

Common Name This species does not have a common name.

Total Body Length Female 9.7–13.6 mm, Male 8.5–10.5 mm

Description The prosoma is dark brown, almost black, with a tan median stripe that looks like two arrows pointing toward the fovea. The edges are mottled with tans, browns, and black. The opisthosoma is mottled with brown, with a pinkish-tan median stripe that forms a series of triangles. White dots are on the basal corners of each triangle. The femurs and patellae are tan and black and strikingly banded. The distal segments are dark brown.

female

Habitat These spiders are often found among debris in coastal plains areas. There are also records of specimens being collected from woodrat nests.

Known records from TX

MITURGIDAE ▶ PROWLING SPIDERS

Genera in This Family *Syspira* (5), *Teminius* (2), and *Zora* (2)

Total Body Length 3.0–17.0 mm

Description These spiders are entelegyne and ecribellate. They have eight eyes in two rows. They lack precoxal triangles. They have a pair of book lungs and a tracheal spiracle near the spinnerets. They have two tarsal claws.

Similar Families Agelenidae, Anyphaenidae, Cheiracanthiidae, Clubionidae, Ctenidae, Lycosidae, Pisauridae, Sparassidae

How to distinguish Miturgidae from other similar families:

Miturgids can be distinguished from agelenids, as they lack the elongated posterior lateral spinnerets that are seen in agelenids. Anyphaenids have an anteriorly placed tracheal spiracle, whereas in miturgids the tracheal spiracle is located near the spinnerets. They can be distinguished from anyphaenids, cheiracanthiids, and clubionids by the lack of precoxal triangles, which are present in anyphaenids, cheiracanthiids, and clubionids. Miturgids have a straight or slightly recurved anterior eye row, and the anterior lateral eyes are not reduced in size, as is seen in ctenids. Lycosids have enlarged posterior median eyes and the appearance of three eye

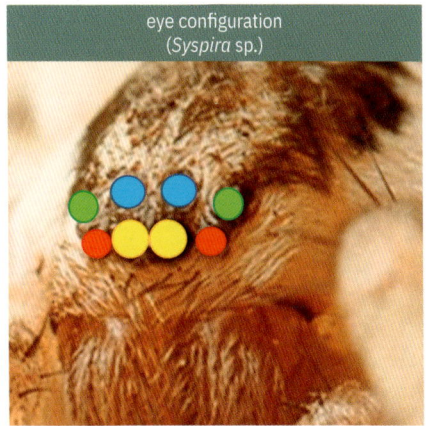
eye configuration
(*Syspira* sp.)

lacking precoxal triangles (*Zora pumila*)

rows, which is very different from the eye configuration in miturgids. Miturgids have two tarsal claws, whereas pisaurids, lycosids, and agelenids have three tarsal claws. Miturgids can be distinguished from sparassids, as they have prograde legs and sparassids have laterigrade legs.

Syspira longipes
Simon, 1895

Common Name
This species does not
have a common name.

Total Body Length
Female 8.5–16.5 mm, Male 7.4–12.5 mm

Description The prosoma is sandy tan. Dark radial lines connect to a paramedian band almost in the shape of a simple flower. The opisthosoma is sandy tan. The heartmark is pale gray. A series of triangles outlined in dark brown runs down the length, and there is additional feathery brown shading next to the median patterned area. The legs are tan, and the metatarsi and tarsi are slightly darker. There are a few faint bands on the legs.

Habitat During the day, these spiders hide under logs, rocks, and other debris. During the night, they can be found wandering hunting for prey.

Known records from AZ, CA, NM, NV, TX

Syspira tigrina
Simon, 1895

Common Name
This species does not
have a common name.

Total Body Length
Female 7.5–11.0 mm, Male 6.2–9.5 mm

Description The prosoma is light brown. There are dark, almost black, paramedian bands. There is also a transverse dark, almost black, band on the posterior of the prosoma. The opisthosoma is mottled with gray and brown, with a brown median stripe. Pale chevron-shaped markings are on the posterior half. The legs are pale brown, with some dark banding.

Habitat During the day, these spiders hide under logs, rocks, and other debris. During the night, they can be found wandering hunting for prey.

Known records from CA, NV

Zora pumila (Hentz, 1850)

Common Name This species does not have a common name.

Total Body Length Female 5.0–6.0 mm, Male 3.5–4.0 mm

Description The prosoma is tan, with a lighter median stripe. There are golden-brown paramedian bands. The edges have brown spots. The opisthosoma is tan to brown. The heartmark is light tan. Two pale stripes are on the opisthosoma, and some pale dots may be on the posterior end. The legs are pale tan and heavily spotted. On the femurs of legs I and II, the spots almost join to make dorsal stripes. The metatarsi and tarsi of legs I and II are often slightly darker.

Habitat These spiders can be found in small shrubs and low vegetation and sometimes on the ground. They can also be found in and around human structures.

Known records from AL, CT, FL, GA, IL, KS, KY, MA, ME, MS, NC, NJ, NY, OH, PA, TN, TX

female

female

female ventral

OXYOPIDAE > LYNX SPIDERS

Genera in This Family *Hamataliwa* (3), *Oxyopes* (13), and *Peucetia* (2)

Total Body Length 3.2–21.5 mm

Description These small to medium-sized spiders are entelegyne and ecribellate. They have eight eyes in a diagnostic hexagonal eye arrangement. They have legs with prominent spines and three tarsal claws. They have a pair of book lungs and a single tracheal spiracle located near the spinnerets. They tend to hunt in tall grass and herbaceous vegetation but can also be found on woody shrubs and trees.

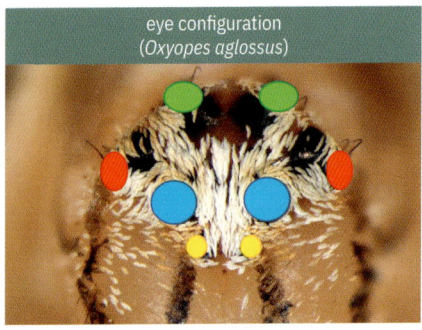

eye configuration
(*Oxyopes aglossus*)

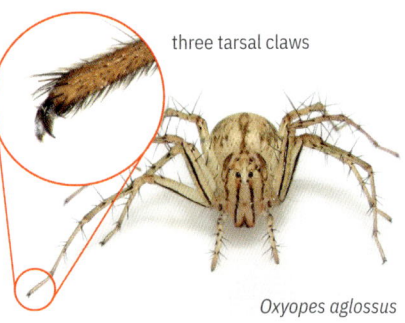

three tarsal claws

Oxyopes aglossus

Similar Families Salticidae

How to distinguish Oxyopidae from other similar families:

The diagnostic eye configuration of oxyopids can distinguish them from all other families. The anterior median eyes are located just above the clypeus, and the other eyes form the shape of a hexagon. Salticids have enlarged anterior median eyes, whereas the anterior median eyes in oxyopids are the smallest.

Hamataliwa grisea Keyserling, 1887

Common Name Bark lynx spider

Total Body Length Female 8.7–10.9 mm, Male 8.4–11.1 mm

Description The prosoma is mottled with black, tan, and browns. The anterior median eyes are smaller than the others and located between and just below the anterior lateral eyes (which are the largest). The posterior eyes are more widely spaced than those of most other Oxyopidae. The color of the opisthosoma is much the same as that of the prosoma. The legs are hairy and colored in the same manner as the rest of the spider. This coloration provides good camouflage on tree bark.

Habitat These spiders are often found in low tree branches and shrubs.

Known records from AZ, CA, FL, GA, MS, NM, TX

female

female

female
male
male
male

Oxyopes aglossus
Chamberlin, 1929

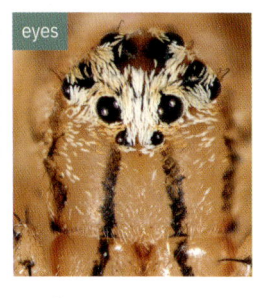

eyes

Common Name
This species does not have a common name.

Total Body Length
Female 4.5–6.7 mm, Male 3.9–4.8 mm

Description The female prosoma is tan, with a dark median stripe that stops well behind the posterior eye row. There are brown paramedian bands. The edges are pale. Thin dark stripes run from the anterior median eyes all the way down the chelicerae. The female opisthosoma is tan, with a light brown heartmark. Three pairs of dark, mottled, almost square markings are on the sides. The legs are tan, with spots and dark contiguous lines on the ventral surfaces of femurs I and II. Dark markings

female

female

female

female with egg sac

male

male

male

may be on the ventral side of femur III, but they do not make a contiguous line. The male prosoma is copper colored, reflective, and iridescent. The opisthosoma is darker, almost green, reflective, and iridescent. The legs are golden and have dark contiguous lines on the ventral surfaces of femurs I and II. Dark markings may be on the ventral surface of femur III, but they do not make a contiguous line. The eyes are in a hexagon pattern, with the anterior median eyes located just below. Thin dark stripes run from the anterior median eyes all the way down the chelicerae, but they are less conspicuous than in the female. The male pedipalps are velvety black.

Habitat These spiders are often found in low herbaceous vegetation.

Known records from AL, AR, FL, GA, LA, MD, MS, NC, OH, SC, TN, TX, VA

Oxyopes apollo Brady, 1964

Common Name This species does not have a common name.

Total Body Length Female 4.2–6.7 mm, Male 3.4–4.4 mm

Description The female prosoma can vary in coloration. It can be cream with pale ginger and light brown markings to golden orange with dark brown to almost black markings. The opisthosoma is pale brown to brown, with a cream median stripe. The ventral surfaces of the legs are dusky. The male prosoma is dark brown, with an orange median stripe and tan edges. The median stripe widens anteriorly and has

undulating edges. The opisthosoma is pale orange, with a light brown to pale gray median stripe. Oblique transverse dark brown bands are on the sides. The legs of both the males and females are pale brown to tan. There are dusky markings at the distal ends of the femurs and proximal ends of the tibiae.

Habitat These spiders are often found in herbaceous vegetation.

Known records from AR, AZ, FL, GA, IL, LA, MO, NC, NM, OK, TN, TX

Oxyopes salticus Hentz, 1845

Common Name Striped lynx spider

Total Body Length Female 4.6–7.4 mm, Male 3.9–5.9 mm

Description The female prosoma is pale yellow. There is a white median stripe, bordered by light brown. There are white paramedian bands, and light brown submarginal bands. A black dash and a dot are on each side of the anterior prosoma below the submarginal band, sometimes referred to as the "cheeks." The dot almost gives the appearance of another eye. Thin black stripes run from the anterior median eyes down the chelicerae. The opisthosoma is cream white, with a golden-brown heartmark. There are stripes in golden and brown, giving an overall striped appearance. The legs are pale yellow, with contiguous black stripes on the ventral surfaces of femurs I, II, and III. The male prosoma is golden brown. There is a brown median stripe, and a light stripe runs through its middle. There are dark submarginal bands. These may be faint to indistinct on some specimens. The cheeks have a black dash-and-dot pattern similar to that seen on the females. Thin dark lines also run from the anterior median eyes down the chelicerae. The opisthosoma is almost green, reflective, and iridescent. A slightly darker heartmark is often visible, as is the pale median stripe.

female

female

female

male

male

male

male

male

subadult
male

The legs are golden in color and have dark contiguous lines on the ventral surfaces of femurs I, II, and III. The male pedipalps are velvety black.

Habitat These spiders are very common in grassy habitats and low herbaceous vegetation.

Known records from ON (Canada); AL, AR, AZ, CA, CO, CT, DC, FL, GA, IA, IL, IN, KS, KY, LA, MA, MI, MO, MS, NC, NE, NH, NJ, NM, NY, OH, OK, OR, PA, SC, TN, TX, VA, WI, WV, WY (United States)

Oxyopes scalaris Hentz, 1845

Common Name Western lynx spider

Total Body Length Female 5.8–9.6 mm, Male 4.7–6.1 mm

Description This species can vary greatly in coloration. The prosoma can be pale gray to golden brown. There may be a pale median stripe. The opisthosoma can be pale gray to golden brown to almost black. There are often stripes and bands, but these can vary. The legs are yellow to golden brown, with dark banding.

Habitat These spiders are often found in forested and sagebrush habitats among the tree branches. They can also be found in grassy and herbaceous vegetation. The egg sacs are attached to vegetation, and the female sits on the egg sac to guard it.

female with egg sac

female

juvenile

female with egg sac

male

male

male

Known records from BC, LB, MB, NS, ON, QC, SK (Canada); AL, AZ, CA, CO, CT, FL, GA, ID, IL, LA, MA, MD, ME, MI, MN, MO, MS, MT, NC, NE, NH, NJ, NM, NV, NY, OH, OK, OR, SD, TN, TX, UT, VT, WA, WI, WY (United States)

Oxyopes tridens
Brady, 1964

Common Name
This species does not have a common name.

Total Body Length
Female 5.5–7.6 mm, Male 4.9–6.4 mm

Description The prosoma is dark brown. There is a tan median stripe. The median stripe has a white border, which runs through the lateral eyes, turns back to the posterior median eyes, and then passes between the posterior median eyes to the front of the prosoma, creating the shape of a trident. The opisthosoma is dark brown, with a white to cream median stripe, and there is a pale brown heartmark. The legs are gray brown, with dark brown banding.

Habitat These spiders are often found on rocks or bare ground. They are usually not found hunting on vegetation. They have also been found on and around human structures.

Known records from AZ, CA, NM, NV, TX

female

female

female

Peucetia viridans (Hentz, 1832)

Common Name Green lynx spider

Total Body Length Female 11.8–21.6 mm, Male 8.3–14.5 mm

Description The prosoma is vivid green. There may be three dark spots around the edge and sometimes a small dark marking near the fovea. The opisthosoma is vivid green, usually with some white chevrons and occasionally with pink markings. The legs are pale, almost white, with black spots. Some specimens are known for having dark pink, almost red, femurs.

male

male

male

Habitat These spiders are common in agricultural fields and low vegetation. The egg sacs can have a slightly spiky appearance and are attached to vegetation. The female sits on the egg sac to guard it.

Known records from AL, AR, AZ, CA, FL, GA, KS, LA, MS, NC, NM, OK, OR, SC, TN, TX, VA

male

male

PHILODROMIDAE › RUNNING CRAB SPIDERS

Other Colloquial Names
Philodromid crab spiders

Genera in This Family *Apollophanes* (3), *Ebo* (8), *Philodromus* (57), *Rhysodromus* (3), *Thanatus* (8), *Tibellus* (7), and *Titanebo* (13)

Total Body Length 3.0–8.2 mm

Description These entelegyne spiders are ecribellate. They have eight eyes in two recurved rows. The sizes of the eyes are usually very similar. They have laterigrade legs. Leg II is the longest (sometimes extremely so), and the spiders sit in a somewhat crab-like pose. They have a pair of book lungs and a single tracheal spiracle located near the spinnerets. They are active predators, chasing prey on a variety of surfaces.

Similar Families Hersiliidae, Selenopidae, Sparassidae, Thomisidae

How to distinguish Philodromidae from other similar families:

Philodromids do not have elongated spinnerets like those seen in hersiliids. Philodromids have their eyes in two recurved rows. This distinguishes them from selenopids, whose posterior eye row is so strongly procurved that it

eye configuration
(*Thantus rubicellus*)

*Philodromus
infuscatus*

laterigrade legs, overall
fairly small in size

appears they have six eyes in the anterior row and only two in the posterior row. Most of the sparassids are quite large (7.0–50.0 mm total body length); in contrast, philodromids tend to be smaller (3.0–8.2 mm total body length). Thomisids have robust legs I and II, the second leg is not longer than the first, and the size of legs III and IV is reduced. These traits distinguish them from philodromids, whose leg II is the longest (sometimes extremely so), and legs III and IV are of similar build to the others.

Apollophanes punctipes (O. Pickard-Cambridge, 1891)

Common Name This species does not have a common name.

Total Body Length Female 5.6–7.4 mm, Male 3.8–7.0 mm

Description The prosoma is yellow orange, with a pale median stripe. The eye region is darker. The lateral areas may be speckled or suffused with black. The prosoma is distinctly longer than it is wide. The eyes of the posterior row are equally spaced. The opisthosoma is cream white, with purple, orange, or black spots and streaks, occasionally forming lateral bands. The opisthosoma is somewhat flattened and angulate at the posterior end. The legs are yellow to orange and speckled with black. They can be quite variable in coloration and markings. *A. punctipes* can look similar to *Apollophanes texanus* but lacks the stout hairs seen on the prosoma of *A. texanus*.

Habitat These spiders can be found in grassy areas. The female attaches her egg sac to plant leaves and will guard the egg sac. They have been found at elevations up to 12,500 ft.

Known records from AZ, NM, TX

female

female

male

male

Ebo pepinensis Gertsch, 1933

Common Name This species does not have a common name.

Total Body Length Female 3.5–4.9 mm, Male 2.2–2.8 mm

Description The prosoma is tan, with various brown markings. The opisthosoma is elongated and tan, with a dark brown heartmark and scattered brown markings. The legs are tan, with a band on the proximal end of the tibia. Leg II is approximately twice as long as leg I. The tibiae and metatarsi I, III, and IV lack a dark dorsal stripe the entire length of the segments; the absence helps distinguish *E. pepinensis* from some other *Ebo* species.

Habitat These spiders are often found in grasslands. They have been found at elevations up to 12,500 ft.

Known records from AB, BC, MB, NB, NS, ON, QC, SK, YT (Canada); CA, CO, ID, IL, IN, KS, MI, MN, MT, NM, NV, OR, TX, UT, WA, WY (United States)

female

Philodromus cespitum (Walckenaer, 1802)

Common Name This species does not have a common name.

Other Colloquial Names Turf running spider

Total Body Length Female 4.5–6.1 mm, Male 4.0–5.0 mm

Description The prosoma is dark orange, with a pale median band. The very edge is lined in off-white. The opisthosoma can range from pale yellow to gray. The heartmark is dark, and there is a very wide pale median stripe. Sometimes a series of transverse bands can be seen within the median stripe. The lateral edges are darker. The opisthosoma is somewhat angulate at the posterior end. Darker dots are on the lateral edges just before the spinnerets. The legs are tan to golden brown and unbanded but may have some speckles.

female

female

Habitat These spiders are often found on low vegetation.

Known records from AB, BC, MB, NB, NS, NT, ON, PE, QC, SK, YT (Canada); AK, AZ, CA, CO, IA, ID, IL, IN, MA, MD, ME, MI, MN, MT, ND, NE, NH, NJ, NM, NV, NY, OR, SD, TX, UT, VT, WA, WI, WY (United States)

Philodromus dispar Walckenaer, 1826

Common Name This species does not have a common name.

Other Colloquial Names Eurasian running crab spider

Total Body Length Female 4.0–6.1 mm, Male 4.0–5.0 mm

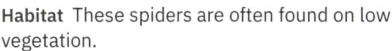

Description The female prosoma is tan with brown hairs, creating a variegated effect. The very edge has a cream stripe. The opisthosoma is also variegated tan with brown, with some pale chevrons in the midline. The lateral edges are tan and cream. There is a distinct line between the darker dorsal and lighter lateral surfaces; according to Tone Killick (personal communication), this is diagnostic for this species. The legs are tan and may have some speckling. The males of this species are easily recognized. The prosoma is dark red,

almost black, and lined at the very edge in white. The opisthosoma is also dark red, almost black, with white lateral edges. The legs are yellow and unbanded.

Habitat These spiders are often found on low vegetation.

Known records from BC (Canada); WA (United States)

This is an introduced species, and its range is probably more extensive than is demonstrated by the scientific literature.

Philodromus infuscatus
Keyserling, 1880

Common Name This species does not have a common name.

Total Body Length
Female 4.2–4.7 mm,
Male 4.0–4.5 mm

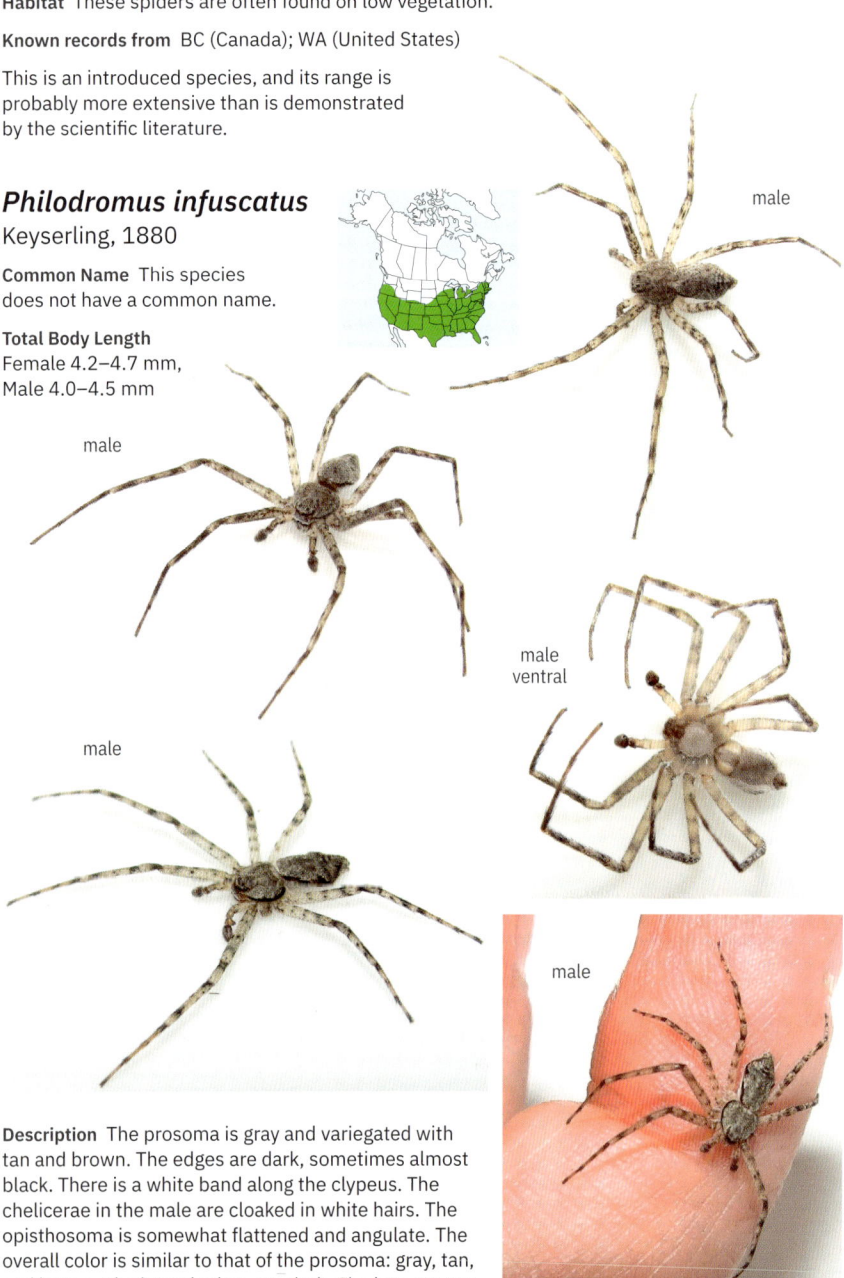

male

male

male

male
ventral

male

Description The prosoma is gray and variegated with tan and brown. The edges are dark, sometimes almost black. There is a white band along the clypeus. The chelicerae in the male are cloaked in white hairs. The opisthosoma is somewhat flattened and angulate. The overall color is similar to that of the prosoma: gray, tan, and brown. The lateral edges are dark. The legs are tan, with dusky leg bands and black spots.

Habitat These spiders are often found in forested habitats and low vegetation.

Known records from ON (Canada) and AL, AR, DC, FL, IA, IL, IN, KS, KY, MA, MD, MI, MN, MO, NC, NY, PA, TX, VA, VT, WI (United States); known records of *Philodromus infuscatus utus* Chamberlin, 1921 from AZ, CA, CO, NM, OR, UT, WY

Philodromus marxi Keyserling, 1884

Common Name This species does not have a common name.

Other Colloquial Names Metallic crab spider

Total Body Length Female 2.7–3.8 mm, Male 2.3–3.0 mm

Description The female prosoma is dark brown, with a cream-white median stripe. The very edge has a thin cream line around it. The opisthosoma is cream colored, with dark lateral edges on the anterior two-thirds. At the ends of these dark marks are lobed markings toward the midline. There is a faint heartmark that is only slightly darker than the surrounding area. Sometimes dots are also near the spinnerets. The legs are cream, with tan bands at the proximal end of each segment. In contrast, the male prosoma and opisthosoma are shiny and metallic, somewhat copper-like in appearance. The legs are yellow without any banding. *Philodromus placidus* can look very similar (both the males and females). To confirm an identification, one would need to look at the genitalia.

Habitat These spiders are often found in tall grasses and other low vegetation.

Known records from AR, CO, CT, FL, GA, IL, IN, MA, MD, ME, MI, MN, MO, MS, NC, NE, NJ, NY, OK, SC, TX, VA, WI, WV

female

female ventral

female with egg sac

male

Philodromus minutus Banks, 1892

Common Name This species does not have a common name.

Total Body Length Female 3.4 mm, Male 2.5 mm

Description The female prosoma is light brown, with a pale median stripe. The opisthosoma is cream, with a brown heartmark and splotchy brown spots down the midline. The legs are tan, and prolateroventral longitudinal stripes are on legs III and IV. The male prosoma is dark brown, with a slightly paler median stripe. There is an overall metallic sheen to it. The opisthosoma is brown to gray and also has somewhat of a metallic sheen. The legs are yellow and have prolateroventral longitudinal stripes on legs III and IV.

Habitat These spiders are often found in low herbaceous vegetation.

Known records from ON, QC (Canada); AR, CO, CT, DC, GA, IA, IN, LA, MA, ME, MI, MN, MO, NM, NY, OH, PA, TX, UT, VA, WI (United States)

female

female

male

female

male

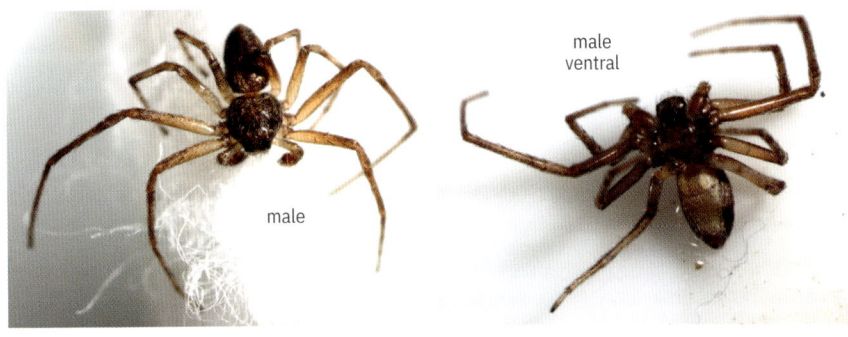

male

male
ventral

Philodromus placidus
Banks, 1892

Common Name This species
does not have a common name.

female

Total Body Length
Female 3.5–5.0 mm, Male 3.0–3.5 mm

Description The female prosoma is dark and mottled with
brown, with a cream-white median stripe. The very edge
has a thin cream line around it. The opisthosoma
is cream colored, with dark lateral edges
on the anterior two-thirds. At the ends of
these dark markings are lobed markings
toward the midline. This darker area
encroaches on the dorsal surface to a
greater extent than is seen in *Philodromus
marxi*. The dorsal opisthosoma is also

female

female
ventral

somewhat shaded with tan and light brown. The legs are pale cream white. There are bands on the proximal ends of each segment. Dark markings are also on the middle of the prolateral surface of femurs III and IV. The male prosoma is golden brown to brown, and an overall metallic sheen is often seen. The opisthosoma is dark gray to purple brown, and it also has a metallic sheen. The legs are yellow and have markings similar to those of the female. *P. marxi* can look very similar. To confirm a species-level identification, one would need to look at the genitalia.

Habitat These spiders are often found in forested habitats and seem to have an affinity for coniferous trees.

Known records from AB, BC, LB, MB, NB, NL, NS, NT, ON, PE, QC, SK, YT (Canada); AK, AL, AR, CO, CT, FL, GA, IL, IN, MA, ME, MI, MN, MS, NC, ND, NE, NH, NJ, NY, OH, OR, PA, RI, SD, TX, VA, VT, WI, WV, WY (United States)

male

Philodromus rufus
Walckenaer, 1826

Common Name This species does not have a common name.

Total Body Length
Female 2.7–4.5 mm, Male 2.5–3.5 mm

Description The prosoma is reddish orange, with a broad yellow median stripe. The opisthosoma is elongated. It is reddish orange, with a faint yellow-orange median stripe and red heartmark. The legs are yellow, with red speckling on legs I and II. The males are often somewhat darker than the females. There are subspecies of *P. rufus*; they all have a similar general appearance, with only minor differences.

Habitat These spiders are often found in grassy and shrubby habitats.

Known records from AB, BC, LB, MB, NB, NL, NS, NT, ON, PE, QC, SK, YT (Canada); AK, AR, AZ, CA, CO, CT, GA, ID, IL, KS, MA, ME, MI, MN, MS, MT, NC, ND, NE, NH, NJ, NM, NY, OH, OR, PA, RI, SD, TX, UT, VA, VT, WA, WI, WY (United States)

female

female
ventral

female

female with egg sac
(young emerging)

all images *Philodromus rufus*

male
ventral

male

male

male

Thanatus rubicellus Mello-Leitão, 1929

Common Name This species does not have a common name.

Total Body Length Female 5.0–7.0 mm, Male 4.0–6.0 mm

Description The prosoma is cream to off-white. A brown to reddish-brown median stripe starts out thin at the posterior and widens and splits as it approaches the eye region, giving the appearance of a very tall V. There are brown to reddish-brown submarginal bands. The opisthosoma is cream, with scattered darker hairs. A very well-defined heartmark in dark brown is outlined in off-white. A pair of brown to reddish-brown stripes on either side of the heartmark run to the spinnerets, sometimes only on the posterior. The legs are cream, with a pair of dull brown to reddish-brown stripes on the dorsal surface of the femurs and brown on the ventral surface of the femurs. Short dark marks are on the proximal prolateral and retrolateral surfaces of the tibiae. *Thanatus* species are difficult to identify to species without viewing a specimen under the microscope.

Habitat Not much has been published on the habitat preferences of this species. The specimens shown here were collected from a prairie, and others were observed in grassy areas.

Known records from AB, LB, MB, QC, SK (Canada); AL, AR, CO, CT, GA, IL, MA, ME, MN, MO, NC, NH, NJ, NM, NY, OK, PA, SC, SD, TX, VA, WY (United States)

female

female

female
ventral

female

subadult
male

subadult
male

Tibellus oblongus (Walckenaer, 1802)

Common Name The genus *Tibellus* has the
common name "slender crab spiders."

Other Colloquial Names Oblong running spider

Total Body Length Female 6.0–10.0 mm, Male 5.0–7.0 mm

Description The prosoma is tan yellow, with a
dark brown median stripe that forks slightly in
the eye region. There are parallel brown lateral
stripes. Faint lines may be between the median
and lateral stripes. Sometimes the lines are
visible only as a dot toward the posterior of
the prosoma. The opisthosoma is tan yellow,
with a brown median stripe. Two thin stripes
that are often dashed are on either side of the
median stripe. There are two spots toward the
posterior on the first set of thin stripes. The
legs are tan yellow and speckled.

Habitat These spiders are often found in
grassy habitats. When sitting, they extend the
first two pairs of legs forward and the last two
pairs backward (or with the third pair wrapped

female

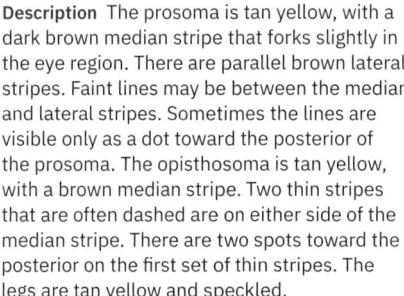

female

female

469

female ventral

juvenile

around the edges of the leaf or branch on which they are sitting), enabling them to sit easily on thin blades of grass.

Known records from AB, BC, MB, NB, NS, NT, ON, QC, SK, YT (Canada); AK, AZ, CA, CO, GA, IL, IN, MA, ME, MI, MN, MT, NH, NM, NV, OH, PA, TX, UT, WA, WI, WY (United States)

PISAURIDAE > NURSERYWEB SPIDERS

Other Colloquial Names Fishing spiders

Genera in This Family *Dolomedes* (8), *Pisaurina* (4), and *Tinus* (1)

Total Body Length 5.0–37.0 mm

Description These medium-sized to large spiders are entelegyne and ecribellate. They have eight eyes in two rows. Their chelicerae do not have a boss or condyle. They have a pair of book lungs and a single tracheal spiracle near the spinnerets. The egg sacs are large spheres, usually white to cream, that are attached to the spinnerets but also held in the chelicerae until the young are ready to emerge. At that time, the female builds a nursery web for them and guards them until they disperse.

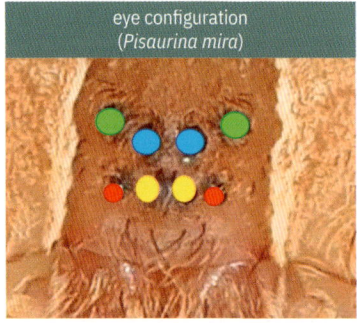

eye configuration
(*Pisaurina mira*)

female carries egg sac in chelicerae
(*Dolomedes albineus*)

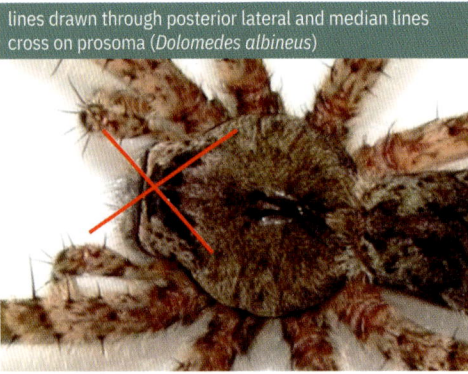

lines drawn through posterior lateral and median lines cross on prosoma (*Dolomedes albineus*)

Similar Families Ctenidae, Deinopidae, Homalonychidae, Lycosidae, Miturgidae, Trechaleidae

How to distinguish Pisauridae from other similar families:

Pisaurids can be distinguished from ctenids by their eye configuration. In ctenids, the anterior lateral eyes are smaller than the other eyes and located close to the posterior lateral eyes, and the eye rows are strongly recurved, giving the appearance of three eye rows. Deinopids have extremely large posterior median eyes, are cribellate, and build a netlike web that is suspended from the first two pairs of legs and used to scoop up prey; these traits distinguish them from pisaurids. In homalonychids, the prosoma narrows anteriorly to a greater extent than is seen in pisaurids. Homalonychids also use their stiff hairs on their prosoma, legs, and opisthosoma to trap dirt and sand to help them camouflage, which pisaurids do not do. Pisaurids can be distinguished from lycosids by their eye configuration. The posterior lateral eyes are further back on the prosoma in lycosids than they are in pisaurids. One way to visualize the difference is to draw lines between the posterior lateral and posterior median eyes on each side and then to extend those lines forward. They would cross in front of the spider in lycosids and on the prosoma in pisaurids. Additionally, pisaurids lack the boss or condyle on the basal chelicerae that is often seen in lycosids. Pisaurids have three tarsal claws; in contrast, miturgids have only two. Pisaurids can be distinguished from trechaleids by their prograde legs; trechaleids have somewhat laterigrade legs. Additionally, trechaleids do not carry their egg sacs in their chelicerae (they are attached to the spinnerets), and they do not build a nursery web (the young ride on the egg sac for a period before dispersing). Trechaleids are also represented in North America north of Mexico by only one species, and it is found only in Arizona.

Dolomedes albineus Hentz, 1845

Common Name Whitebanded fishing spider

Total Body Length Female 23.0 mm, Male 18.0 mm

Description The prosoma is variegated white to gray green, with a dark V mark at the fovea and faint radial lines. The eye region is elevated, and the area around the eyes is usually darker. A white band crosses the clypeus. White hairs are on the chelicerae. The sternum is light in color, occasionally with a slightly darker border. The very edge of the prosoma is lined with black. The opisthosoma is gray green, occasionally yellow, and somewhat variegated, with a dark brown median stripe made up of chevron shapes. White dots may be at the base corner of each chevron. The legs are cream to gray, with greenish-gray banding. There are tufts of white hairs throughout, giving the spider a fuzzy appearance. The legs of the males are proportionally longer than those of the females. *D. albineus* can be distinguished from the other species of *Dolomedes* by the

female

female with
egg sac

471

female with
egg sac

male

male

juvenile

juvenile

elevated eye region of the prosoma, the lack of longitudinal dark stripes down the clypeus and chelicerae, the white band across the clypeus, and the overall fuzzy appearance.

Habitat These spiders are usually found on tree trunks, where they sit head down. They are often found in swamps or near streams, lakes, and ponds.

Known records from AL, AR, FL, GA, IN, KY, LA, MO, MS, NC, OH, SC, TN, TX, VA

Dolomedes scriptus Hentz, 1845

Common Name The genus *Dolomedes* has the common name "fishing spiders."

Other Colloquial Names Striped fishing spider

Total Body Length Female 17.0–24.0 mm, Male 13.0–16.0 mm

Description The prosoma is light brown to brown, with a thin pale cream median stripe. There is sometimes a pair of dark triangles, one on either side

472

of the median stripe anterior of the fovea. A pair of curved lines are also on either side of the median stripe posterior of the eyes. There are pale radial lines. The lateral areas are tan to cream, occasionally white. The chelicerae are covered with tan hairs and may have a gray longitudinal stripe on them. The sternum is brown, with a pale median stripe. The opisthosoma is light brown to brown. The heartmark is pale tan to gray. Tan to cream markings are on the anterior: a V and a pair of lobed markings. A series of chevrons that run down the median of the opisthosoma is highlighted in tan to cream, giving the appearance of W shapes. There is shading within the chevrons, giving them a sculpted appearance. The legs are tan and banded in brown. The legs of the males are proportionally longer than those of the females. *D. scriptus* can be distinguished from the other *Dolomedes* species by the W markings on the opisthosoma, the curved cream lines on the prosoma posterior of the eyes, and the lack of a cluster of hairs on the male ventral femur IV.

Habitat These spiders are commonly found near fast-moving bodies of water, such as streams and rivers.

Known records from MB, NB, NS, ON, QC (Canada); AL, AR, CO, CT, DE, FL, GA, IA, IL, IN, KS, MA, MD, ME, MI, MN, MO, MS, NC, NE, NH, NJ, NY, OH, OK, PA, RI, SC, TN, TX, VT, WI, WV (United States)

Dolomedes tenebrosus Hentz, 1844

Common Name The genus *Dolomedes* has the common name "fishing spiders."

Other Colloquial Names Dark fishing spider

Total Body Length Female 15.0–26.0 mm, Male 7.0–13.0 mm

Description The prosoma is variegated light to dark brown, with a pale median stripe, which may be indistinct. There is usually a pair of dark triangles, one on either side of the median stripe just anterior of the fovea. There is a broken submarginal band that is dark brown, almost black, highlighted with cream. A pair of dark longitudinal stripes start in the eye region

female

female

female

juvenile

male

male

and run the length of the chelicerae. The sternum is brown, with a pale median stripe. The opisthosoma is light to dark brown. The heartmark is gray brown. A series of chevrons is on the posterior half. These may be highlighted by cream-colored dots at the base corners. The sides are tan to light brown. The legs are tan and banded in brown. The legs of the males are proportionally longer than those of the females. *D. tenebrosus* can be distinguished from the other *Dolomedes* species by the distinct dark lines that run from the eye region to the ends of the chelicerae, the lack of distinct W-shaped markings on the opisthosoma, and the overall rich brown color.

Habitat This species does not seem to be as closely associated with water as the other *Dolomedes* species and is often found in and around human structures. It can also be found in forested habitats.

Known records from MB, NB, NS, ON, QC (Canada); AL, AR, CT, DC, FL, GA, IL, IN, KS, KY, LA, MA, MD, ME, MI, MN, MO, MS, NC, ND, NE, NH, NJ, NY, OH, PA, RI, SC, TN, TX, VA, VT, WI, WV (United States)

Dolomedes triton (Walckenaer, 1837)

Common Name The genus *Dolomedes* has the common name "fishing spiders."

Other Colloquial Names Sixspotted fishing spider

Total Body Length Female 17.0–20.0 mm, Male 9.0–13.0 mm

Description The prosoma is green to brownish green (occasionally dark brown), with pale cream to white submarginal bands. Sometimes a pale stripe or dots cross the clypeus. The chelicerae are covered with tan to golden hairs.

female

female

female

female on water surface

juvenile

female
ventral

male

male

male

The sternum is pale yellow to pale green, with six spots on the lateral edges (leading to the common name "sixspotted fishing spider"). Sometimes these spots are fused. The opisthosoma is tan to pale green and pale cream on the lateral edges. Pairs of white dots often run down the midline. The legs are pale yellow to dark green. Some have dusky markings and longitudinal stripes on the femurs. The legs of the males are proportionally longer than

male

male
ventral

those of the females. The males have a tubercle on the ventral side of femur IV. *D. triton* can be distinguished from the other species of *Dolomedes* by the six spots on the sternum, the tubercle on the ventral surface of the male femur IV, and the fairly plain prosoma.

Habitat These spiders usually seem to be found near slow-moving water. They are often found in the emergent aquatic vegetation or sitting on the surface of the water. When startled, they will quickly dive below the water's surface, where they can stay submerged for some time.

Known records from AB, BC, MB, NT, ON, PE, QC, SK, YT (Canada); AK, AL, AR, CA, CO, CT, DE, FL, GA, ID, IL, IN, KS, KY, LA, MA, MD, ME, MI, MN, MO, MT, NC, ND, NE, NH, NJ, NM, NY, OH, OK, OR, PA, RI, SC, TN, TX, UT, VA, WA, WI, WV (United States)

Dolomedes vittatus Walckenaer, 1837

Common Name The genus *Dolomedes* has the common name "fishing spiders."

Other Colloquial Names Banded fishing spider

Total Body Length Female 19.0–28.0 mm, Male 10.0–25.0 mm

Description The prosoma is dark brown. There is a pair of dark triangles, one on each side of the midline just anterior of the fovea. These triangles are usually very pronounced in this species, more so than in the others with this characteristic. The prosoma in the males often has white to cream lateral edges. The sternum is gray, with an indistinct median stripe. The opisthosoma is dark brown. Paired cream

female

juvenile

spots are often on the posterior half. The legs are dark brown and indistinctly banded. The males have a cluster of stiff hairs on the ventral surface of femur IV. *D. vittatus* is most easily confused with *Dolomedes scriptus*, as both species can have somewhat variable markings, but can be distinguished by the very pronounced dark triangles on the prosoma, the lack of W markings on the opisthosoma, and the cluster of hairs on male femur IV.

juvenile

Habitat These spiders are usually found in shaded areas near streams.

Known records from ON (Canada); AL, AR, CT, DE, FL, GA, IL, IN, KS, KY, MA, MD, MO, MS, NC, NH, NJ, NY, OH, OK, PA, SC, TN, TX, VA, WI, WV (United States)

Pisaurina dubia (Hentz, 1847)

Common Name This species does not have a common name.

Total Body Length Female 8.0–10.0 mm, Male 8.0–9.0 mm

Description The prosoma is tan to yellow, with a series of longitudinal stripes in browns and yellows. This striped pattern extends the length of the opisthosoma. A pair of dark dots is usually on either side of the heartmark. A tuft of straight hairs projects in front of the spider from between the posterior median eyes. The legs are tan to yellow, and a pale longitudinal stripe can often be seen on the dorsal surface of the femurs. *P. dubia* looks similar to *Pisaurina undulata* but can be distinguished by the straight hairs that project forward from the prosoma. In *P. undulata*, curved spines project forward from a similar position, and the lateral opisthosoma stripes are often broken. *P. dubia* can easily be

female

female with egg sac

female

mistaken for a *Tibellus* species (Philodromidae), but note the eye configuration, the placement of the dark opisthosoma spots, and the hair tufts that are present on *P. dubia* but absent in *Tibellus* species.

Habitat This species has been found hunting on and around pitcher plants, in sugarcane fields, and in wetlands.

Known records from AR, FL, GA, IN, KS, LA, MO, MS, NC, OH, OK, TN, TX

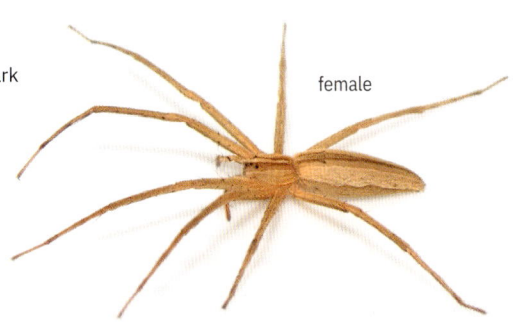

female

Pisaurina mira (Walckenaer, 1837)

Common Name Nurseryweb spider

Other Colloquial Names American nurseryweb spider

Total Body Length Female 12.5–16.5 mm, Male 10.5–15.0 mm

Description The appearance of this species varies greatly. The prosoma is usually yellow to tan to light brown. There may be a broad slightly darker median stripe, a thin very dark median stripe, or no median stripe. The opisthosoma can range from yellow to tan to light brown. There may be a broad median stripe with undulating edges of cream, a thin dark median stripe, an indistinct median stripe with undulating broken white markings on either side toward the posterior, or very little to no markings. The legs range from yellow to light brown. There may be distinct broad dark brown bands, indistinct pale bands, or no visible banding.

female

female

female

Habitat These spiders are often found in forest understories, fields, and shrubs. They often hold the first two pairs of legs together while sitting.

Known records from NB, NS, ON, QC (Canada); AL, AR, CT, FL, GA, IA, IL, IN, KS, KY, LA, MA, MD, ME, MI, MN, MO, MS, NC, NH, NJ, NY, OH, OK, PA, SC, TN, TX, VA, VT, WI, WV (United States)

female

female

female

female with egg sac

nursery web full of spiderlings

juvenile

juvenile

juvenile

Tinus peregrinus (Bishop, 1924)

Common Name This species does not have a common name.

Total Body Length Female 5.0–10.0 mm, Male 4.3–10.0 mm

Description The prosoma is light to medium brown, with lighter submarginal bands. The opisthosoma is tan to light brown, with a broad brown median stripe that narrows just over halfway down, expands again, and then tapers to the spinnerets. The legs are tan to brown, with indistinct banding. *T. peregrinus* looks very similar to some *Dolomedes* species. To distinguish them, note the ranges of the species and the opisthosoma markings.

female

Habitat These spiders are often found on rocks and low vegetation near water.

Known records from AR, AZ, CA, NV, TX

SALTICIDAE ▶ JUMPING SPIDERS

Genera in This Family *Admestina* (3), *Anasaitis* (1), *Attidops* (4), *Attinella* (3), *Attulus* (8), *Bagheera* (1), *Beata* (1), *Bellota* (2), *Chalcoscirtus* (4), *Cheliferoides* (2), *Chinattus* (1), *Colonus* (3), *Corythalia* (2), *Dendryphantes* (1), *Eris* (4), *Euophrys* (1), *Evarcha* (3), *Ghelna* (4), *Habronattus* (74), *Hakka* (1), *Hasarius* (1), *Heliophanus* (1), *Hentzia* (8), *Hyetussa* (2), *Leptofreya* (1), *Lyssomanes* (1), *Maevia* (3), *Marchena* (1), *Marpissa* (10), *Menemerus* (2), *Messua* (2), *Metacyrba* (4), *Metaphidippus* (10), *Mexigonus* (3), *Myrmarachne* (2), *Naphrys* (4), *Neon* (6), *Neonella* (2), *Paradamoetas* (2), *Paramaevia* (3), *Paramarpissa* (3), *Paraphidippus* (3), *Peckhamia* (4), *Pelegrina* (26), *Pellenes* (12), *Phanias* (8), *Phidippus* (48), *Phlegra* (1), *Platycryptus* (3), *Plexippus* (1), *Poultonella* (2), *Pseudeuophrys* (2), *Rhetenor* (1), *Salticus* (5), *Sarinda* (2), *Sassacus* (4), *Sibianor* (1), *Sittisax* (1), *Synageles* (8), *Synemosyna* (2), *Talavera* (1), *Terralonus* (7), *Tomis* (1), *Tutelina* (4), and *Zygoballus* (4)

Total Body Length 1.0–22.0 mm

Description These diverse spiders are ecribellate and entelegyne. They have eight eyes in three rows, and the anterior median eyes are greatly enlarged. They have excellent vision. They have a pair of book lungs and a single tracheal spiracle located near the spinnerets. They tend to have stocky legs, and the tarsi have two claws. They are active diurnal hunters and pounce on their prey.

Similar Families Corinnidae, Lycosidae, Oxyopidae, Thomisidae, Trachelidae

How to distinguish Salticidae from other similar families:

The eye configuration of salticids distinguishes them from all similar families. Salticids have eight eyes in three rows, and the anterior median eyes are greatly enlarged. This is unlike all other similar families. Corinnids have eight eyes in two rows; usually the anterior eye row is recurved, and the posterior eye row is procurved. Lycosids have four small anterior eyes in a row across the front of the prosoma; the posterior median eyes are usually large, and the posterior lateral eyes are placed back on the prosoma, giving the appearance of three eye rows. Oxyopids have small anterior median eyes located just above the clypeus, and the other eyes form the shape of a hexagon. Both thomisids and trachelids have eight eyes in two rows.

eye configuration
(*Colonus sylvanus*)

Admestina wheeleri Peckham & Peckham, 1888

Common Name This species does not have a common name.

Total Body Length Female 4.3 mm, Male 3.3 mm

Description The prosoma is elongated and somewhat flattened and variegated tan and brown. The opisthosoma is cream, with a brown median stripe in a feather-like pattern. Tibia I is more robust than the other legs in both the males and the females and is brown with dark banding. The other legs are tan with brown banding. Three species of *Admestina* are known in North America. All look similar but can easily be distinguished by looking at the genitalia under a microscope. There is some minor overlap in their known ranges, but one can often make an educated guess about species identification based on the geographic

female

female

female
ventral

male

male with quarter
for scale

male

male

male

male
ventral

location where the specimen was found.
This is a small spider and often overlooked.

Habitat Little is known about the habitat
preferences of these spiders, but they are often found on tree branches. They are
somewhat flat, and it is thought they like to hide in the cracks and crevices of bark.

Known records from MB, ON (Canada); AR, CT, IN, KS, MA, ME, MI, MN, NY, OH,
OK, SD, WI (United States)

Anasaitis canosa (Walckenaer, 1837)

Common Name Twinflagged jumper

Total Body Length Female 5.0–6.0 mm, Male 4.0–5.2 mm

Description The prosoma is black and shiny. It has a pair of white dots
or dashes on the posterior end. The opisthosoma is light brown to brown.
A white marking is at the posterior end of the heartmark, beyond which is
a light brown median stripe. The edges are tan to white. About halfway down, the lateral markings
extend toward the midline (but do not meet). The first two pairs of legs are the darkest, with some

light banding. The other legs are tan to brown, with banding. The femurs and patellae of the pedipalps are white.

Habitat These spiders are often found on tree trunks, low vegetation, and tall grasses. They are also found in and around human structures.

Known records from AL, AR, FL, GA, LA, MD, MS, OK, SC, TX

female female

Attidops youngi
(Peckham & Peckham, 1888)

Common Name This species does not have a common name.

Total Body Length
Female 2.4–2.7 mm,
Male 2.4–2.8 mm

Description The prosoma is dark brown, with scattered yellow to white hairs and four indistinct yellow to white dots on the dorsal surface. The opisthosoma is brown, with scattered yellow to white hairs. Three transverse golden bands are on the posterior; the first two bend slightly in the middle toward the anterior. The legs are brown and banded in golden.

female

female ventral

female on penny for scale

Habitat These spiders are often found on tree trunks and other vertical surfaces.

Known records from ON, QC (Canada); CT, IL, KS, LA, MD, MI, NY, OH, PA, TX, VA, WI (United States)

Attulus ammophilus (Thorell, 1875)

Common Name This species does not have a common name.

Total Body Length Female 4.5–6.0 mm, Male 3.7–5.0 mm

Description The female prosoma is variegated brown. The female opisthosoma is also variegated brown, but indistinct pale chevrons run down the median. The male prosoma is brown and darker posteriorly and lightest on the anterior. A white median stripe does not reach the anterior. White bands are on the lateral edges. The male opisthosoma is brown, with pale lateral edges. A series of pale chevrons

female   female

female ventral   male

male   male

that vary in size run down the midline. The legs are mottled with brown, tan, white, and black, with the vague appearance of banding.

Habitat These spiders are often found in sandy habitats and in and around human structures.

Known records from ON (Canada); PA, UT (United States)
This is an introduced species.

Attulus fasciger (Simon 1880)

Common Name This species does not have a common name.

Other Colloquial Names Asiatic wall jumping spider

Total Body Length Female 4.5–5.3 mm, Male 3.6–4.5 mm

Description The prosoma is tan and cream, with a pale median stripe.
There is a pale triangle on the posterior end. The opisthosoma is tan, cream, and brown. There is a dark median stripe. The heartmark is tan. There are four white dots, two on each side of the heartmark. A large lobed cream marking on the posterior half of the opisthosoma covers the median stripe in that area. The legs are tan and banded in light brown. The males tend to be slightly darker. *A. fasciger* can look superficially similar to *Naphrys pulex*. The males can easily be distinguished by the size of the pedipalps (much smaller in *N. pulex*) and the height of the clypeus (much higher in *N. pulex*). The females can be distinguished by the dark heartmark and overall markings.

Habitat These spiders are usually found in and around human structures.

Known records from BC, MB, ON, QC (Canada); CO, IL, IN, KS, MI, MN, NH, NJ, NY, OH, PA, WI (United States)
This is an introduced species.

female
ventral

female

juvenile

juvenile
ventral

male

male

male

male
ventral

Bagheera prosper (Peckham & Peckham, 1901)

Common Name This species does not have a common name.

Total Body Length Female 6.0–8.0 mm, Male 5.0–6.5 mm

Description The female prosoma is silvery gray, with a light band between the posterior lateral eyes and a white spot toward the anterior. The opisthosoma is shiny golden. There are paired white spots or dashes on either side of the midline. The male prosoma is dark brown and has an iridescent sheen. There is a white band between the posterior lateral eyes and a dot toward the anterior. There are white

female

female

female

female
ventral

male

male

male
ventral

male

bands on the lateral edges. The chelicerae and fangs are extremely elongated. The opisthosoma is tan with faint, paired, light-colored spots on either side of the midline, and a white to tan band that runs across the anterior and down the lateral edges. The first pair of legs is darker, with some shiny hairs and banding. The other legs are cream, with tan banding and light-colored hairs.

Habitat These spiders are often found in woody habitats and riparian vegetation.

Known records from AR, OK, TX

Chalcoscirtus diminutus (Banks, 1896)

Common Name This species does not have a common name.

Total Body Length Female 2.6 mm, Male 2.6 mm

Description The prosoma is dark brown and shiny. The male has a white clypeus. The opisthosoma is dark brown, with a thin white anterior band.

female

female

female

female

A transverse pale brown band is about halfway down but often does not meet in the middle. A pair of parallel lines that are often broken at the anterior end are on either side of the midline above the spinnerets.

Habitat These spiders are found near the ground, often in leaf litter or other debris.

Known records from AZ, CA, FL, GA, IL, KS, MI, MN, MO, NE, NM, NY, SC, TX

Chinattus parvulus (Banks, 1895)

Common Name This species does not have a common name.

Other Colloquial Names Little Mountain jumping spider

Total Body Length Female 4.0–5.5 mm, Male 4.5–5.5 mm

Description The prosoma is reddish brown, with a dark border and a dark area between the posterior eyes. The males have white spots just posterior to the posterior lateral eyes and, occasionally, a white spot where the prosoma starts to decline toward the pedicle. The female opisthosoma is reddish brown, with a mottled tan anterior stripe. A series of tan markings makes a broken median stripe: chevrons toward the posterior end and stripes posteriorly from the anterior band. The male opisthosoma is reddish brown,

female

female

female

male

with a white anterior band. A pair of white spots are on each side about halfway down. There is also a white spot just above the spinnerets. The legs are yellow to brown. Often the males have darker legs I and sometimes darker legs II as well.

Habitat These spiders are often found in moist leaf litter and swampy areas.

Known records from ON, QC (Canada); AL, AR, CO, DC, FL, GA, IL, IN, KY, MS, NC, NM, NY, OH, TN, VA (United States)

Colonus hesperus (Richman & Vetter, 2004)

Common Name This species does not have a common name.

Total Body Length Female 7.0 mm, Male 5.6 mm

Description The female prosoma is tan to white, with four large black spots on the dorsal surface. A pair of dark dashes that often are just posterior to these marks remind me of Groucho Marx's eyebrows. The opisthosoma is tan to white, with black dots, sometimes with two wide tan stripes on either side of the midline. The anterior end is somewhat darker than the posterior. The legs are tan to white. The male prosoma is dark brown, with a white spot in the middle of the dorsal surface that widens toward the anterior eyes. White lateral bands are on the anterior half, and a pair of white lines is on the posterior end. The opisthosoma is golden brown, with a white band around the anterior. The legs are golden brown and banded in black.

female

female

male

Habitat These spiders are often found on trees, including those in orchards. They can also be found in and around human structures.

Known records from AZ, CA, NM, TX

Colonus puerperus (Hentz, 1846)

Common Name This species does not have a common name.

Total Body Length Female 7.0–10.0 mm, Male 7.0–7.2 mm

Description The female prosoma is yellow to golden. There is a white marking on the anterior. The posterior eyes are ringed in black. The opisthosoma is tan to yellow, with scattered small black dots, and sometimes a pair of widely spaced tan to light brown stripes are on the dorsal surface. The ventral opisthosoma has no distinct markings. The legs are tan to yellow and unbanded. The male prosoma is golden to dark brown,

female

female

female

491

female
ventral

male

sometimes almost black. A white median line is confined to the middle of the dorsal surface. White lines are on the anterior half of the lateral edges, and two white dashes are on the posterior end. The male opisthosoma is yellow to brown. A pair of widely spaced white longitudinal lines are on the dorsal surface. The legs are golden to dark brown. *C. puerperus* looks very similar to *Colonus sylvanus*. Note that, in the males, the dorsal prosoma marking is usually a line in *C. puerperus* and usually a circular dot in *C. sylvanus*. Note that, in the females, *C. sylvanus* often has orange to red coloration near the eyes that is missing in *C. puerperus*, and that *C. sylvanus* has a larger black spot on the ventral opisthosoma near the spinnerets than the small spot that may or may not be seen in *C. puerperus*.

Habitat These spiders are usually found in grassy areas or the understory vegetation of wooded areas.

Known records from AL, AR, FL, GA, IL, IN, KS, MD, MN, MO, NC, NJ, OH, OK, PA, SC, TX

Colonus sylvanus (Hentz, 1846)

Common Name This species does not have a common name.

Total Body Length Female 8.0–10.0 mm, Male 7.0–9.0 mm

Description The female prosoma is pale yellow, with a darker yellow to golden area around the eyes. Orange and even reddish coloration may be near the eyes. Occasionally, a large black spot is in the middle between the posterior eyes. The eyes are ringed in black. A white marking crosses the anterior. The opisthosoma is pale yellow to tan, with white around the lateral edges and a white median stripe. Black dots are scattered throughout. On the ventral opisthosoma, a black spot is just before the spinnerets. The legs are

female

female

female

female ventral

female

male

pale and have indistinct banding at the ends of each segment. The male prosoma is reddish brown to almost black. There are red stripes between the posterior lateral and median eyes and a circular or square white dot on the dorsal surface. Two to three white dashes are on the anterior lateral edges. The opisthosoma is dark brown, with a pair of widely spaced longitudinal white stripes. The edges are somewhat paler than the center. The legs are predominantly black, with some yellow bands. *C. sylvanus* looks very similar to *Colonus puerperus*. Note that, in the males, the dorsal prosoma marking is usually a line in *C. puerperus* and usually a circular dot in *C. sylvanus*. Note that, in the females, *C. sylvanus* often has orange to red coloration near the eyes that is missing in *C. puerperus,* and that *C. sylvanus* has a larger black spot on the ventral opisthosoma near the spinnerets than the small spot that may or may not be seen in *C. puerperus*.

Habitat These spiders tend to be found in forested habitats.

Known records from AL, AR, AZ, DE, FL, GA, IL, IN, KS, LA, MD, MO, MS, NC, NJ, NY, OH, OK, SC, TN, TX, VA

male

male

Dendryphantes nigromaculatus (Keyserling, 1885)

Common Name This species does not have a common name.

Other Colloquial Names Black-marked jumping spider

Total Body Length Female 5.6–7.2 mm, Male 4.6–5.5 mm

Description The prosoma is black. White bands on the lateral edges continue around above the anterior eyes. The females often have white hairs on the dorsal surface of the prosoma as well. The opisthosoma is black, with a white anterior band. Four pairs of spots run down the midline. Short transverse dashes are on the lateral edges. The legs are black, with some white banding, and the tarsi are paler. The males have a tuft of white hair on the prolateral distal femurs and patellae of leg I.

Habitat Not much is published about the habitat preferences of this species. Some specimens have been collected from mossy areas.

Known records from AB, BC, NL, NT, QC, YT (Canada); AK, AZ, CO, ID, ME, MN, MT, NH, NM, TX, UT, WI, WY (United States)

Eris militaris
(Hentz, 1845)

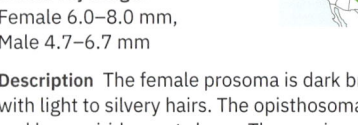

Common Name Bronze jumper

Total Body Length
Female 6.0–8.0 mm,
Male 4.7–6.7 mm

female

Description The female prosoma is dark brown and covered with light to silvery hairs. The opisthosoma is medium brown and has an iridescent sheen. Three pairs of white dashes are on either side of the midline. Additionally, there are dashes on the lateral edges. The legs are brown, with minimal banding. The male prosoma is dark brown. There is a transverse white

female

female
ventral

juvenile

male

male

male

male

male
ventral

band anteriorly, white bands are on the lateral edges, and a small white dash is on the posterior. The male opisthosoma is dark brown and has an iridescent sheen. A white band across the anterior continues down the lateral edges to the spinnerets. The legs have pale femurs and tarsi but are otherwise dark brown, with minimal banding. The male prosoma has neither the white dot and dash on the dorsal surface nor the white band on the clypeus that is seen in the similar *Eris flava*.

Habitat These spiders are often found in sunny areas on low vegetation and trees.

Known records from AB, BC, MB, NB, NS, NT, ON, PE, QC, SK, YT (Canada); AK, AL, AR, CA, CO, CT, FL, GA, IA, ID, IL, IN, KS, LA, MA, ME, MI, MN, MO, MS, MT, NC, ND, NE, NH, NJ, NM, NY, OH, OK, OR, PA, SC, SD, TN, TX, UT, WA, WI, WV, WY (United States)

Euophrys monadnock Emerton, 1891

Common Name This species does not have a common name.

Other Colloquial Names Contrasting jumping spider

Total Body Length Female 3.6–5.0 mm, Male 3.6–4.0 mm

Description The female prosoma is dark brown, almost black. The opisthosoma is dark brown, almost black, with scattered light-colored hairs. The first pair of legs are dark, and the other legs are reddish brown, with occasional black bands. The male prosoma is black. The male opisthosoma is black. The male pedipalps are bright yellow. The legs are black except for both the tarsi, which are white to yellow, and the femurs of legs III and IV, which are brilliant red.

Habitat These spiders can often be found in coniferous forests. The female is often overlooked, due to her drab appearance, whereas the male is easily recognizable.

Known records from AB, BC, MB, NB, NS, ON, SK (Canada); CA, CO, ME, MI, MN, NH, SD, WI (United States)

Habronattus americanus (Keyserling, 1885)

Common Name The genus *Habronattus* has the common name of "paradise spiders".

Total Body Length Female 5.5 mm, Male 4.2–4.8 mm

Description The female prosoma is tan to light gray brown. There is a reddish-brown area in the eye region, with a darker marking just posterior, sometimes all the way to the posterior edge. Black bands are on the sides of the prosoma near the eyes. The clypeus is white to cream. The opisthosoma is light gray brown. There is a tan median stripe made up of chevrons or diamond shapes; lateral to this are light reddish-brown markings. A few indistinct transverse dark bands are on the lateral edges. The legs are light gray brown, with indistinct banding. The male prosoma is dark brown, with a few scattered light hairs. A thin white line is around the edge. The anterior eye row is surrounded by long hairs that can be black, gray, or golden, in some cases creating crests. The clypeus is covered with iridescent white scales. The chelicerae are covered with white, blue, or pink hairs in seven longitudinal bundles. The male opisthosoma is black, with a white median stripe that is sometimes broken and white lateral edges. The male pedipalps are clothed in red and white hairs. Leg I is dark on the dorsal surface and usually has black and white ventral longitudinal stripes. There is a fringe of red hairs on femur I. The other legs are similarly marked but lack the red fringe. Iridescent scales are on the patella of leg III.

female

male

Habitat These spiders are usually found in open habitats.

Known records from AB, BC, MB, ON, SK (Canada); CA, CO, ID, MN, MT, NV, OR, UT, WY (United States)

Habronattus coecatus
(Hentz, 1846)

Common Name The genus *Habronattus* has the common name of "paradise spiders".

Total Body Length
Female 5.5 mm, Male 4.3–4.7 mm

Habronattus coecatus

female

female

Description The female prosoma is tan. The opisthosoma is tan, with a pale teardrop at the posterior end and a pair of transverse dashes on the lateral edges toward the midline about halfway down. These markings are sometimes highlighted by a dark border. The legs are tan, with occasional black bands. The male prosoma is dark brown to almost black. The clypeus is red only to the width of the anterior lateral eyes. A patch of tan to white in a transverse band behind the

female

male

male

male

male

male

male
ventral

posterior lateral eyes forms a pi shape. Just posterior of this mark is a pair of parallel lines that run from this marking to the posterior edge of the prosoma. The opisthosoma is tan to dark brown. The heartmark is tan. There is a tan to cream anterior band, and there may be a transverse tan to cream band about halfway down, which may or may not meet in the middle. A tan to cream teardrop marking is toward the posterior end, and then there are two white dots on either side of the spinnerets. The legs are tan to light brown. The ventral surface of leg I is black. Leg III is the longest, and the lateral surfaces of the tibia on leg III are green. A modification to the femurs of leg III creates an acute point. Superficially, the males look similar to *Habronattus pyrrithrix*, but the prosoma and opisthosoma markings are different and the ranges of the two species do not overlap.

Habitat These spiders are often found in grassy areas, including lawns, and they are often near the ground.

Known records from AL, AR, AZ, DC, FL, GA, IL, IN, KS, KY, LA, MA, MO, MS, NC, NJ, NM, NY, OH, OK, PA, SC, SD, TN, TX, VA

Habronattus decorus (Blackwall, 1846)

Common Name The genus *Habronattus* has the common name of "paradise spiders".

Total Body Length Female 5.0–7.5 mm, Male 4.0–5.0 mm

Description The female prosoma is tan to white and sometimes somewhat variegated. The opisthosoma is white to tan. There is a transverse light brown band toward the anterior. Two large black spots are about halfway down. The posterior has a large black marking that contains two small white dots. The legs are yellow, with scattered light and dark hairs. The male prosoma is black and sometimes covered in white hairs. The anterior of the opisthosoma is usually pale to white, and the remaining area is covered in iridescent red hairs. The legs are brown to dark brown. Leg III is the longest.

Habitat These spiders are often found near the ground, sometimes in the litter of grassy habitats and fields.

female

male

male

male

male

Known records from AB, BC, MB, NB, NS, ON, QC, SK (Canada); CT, FL, IA, KS, KY, MA, ME, MI, MN, NE, NH, NJ, NY, OH, OR, PA, SC, TX, UT, VT, WI (United States)

Habronattus geronimoi Griswold, 1987

Common Name The genus *Habronattus* has the common name of "paradise spiders".

Total Body Length Female 4.6–5.1 mm, Male 3.4–4.4 mm

Description The prosoma is tan to pale gray brown. A dark band is between the posterior lateral eyes. At times, the band is confined to just the middle

female

female

female

male

male

male

and does not reach the eyes. A dark marking is on the midline at the posterior. The anterior median eyes are surrounded with gold or pale brown. The clypeus has a fringe of white or golden-brown hairs. The opisthosoma is variegated with browns, tans, grays, white, and black. There is a dark W marking about halfway down. The legs are variegated in colors similar to the opisthosoma.

Habitat These spiders are often found in wooded habitats.

Known records from AZ, NM

Habronattus hallani (Richman, 1973)

Common Name The genus *Habronattus* has the common name of "paradise spiders".

Total Body Length Female 5.5–6.9 mm, Male 4.9–5.3 mm

Description The female prosoma is light brown. A dark band is between the posterior lateral eyes, and there is a dark marking along the midline at the posterior. The very edge has a dark line. The clypeus is pale yellow. The opisthosoma is gray brown, with a tan anterior band. The legs are dark brown, with tan and gray hairs. The male prosoma is black, with iridescent scales. Pale markings are on either side of the midline toward the posterior. The clypeus is iridescent

female

male

male

male

male

white, with golden hairs over the chelicerae. The opisthosoma is black, with a white anterior band. There is a tan heartmark, followed by an off-white median stripe. The lateral edges have tan bands. The legs are black. Leg I has pale stripes on the prolateral and retrolateral surfaces. The femur and patella of leg I are adorned with iridescent scales. The male pedipalps are clothed in golden hairs.

Habitat These spiders have been found in grassy habitats and leaf litter.

Known records from AZ, CA, TX

Habronattus oregonensis (Peckham & Peckham, 1888)

Common Name The genus *Habronattus* has the common name of "paradise spiders".

Other Colloquial Names Oregon paradise spider

Total Body Length Female 4.8–7.6 mm, Male 4.2–6.0 mm

Description The female prosoma is mottled with tan and brown. A dark marking is on the posterior, and dark bands are on the sides. The clypeus is off-white to tan. The opisthosoma is brown, with a tan anterior band. A faint light brown marking is toward the posterior, and a pair of white spots are above the spinnerets. Faint tan transverse bands on the lateral surfaces do not meet in the middle. The legs are brown, with pale hairs. The male prosoma is dark brown. There is sometimes a hint of rust-colored bands near the posterior eyes. White to pale gray hairs are above the anterior

female

female

female

male

male

median eyes. The clypeus is brown, with iridescent scales. The chelicerae are coved in rust-colored hairs. The opisthosoma is dark brown, sometimes with a tan to rust-colored anterior band that reaches all the way around to the spinnerets and a tan to rust-colored median stripe. Leg I has swollen tibiae that are lined with a brush of dark hairs. Legs I and II have dark brown femurs, and the distal segments are dark brown to black, with light-colored hairs. Legs III and IV are brown, with light-colored hairs. The male pedipalps are covered in long dark rust-colored hairs.

Habitat These spiders are usually found in open habitats.

Known records from AB, BC (Canada); AZ, CA, MT, NV, OR, UT, WA (United States)

Habronattus pugillis Griswold, 1987

Common Name The genus *Habronattus* has the common name of "paradise spiders".

Total Body Length Female 4.6–5.6 mm, Male 4.6–5.4 mm

Description The female prosoma is mottled with light brown. There is a thin dark line around the edge. The clypeus is cream to tan. The opisthosoma is mottled with light brown to tan. A pair of dark markings that can be undulating or straight are on either side of the midline. The legs are brown, with light-colored hairs. The male prosoma is dark golden brown. There is often a paler area behind the posterior eyes. Scattered white hairs are in line with the posterior eyes. Sometimes a pair of white lines go in a curve from the anterior median eyes under the anterior lateral eyes. The clypeus can be completely white, or it can be brown above and white below. There are iridescent scales just above the chelicerae. The chelicerae are covered in white hairs. The opisthosoma is mottled with tan brown. The legs are dark brown, with light hairs. The male pedipalps are dark brown with light hairs and sometimes golden brown with golden hairs.

female

female

female

male

male male

male male

Habitat These spiders are usually found in montane habitats at elevations between 4,000 ft and 8,500 ft. They tend to be found in leaf litter or other ground debris in oak or pine woodlands.

Known records from AZ

Habronattus pyrrithrix (Chamberlin, 1924)

Common Name The genus *Habronattus* has the common name of "paradise spiders".

Total Body Length Female 5.0–8.0 mm, Male 4.0–5.0 mm

Description The female prosoma is dark gray and has an iridescent sheen. There is a pale line around the edge. Scattered light hairs are above the anterior eyes. The clypeus is white.

female

female

male

male

male

male

male

male

male

The opisthosoma is dark gray to copper colored, and a pale tan anterior band continues down the sides to the spinnerets. The median stripe is a series of tan chevrons and diamonds. The legs are dark brown, with light-colored hairs. The male prosoma is dark brown. A pale line runs through the eye row and back to the posterior of the prosoma. The eye region can be dark gray or pale tan. The clypeus can be entirely red or black in the middle with red at the sides. The opisthosoma is dark brown, and a tan anterior band continues on the sides to the spinnerets. There is pale tan median stripe that is made up of dots, diamonds, and/or chevrons. The first pair of legs have green femurs and tibiae with long white hairs, and the metatarsi and tarsi are dark. There are dark longitudinal lines on the proventral surface. Legs II and IV are tan. Leg III has green tibiae and modified patellae with white spots. The male pedipalps are cream colored. Superficially, the males look similar to *Habronattus coecatus*, but the prosoma and opisthosoma markings are different, and the ranges of the two species do not overlap.

Habitat These spiders can be found in agricultural fields, leaf litter, and lawns.

Known records from AZ, CA

Habronattus ustulatus (Griswold, 1979)

Common Name The genus *Habronattus* has the common name of "paradise spiders".

Total Body Length Female 4.2–4.8 mm, Male 3.4–3.9 mm

Description The female prosoma is light brown to tan, with a white clypeus. The opisthosoma is light brown to tan. The legs are light brown to tan, with bands at the proximal and distal ends of each segment. The male prosoma is golden to dark brown and lighter on the sides. The clypeus is reddish to yellow brown. Reddish hairs are sometimes above the eyes. The opisthosoma is tan to reddish brown and may have some chevron markings. A fringe of black hairs is on the ventral surface of leg I, and in some cases the entire femur, patella, and tibia are black. The other legs are usually yellow to brown, with some heavy banding. The appearance of this species varies quite a bit based on the environment of the habitat it is living in.

Habitat These spiders are often found on bare ground, and matching the environment allows them to camouflage well.

Known records from AZ, CA, NM, NV, OR

female with prey

male

Hentzia mitrata (Hentz, 1846)

Common Name This species does not have a common name.

Other Colloquial Names White-jawed jumping spider

Total Body Length Female 2.9–4.3 mm, Male 3.5–4.1 mm

Description The female prosoma is yellow with white hairs, giving it a mottled appearance. There is a thin dark line around the edge. Two black hairs stick out laterally from beside the anterior lateral eyes on each side. The opisthosoma is pale yellow to reddish

female

female

female

male

male

male ventral

subadult male

brown, with white hairs. Indistinct transverse bands cross the opisthosoma, sometimes giving the appearance of tan rectangular shapes, the first of which is divided in two. All the legs are pale yellow; the first pair is most robust. The male prosoma is golden on top and white around the edges. The chelicerae are covered with white hairs. The opisthosoma has a similar color pattern: golden on top and white around the edges. All the legs are white, and the first pair is significantly longer than the others. The female looks very similar to *Hentzia palmarum*. The first pair of legs on female *H. mitrata* tends to be paler, but this may not be a consistent trait. The best way to confirm an identification is to look at the genitalia with a microscope. The males are not easily confused.

Habitat These spiders are often found on understory vegetation of forested habitats.

Known records from ON, QC (Canada); AL, AR, CT, DC, FL, GA, IA, IL, IN, KS, LA, MA, MD, ME, MI, MN, MO, MS, NC, NJ, NY, OH, PA, SC, TX, VA, WI, WV (United States)

Hentzia palmarum (Hentz, 1832)

Common Name This species does not have a common name.

Other Colloquial Names Common Hentz jumping spider

Total Body Length Female 4.7–6.1 mm, Male 4.0–5.3 mm

Description The female prosoma is yellow with white hairs, giving it a mottled appearance. There is a thin dark line around the edge. Two black hairs stick out laterally from beside the anterior lateral eyes on each side. The opisthosoma is pale yellow to reddish brown, with white hairs. Indistinct transverse bands cross the opisthosoma, sometimes giving the appearance of tan rectangular shapes, the first of which is divided in two. All the legs are pale yellow; the first pair is most robust and usually darker. The male prosoma is dark brown and has an iridescent sheen. A white band around the lateral edges continues across the clypeus. The chelicerae are elongated and have a row of long white hairs. The male opisthosoma is dark brown, with an iridescent sheen and a white band down the lateral edges. The first pair of legs is elongated and dark. The other legs are yellow. The female looks very similar to *Hentzia mitrata*. The first pair of legs on

female ventral

female

female

male

male

male

male on penny for scale

female *H. mitrata* tends to be paler, but this may not be a consistent trait. The best way to confirm an identification is to look at the genitalia with a microscope. The males are not easily confused.

Habitat   These spiders tend to be found near the ground in herbaceous vegetation.

Known records from   MB, NB, NS, ON, QC (Canada); AL, AR, CO, CT, DC, DE, FL, GA, IA, IL, IN, KS, KY, LA, MA, MD, MI, MN, MO, MS, MT, NC, NE, NJ, NY, OH, OK, PA, SC, TX, VA, WI (United States)

Lyssomanes viridis (Walckenaer, 1837)

Common Name   Magnolia green jumper

Total Body Length   Female 7.0–8.0 mm, Male 5.0–6.0 mm

Description   The prosoma is vivid green. There is usually white and/or orange to red around the eyes, especially in the males. The opisthosoma is vivid green, often with four pairs of black dots on either side of the midline.

male

male

male

(In the males, the dots are usually connected by a pair of dark longitudinal lines.) The legs are yellow to green, and the males have conspicuous black bands at the patellae, the distal end of the tibiae, the distal end of the metatarsi, and the tarsi. The male chelicerae are elongated and orange red.

Habitat These spiders are often found in wooded areas on herbaceous vegetation.

Known records from AL, AR, FL, GA, LA, NC, SC, TX, VA

male

juvenile

juvenile

Maevia inclemens (Walckenaer, 1837)

Common Name Dimorphic jumper

Total Body Length Female 7.0–10.0 mm, Male 5.3–7.0 mm

Description The female opisthosoma is tan to pale orange. The eyes are ringed in gold. The anterior is usually darker, especially between the eyes. A white band crosses the clypeus. The opisthosoma is tan to pale yellow. There is usually a pair of orange-red longitudinal stripes with a series of chevron shapes between them. The legs are pale yellow to pale orange and unbanded. The males come in two different morphologies or color forms. The tufted male has a black prosoma, with three erect tufts of hairs. The opisthosoma is black. The legs are pale yellow. The pedipalps are black but lined with yellow hairs. The gray male has a black and white variegated prosoma and opisthosoma. A series of chevron markings often runs down the opisthosoma;

female

female

female

female

gray male

gray male

gray male ventral

gray male

tufted male

in some individuals, the markings are reddish. The legs are banded in black and white. The pedipalps are covered with yellow hairs. The mating performances of the males are as different as their appearances. The tufted males wave their front legs high while standing on their tiptoes, whereas the gray males do a close-to-the-ground shuffle display. Regardless of which male successfully mates with the female, her male offspring will have both color forms.

Habitat These spiders are often found in trees, shrubs, and herbaceous vegetation. They are also often found on and around human structures.

Known records from MB, ON, QC (Canada); AL, AR, CT, FL, GA, IA, IL, IN, KS, KY, LA, MA, MD, ME, MI, MN, MO, NE, NJ, NY, OK, PA, SC, TN, TX, VA, WI, WV (United States)

tufted male

tufted male

Marpissa formosa (Banks, 1892)

Common Name This species does not have a common name.

Other Colloquial Names Short-bellied slender jumping spider

Total Body Length Female 6.5–9.0 mm, Male 5.2–6.0 mm

Description The female prosoma is dark brown and clothed in light-colored hairs. There is a thin pale line around the rim. The chelicerae are clothed in long white hairs. There is a brown band around the anterior eyes. The opisthosoma is cream white, with a pair of broad reddish-brown longitudinal stripes. The legs are pale yellow, with faint dusky bands. The first pair of legs is robust in comparison to the others. The male prosoma is jet-black, with a thin white line

female

female

around the edge. Two rows of white spots are on the dorsal surface; these are made of specialized hairs and are sometimes rubbed off. The opisthosoma is jet-black, with a white anterior band and a series of paired white spots down the dorsal surface. The legs are black, with some yellow banding. The first pair of legs is robust and completely black.

Habitat These spiders are often found in herbaceous vegetation and understory shrubs and seem to like humid environments.

Known records from ON, QC (Canada); AL, AR, IA, IL, IN, KS, LA, ME, MN, MS, NC, NE, NJ, NY, OH, OK, PA, TN, TX (United States)

female ventral

juvenile

female

male

male ventral

male

male

Marpissa lineata (C. L. Koch, 1846)

Common Name This species does not have a common name.

Other Colloquial Names Four-lined slender jumping spider

Total Body Length Female 4.0–5.3 mm, Male 3.0–4.0 mm

Description The prosoma is light brown, with a brown submarginal band, and it is darker in the eye region. A dark median stripe on the posterior half forks at the fovea into a large U shape. The opisthosoma is covered with a series of longitudinal stripes. The midline stripe is reddish brown. On each side is a white stripe, a dark brown stripe, and then another white stripe. The legs are yellow, with brown banding. The male tibiae I are black, with a bristle of hairs.

female

Habitat These spiders are usually found in the leaf litter of forested areas and prairies.

Known records from ON (Canada); CO, GA, IA, IL, IN, KS, MA, MI, MN, MO, MS, NJ, NM, NY, OH, PA, SD, TN, TX, WI (United States)

female

female ventral

male

male

male

male ventral

male

Marpissa pikei (Peckham & Peckham, 1888)

Common Name Pike slender jumper

Total Body Length Female 6.5–9.5 mm, Male 6.0–8.2 mm

Description The female prosoma and opisthosoma are tan to white, and both segments are elongated. The legs are yellow, and the first pair is robust and somewhat darker. The male prosoma and opisthosoma are dark brown, with white bands on the lateral edges, and both segments are elongated. The legs are yellow, and the first pair is robust, elongated, and darker.

female

female ventral

female

male

male

male

Habitat These spiders are often found in grassy habitats, prairies, and fields.

Known records from ON (Canada); AL, AR, AZ, CO, CT, FL, GA, IA, IL, IN, KS, LA, MA, MI, MN, MO, MS, NJ, NM, NY, OH, OK, SC, TX, VA, WI (United States)

Menemerus bivittatus (Dufour, 1831)

Common Name Gray wall jumper

Other Colloquial Names Gray wall jumping spider

Total Body Length Female 7.0–10.2 mm, Male 5.5–7.3 mm

Description The prosoma is dark brown, almost black. A broad cream-white median stripe widens in the eye region. There is a thin cream line around the edge. The clypeus and the area between the anterior eyes are orange to light brown. The opisthosoma is mottled with cream on the dorsal surface. A pair of dark brown, almost black, stripes on the lateral edges

515

female

female

female

male

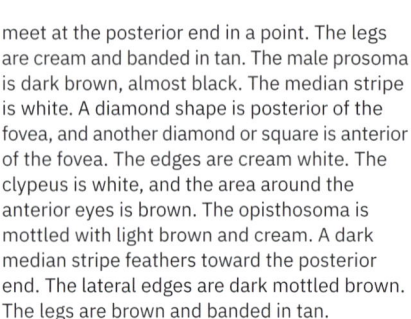

male

meet at the posterior end in a point. The legs are cream and banded in tan. The male prosoma is dark brown, almost black. The median stripe is white. A diamond shape is posterior of the fovea, and another diamond or square is anterior of the fovea. The edges are cream white. The clypeus is white, and the area around the anterior eyes is brown. The opisthosoma is mottled with light brown and cream. A dark median stripe feathers toward the posterior end. The lateral edges are dark mottled brown. The legs are brown and banded in tan.

Habitat In North America, this species seems to be associated only with humans and is often found on walls or other human structures.

Known records from CA, FL, LA, MN, NC, TX, WV
This is an introduced species.

Messua limbata (Banks, 1898)

Common Name This species does not have a common name.

Total Body Length Female 5.0–5.5 mm, Male 4.5–5.0 mm

Description The prosoma is black and iridescent. In the males, a white band runs from the anterior lateral eyes to the posterior of the prosoma. The males have elongated chelicerae. The opisthosoma is black and covered with iridescent scales. There is a white to yellow anterior band.

female

female

Habitat There is not much published about the habitat preferences of these spiders, but they have been seen and collected in herbaceous vegetation.

Known records from AZ, TX

male

male

Metacyrba floridana Gertsch, 1934

Common Name This species does not have a common name.

Total Body Length Female 5.1–6.6 mm, Male 4.5–4.6 mm

Description The prosoma is black and somewhat flattened. The edge is lined in a thin white band. The clypeus of the male is covered in yellow to white hairs. The opisthosoma is black and somewhat narrow and elongated, and a series of yellow markings that may or may not be contiguous run down each lateral edge.

female

male

A white dot is on the dorsal side of the spinnerets. The first pair of legs is robust and darker than the others, almost black. The other legs are light brown.

Habitat These spiders are usually found in dry habitats, often under bark.

Known records from AZ, FL, GA, LA, MS, TX

Metacyrba taeniola (Hentz, 1846) *and*
Metacyrba taeniola similis Banks, 1904

Common Name This species does not have a common name.

Other Colloquial Names Ribbon jumping spider

Total Body Length Female 5.0–7.2 mm, Male 4.4–6.0 mm

Description The prosoma is dark brown, almost black. There is a thin, sometimes broken, white line around the edge. The male clypeus has white hairs. The opisthosoma is reddish brown to dark brown, almost black. A pair of thin, often broken, yellow to white lines run on either side of the midline. The legs are reddish brown, and the femurs and leg I are usually the darkest. In some specimens, femur IV is more red than brown. There are scattered light hairs on the femurs. Many online resources also list *Metacyrba taeniola taeniola* as a subspecies, but this is not recognized by the World Spider Catalog.

Habitat These spiders are often found in dry habitats under rocks or in deciduous forests under bark.

male

male

male

male

male
ventral

(*similis*) female

(*similis*) female

(*similis*) male

(*similis*) male

Known records of *Metacyrba taeniola* from AL, AR, AZ, CA, CO, DC, DE, FL, GA, KS, LA, MD, MS, NC, NM, NV, OK, PA, SC, TN, TX, UT, VA; known records of *Metacyrba taeniola similis* from AZ, CA, CO, NV, OK, TX, UT

Metaphidippus chera (Chamberlin, 1924)

Common Name This species does not have a common name.

Total Body Length Female 3.4–5.0 mm, Male 3.4–4.8 mm

Description The female prosoma is cream white and mottled with hints of pale brown. The opisthosoma is light golden brown, with a white median stripe. Four to five angled white lines sometimes extend from the median stripe. The legs are cream, and some have light brown banding. The male prosoma is white, with a golden-brown median stripe that widens at the posterior eye row. A white spot is in the median stripe between the posterior median eyes. Bright white lines run down the chelicerae.

519

The opisthosoma is white, with a golden-brown median stripe. There are often dark spots at the edges of the median stripe. The legs are white and banded in light brown.

Habitat These spiders are often found on desert vegetation and have been found at elevations up to 7,000 ft. It has been noted that this is one of the most commonly encountered Salticidae species in the southwestern United States.

Known records from AZ, CA, NM, NV, OK, TX, UT

Myrmarachne formicaria (De Geer, 1778)

Common Name This species does not have a common name.

Total Body Length Female 5.0–6.0 mm, Male 5.5–6.5 mm

Description The prosoma is reddish brown and darker and elevated in the eye region. The males have extremely elongated chelicerae that have an iridescent sheen. (The length seems to vary quite a bit.) The opisthosoma is dark brown to golden brown, sometimes lighter anteriorly. There may be a pale chevron-shaped marking about halfway down. The legs are pale orange; the metatarsi and prolateral surface of the tibiae and patellae of leg I are black; and the tarsi are extremely pale. These spiders often hold the first pair of legs up like antennae.

female
ventral

female

male

male

male
ventral

male

juvenile

Habitat In North America, this species seems to be associated with human structures.

Known records from ON (Canada); NY, OH, PA (United States) This is an introduced species.

Naphrys pulex (Hentz, 1846)

Common Name This species does not have a common name.

Other Colloquial Names Flea jumping spider

Total Body Length Female 4.5–5.5 mm, Male 4.0–5.5 mm

Description The prosoma is dark brown, with a pale median stripe that is almost triangular and points posteriorly. In the females, the widest area is

female

male

in the eye region; in the males, the widest area is just posterior of the eye region, as the eye region is dark. The male has a high, pale yellow clypeus and faint dusky longitudinal markings on the chelicerae. The opisthosoma of both the males and the females is dark brown. There is a tan to white heartmark. A transverse tan to white band is just posterior of the heartmark, and it may have dark markings within it posteriorly. The lateral edges are also tan. The legs are banded. The contrast between colors is more dramatic on the males than the females. *N. pulex* can look superficially similar to *Attulus fasciger*, but the males can easily be distinguished by

male

male

male and female mating

male and female mating

the size of the pedipalps (much smaller in *N. pulex*) and the height of the clypeus (much higher in *N. pulex*). The females can be distinguished by the light-colored heartmark and overall markings.

Habitat These spiders are often found on tree trunks, grass, and the ground, and in and around human structures. This species is known to feed on ants.

Known records from MB, NL, NS, ON, QC (Canada); AL, AR, CT, DC, FL, GA, IA, IL, IN, KS, KY, LA, MA, MD, ME, MI, MN, MO, MS, NC, NE, NH, NJ, NY, OH, OK, PA, SC, TN, TX, VA, VT, WI, WV (United States)

Neon nelli Peckham & Peckham, 1888

Common Name This species does not have a common name.

Other Colloquial Names Nell's tiny jumping spider

Total Body Length Female 1.8–3.0 mm, Male 2.0–2.6 mm

female

female
ventral

male

male
ventral

Description The prosoma is dark brown, with a tan area on the posterior end. The opisthosoma is golden brown, with dark brown markings. The heartmark is dark brown; on either side is a stripe that may be broken, approximately the length of the heartmark. Posterior of the heartmark are transverse chevron-shaped bands. The legs are tan, with dusky bands. This species looks very similar to *Neon reticulatus*, which is found west of the Rockies (*N. nelli* is found east of the Rockies), and can be distinguished only by looking at the genitalia.

Habitat These spiders are often found near the ground in leaf litter or under rocks.

Known records from AB, BC, MB, NB, NL, NS, ON, PE, QC (Canada); CA, CT, FL, GA, IA, IL, MA, MD, ME, MI, MN, NC, NH, NJ, NM, NY, OH, OR, PA, SD, TN, TX, VA, WI (United States)

male near penny for scale

Paraphidippus aurantius (Lucas, 1833)

Common Name Emerald jumper

Other Colloquial Names Emerald jumping spider, golden jumping spider

Total Body Length Female 8.0–12.0 mm, Male 7.0–10.0 mm

Description The prosoma is dark reddish brown to black, usually darker on the males, with an iridescent sheen. White to pale orange bands are on either side (usually white on the males and variable on the females). The female chelicerae have long white hairs on the top portion; the male chelicerae are completely black.

The opisthosoma is reddish brown to black, with an iridescent sheen that often looks purple, golden, or green. A white to orange band (usually white on the males and variable on the females) crosses the anterior, and there are three paired transverse dashes on the lateral edges posteriorly. Three to four pairs of white spots are on either side of the midline. The female legs are pale brown, with broad dark bands on femurs I and II and tibiae I and IV. In the males, leg I is black, and the proximal segments of the other legs are dark, with paler distal segments.

female

female

female

female

female

juvenile

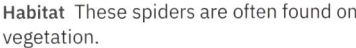

male

male

Habitat These spiders are often found on vegetation.

Known records from AZ, CO, DE, FL, IL, KS, MD, MO, MT, NC, NE, NM, OH, OK, SC, TN, TX, VA

Peckhamia americana
(Peckham & Peckham, 1892)

Common Name This species does not have a common name.

Total Body Length
Female 5.0 mm, Male 4.0 mm

Description The prosoma is dark golden brown. There are a few white hairs between the posterior lateral eyes. The eye region occupies approximately half the prosoma. The prosoma slopes steeply behind the posterior lateral eyes. The opisthosoma is dark golden brown and slightly lighter anteriorly. A pair of white dots are on either side less than halfway down and almost on the ventral surface, and there is a constriction (more pronounced in the males) in

female

male

female

the opisthosoma at this point. The legs are golden brown, with a somewhat darker line on the prolateral surfaces of the tibiae and patellae. These spiders often raise their second pair of legs in an antenna-like pose. Spiders in the genus *Peckhamia* look very similar to those in *Synageles*, and both use the second pair of legs to mimic antennae. The slope behind the posterior lateral eyes is steeper in *Peckhamia* species, and the constriction on the opisthosoma is usually less pronounced in *Synageles*. It is likely that the specimens from west of the Great Plains are a different species, but this has not been resolved at this time.

Habitat These spiders are often found on tree trunks, on understory vegetation, and on and around human structures.

Known records from AL, AR, AZ, CA, CO, FL, GA, IL, IN, KS, LA, MD, MO, NC, NH, NJ, NM, NY, OH, OK, PA, SC, TN, TX, VA

Pelegrina galathea (Walckenaer, 1837)

Common Name Peppered jumper

Other Colloquial Names Peppered jumping spider

Total Body Length Female 3.6–5.8 mm, Male 2.7–4.4 mm

Description The female prosoma is cream to pale yellow, with dark mottling throughout. The female opisthosoma is cream to pale yellow, with mottled dark markings in the general appearance of three pairs of spots on either side of the midline. The male prosoma is dark reddish brown. There are white submarginal bands and a white marking between the posterior median eyes. A black angled dash reaches from the outer edge of the anterior median eyes to the

female

female ventral

male

female

male

male

posterior lateral eyes. (I often hear people refer to these marks as "angry eyebrows.") The clypeus is dark reddish brown. The male opisthosoma is dark reddish brown, and a band crosses the anterior and continues down the lateral sides to the spinnerets; it is broken at the posterior end. Paired white spots may be on the dorsal surface, but these often get rubbed off and may be missing. The legs of both the males and the females are cream and banded in black. *P. galathea* looks similar to other *Pelegrina* species. The females can be distinguished by the pattern on the opisthosoma, while the males have a wider prosoma than the other species and do not have a white marking on the clypeus.

Habitat These spiders are usually found in herbaceous vegetation and can often be found in agricultural fields.

Known records from NS, ON (Canada); AL, AR, CO, CT, DC, DE, FL, GA, IL, IN, KS, KY, LA, MA, ME, MI, MO, MS, NC, NE, NH, NJ, NM, NY, OH, OK, PA, RI, SC, TN, TX, VA, VT, WV, WY (United States)

Pelegrina proterva (Walckenaer, 1837)

Common Name This species does not have a common name.

Other Colloquial Names Common white-cheeked jumping spider

Total Body Length Female 4.4–5.6 mm, Male 3.3–4.2 mm

Description The female prosoma is pale tan to cream colored, with a somewhat mottled appearance. The opisthosoma is pale tan to cream colored. There is a pale anterior band. Four pairs of dark spots are on either side of

female

female

527

female ventral

juvenile

male

male

male
ventral

the midline; the more posterior the spots are, the larger they are. The male prosoma is dark reddish brown. White submarginal bands and a white marking are between the posterior median eyes. A black angled dash reaches from the outer edge of the anterior median eyes to the posterior lateral eyes. (I often hear people refer to these marks as "angry eyebrows.") The clypeus has a central white marking, and there are white cheek markings. The male opisthosoma is dark reddish brown, and a band crosses the anterior and continues down the lateral sides to the spinnerets. The legs are tan and banded in dark brown. *P. proterva* looks similar to other *Pelegrina* species. The females can be distinguished by the pattern on the opisthosoma, while the males have a narrower prosoma than the other species and have a white marking on the clypeus.

Habitat These spiders are often found in forested habitats.

Known records from AB, BC, MB, NB, NS, ON, PE, QC, SK (Canada); AL, CO, CT, GA, IA, IL, IN, KS, KY, MA, MD, ME, MI, MN, MO, MT, NC, ND, NE, NH, NJ, NM, NY, OH, PA, RI, SC, SD, TX, VA, VT, WI, WV, WY (United States)

Phidippus apacheanus Chamberlin & Gertsch, 1929

Common Name This species does not have a common name.

Other Colloquial Names Apache jumping spider

Total Body Length Female 7.1–13.4 mm, Male 5.2–10.6 mm

Description The prosoma is yellow orange to red, with dark sides and a dark clypeus. The females have a fringe of white hairs at the base of the clypeus, and the pedipalps are covered in long white hairs. The males have a black clypeus and black pedipalps.

female with prey

female with prey

male

male

The chelicerae are metallic green to blue. Tufts of black hairs are above the posterior median eyes. The opisthosoma is yellow orange to red, with black on the lateral edges and ventral surface. A black median stripe may or may not be on the posterior half of the dorsal surface. The legs are black, with some light-colored bands. The males tend to be darker.

Habitat These spiders are often found in dry fields and grasslands and on shrubs and cacti.

Known records from AL, AR, AZ, CA, CO, DC, FL, GA, ID, KS, LA, MA, MN, MO, MS, MT, NC, NE, NM, NV, NY, OK, SC, TX, UT, WI, WY

Phidippus ardens Peckham & Peckham, 1901

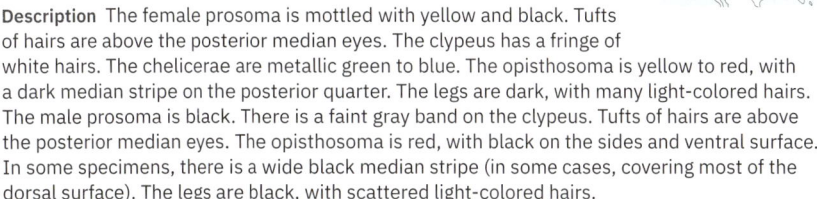

Common Name This species does not have a common name.

Other Colloquial Names Desert red jumping spider

Total Body Length Female 11.5–15.2 mm, Male 8.3–12.7 mm

Description The female prosoma is mottled with yellow and black. Tufts of hairs are above the posterior median eyes. The clypeus has a fringe of white hairs. The chelicerae are metallic green to blue. The opisthosoma is yellow to red, with a dark median stripe on the posterior quarter. The legs are dark, with many light-colored hairs. The male prosoma is black. There is a faint gray band on the clypeus. Tufts of hairs are above the posterior median eyes. The opisthosoma is red, with black on the sides and ventral surface. In some specimens, there is a wide black median stripe (in some cases, covering most of the dorsal surface). The legs are black, with scattered light-colored hairs.

Habitat These spiders are often found on shrubs or grasses and can be found at elevations up to 8,000 ft.

Known records from AZ, CA, CO, ID, KS, NM, NV, OK, TX, UT, WA, WY

female

female

male

male

Phidippus asotus
Chamberlin & Ivie, 1933

Common Name
This species does not
have a common name.

Total Body Length
Female 7.2–11.4 mm, Male 6.7–9.0 mm

Description The female prosoma is mottled with
dark gray and white. There is a white area around
the anterior median eyes (sometimes broken into
three spots), and white spots are usually below
the posterior median eyes. There is also usually a
white spot on the top of the prosoma between the
posterior eyes. Tufts of black hairs are above the
anterior lateral eyes. The clypeus has a fringe of
long white hairs. The chelicerae are metallic blue to
green. The female opisthosoma is variegated with
gray, white, black, and pink. Paired light-colored
oblique dashes are on either side of the midline.
The ventral surface is gray. The male prosoma is

female

male

male

male courting female

dark brown, almost black. There are pale red median stripes and three pale red dots above the anterior eyes. Tufts of black hairs are above the anterior lateral eyes. The clypeus is pale red to gray, and there are long cream to pale red hairs on the chelicerae. The chelicerae are metallic blue to green. The opisthosoma is marked like the female but darker. The legs of both males and females are dark brown, with rings of light-colored hairs.

Habitat These spiders have been found at elevations up to 9,200 ft. They are most often found on oak, juniper, and other shrubs.

Known records from AZ, CA, CO, NM, NV, OK, TX, UT

Phidippus audax (Hentz, 1845)

Common Name Bold jumper

Other Colloquial Names Bold jumping spider

Total Body Length Female 8.0–18.1 mm, Male 6.0–15.3 mm

Description The prosoma is black, and there may be a white marking around the posterior end and a white spot between the posterior lateral eyes. The chelicerae have a metallic green to blue coloration. The

female

female in retreat

female

female

female
ventral

juvenile

male

male

male

male

females have two tufts of hairs that sit erect above the anterior lateral eyes. The opisthosoma is black, with paired matt-black markings on the posterior half and iridescent scales on the anterior half. A band may be around the anterior end. In some individuals, broad white to light tan stripes are on the lateral edges, some with a tan median marking as well. A central marking, often heart shaped, on the dorsal surface can be white to orange. Some people think that the orange is an indicator of immaturity, but it has been documented that many specimens maintain the orange coloration through adulthood. One or two pairs of smaller white markings are on either side of the midline posteriorly. The legs are black, with white hairs that create some banding (in some females and immature specimens, the bands can be golden to brown) and, in the males, white tufts on the prolateral surface of femur I. The palps of the males have white markings on the dorsal femurs and patellae. This species, especially the males, looks similar to *Phidippus regius*. The two are distinguished by the presence of iridescent scales on the anterior opisthosoma, which *P. regius* lacks, and the matt-black posterior opisthosoma markings on *P. audax*.

Habitat These spiders are often found in wooded areas, fields, and prairies, and in and around human structures.

Known records from ON, QC, SK (Canada); AL, AR, AZ, CA, CO, CT, DC, DE, FL, GA, ID, IL, IN, KS, KY, LA, MA, MD, ME, MI, MN, MO, MS, NC, NE, NH, NJ, NM, NY, OH, OK, OR, PA, RI, SC, SD, TN, TX, UT, VA, WA, WI, WV, WY (United States)

Phidippus borealis
Banks, 1895

Common Name This species does not have a common name.

Other Colloquial Names
Boreal tufted jumping spider

Total Body Length
Female 8.5–13.8 mm, Male 4.9–9.1 mm

Description The prosoma is black. The females have white hairs on the lateral and posterior areas. Tufts of hairs are above the anterior lateral eyes. The clypeus is white in the females and black in the males. The chelicerae are metallic green. The opisthosoma is black, with a cream anterior band. A pair of broad cream to pale red bands run down the lateral edges. Small paired white spots are on either side of the midline. This species looks very similar to *Phidippus purpuratus* and is most easily distinguished by its genitalia.

female

female

Habitat These spiders are often found in wooded areas with aspen, pine, willow, or spruce trees. They have been found at elevations up to 9,500 ft.

Known records from AB, BC, MB, NL, NT, ON, QC, SK, YT (Canada); AK, CO, ID, IL, IN, ME, MI, MN, MT, ND, NH, NY, SD, UT, WA, WI, WY (United States)

Phidippus californicus Peckham & Peckham, 1901

Common Name This species does not have a common name.

Total Body Length Female 8.7–15.5 mm, Male 5.5–10.3 mm

Description The prosoma is black, with an iridescent sheen. There is a broad white to gray band around the edges. In the females, the white band continues through the anterior eyes, while in the males, it stops before the eyes. Tufts of black hairs are above the posterior median eyes. The females often have a white spot between the posterior eyes. The clypeus is white, with a fringe of white hairs on the females

and black hairs on the males. The chelicerae are metallic blue to green. The opisthosoma is black and has an iridescent sheen. A white anterior band continues down the edges but stops short of the spinnerets. On either side of the midline is a pair of broad red bands. There are paired white spots within the midline. (The first pair is usually fused.) The legs are dark brown to black; the males have white banding, while the females have yellow. The male pedipalps have white hairs on the femur, patella, and tibia. The cymbium can be black or white.

Habitat These spiders are often found on shrubby plants, especially those that flower.

Known records from AZ, CA, NM, NV, OR, TX, UT

Phidippus carneus Peckham & Peckham, 1896

Common Name This species does not have a common name.

Total Body Length Female 8.4–15.2 mm, Male 7.2–10.4 mm

Description The prosoma is black. The males often have a white band under the posterior eyes, and sometimes it runs through the anterior eyes. The clypeus is gray to tan. The females have a fringe of white hairs. The chelicerae are metallic blue to green. The opisthosoma is red, with a white anterior band. The females often have a black median stripe that widens posteriorly. Paired white spots are often on either side of the median stripe in the females. The legs are black, with white hairs. The males have tufts of white hairs on the femur, patella, and tibia I. The male pedipalps have white stripes on the femur and patella, with a black cymbium.

Habitat These spiders are often found on oaks, cacti, and shrubs. They have been found at elevations up to 7,000 ft.

Known records from AZ, CO, NM, TX

female

female

juvenile with prey

male

male

male

Phidippus carolinensis Peckham & Peckham, 1909

Common Name This species does not have a common name.

Other Colloquial Names Carolina jumping spider

Total Body Length Female 9.0–13.4 mm, Male 7.0–11.0 mm

Description The female prosoma is covered with pale yellow to orange hair.
There are three pairs of hair tufts: one pair points to the sides near the anterior lateral eyes,
another pair is erect above the anterior lateral eyes, and there is a pair in the middle. The clypeus
is white, and the chelicerae are covered
in pale hairs. The female opisthosoma
is covered in pale yellow to orange
hairs. A white band crosses the anterior,
and four pairs of white markings are on
either side of the midline. The male
prosoma is black, and the dorsal
surface is covered in white to yellow
hair. The area around the eyes is black.
Hair tufts project to the sides near the
anterior lateral eyes. The chelicerae are
gray at the base and then clothed in
golden hair. The male opisthosoma is
black, and the dorsal surface is covered
in pale yellow hairs. There is a white
anterior band, and two transverse white
bands are on the lateral edges on the

female

male

male

male

male
ventral

posterior half. A pale central marking is just visible. The legs in both the males and females are black, with yellow and white hairs.

Habitat These spiders are often found on coniferous trees.

Known records from CO, KS, NC, OK, TX

Phidippus clarus Keyserling, 1885

Common Name This species does not have a common name.

Other Colloquial Names Brilliant jumping spider

Total Body Length Female 4.3–14.2 mm, Male 3.3–10.1 mm

Description The female prosoma is covered with tan to golden hairs and sometimes has a red marking between the posterior lateral eyes. The clypeus is white to pale yellow. The chelicerae are metallic green. The female opisthosoma is pale golden to pale red. There is a white to cream anterior band. The heartmark is pale yellow to pale red. A black median stripe starts about one-quarter of the way down. Four pairs of white markings are within the black stripe. A transverse cream to white line about halfway down on the lateral edges points toward the anterior and heads toward the midline but does not cross the black lines. The legs are yellow and banded in black. The male prosoma is black. The chelicerae are metallic green. The male opisthosoma is black, with a white anterior band. A pair of brilliant red stripes are on either side of the midline. A transverse white band points toward the anterior and the midline but does not cross the red marking completely. The legs are black, with white and yellow bands. The femur, patellae, and cymbium of the pedipalps have white hairs on the dorsal surface.

Habitat These spiders are often found in open areas, prairies, old fields, and grassy habitats. They can also be found in suburban gardens.

female

female

female

juvenile

male

male

male

male
ventral

Known records from ON, QC, SK (Canada); AL, AR, AZ,
CA, CO, CT, DC, FL, GA, IA, ID, IL, IN, KS, KY, LA, MA,
MD, ME, MI, MN, MO, MS, MT, NC, ND, NE, NH, NJ, NM,
NY, OH, OK, OR, PA, RI, SC, SD, TN, TX, UT, VA, VT, WA,
WI, WV, WY (United States)

Phidippus comatus Peckham & Peckham, 1901

Common Name This species does not have a common name.

Other Colloquial Names Hairy tufted jumping spider

Total Body Length Female 7.1–10.2 mm, Male 5.7–8.0 mm

Description The female prosoma is reddish brown, with a pale median
stripe that ends before the anterior eyes. A white mottled band on the edges runs through the
anterior eye area. The clypeus is white, and a fringe of long white hairs covers the metallic blue
to green chelicerae. Two pairs of tufts of black hairs are near the posterior median and anterior
lateral eyes. The opisthosoma is dark brown. A pale anterior band continues down the edges
to the spinnerets. The heartmark is pale to reddish brown and is immediately followed by a
cream-colored marking. Two pairs of white transverse bands that do not meet in the middle are
toward the posterior. The ventral opisthosoma is black, with a white stripe on either side. The legs

are dark brown, with tan bands. The male prosoma is dark brown, and a pair of white triangles are between the posterior eyes. Tufts of black hairs are near the posterior median eyes. A white band runs from the posterior lateral eyes along the sides to the posterior. The clypeus is reddish brown. The chelicerae are covered with long white and dark hairs (giving the impression of stripes). The chelicerae are metallic blue to green. The male opisthosoma is marked like the female opisthosoma. The legs are dark brown, with yellow bands and stripes. Femur I is enlarged and black, with a metallic blue sheen and gray tufts. The male pedipalps are banded in yellow and white.

Habitat These spiders are often found in the understory of wooded habitats, including apple and pear orchards. They have been found at elevations up to 8,200 ft.

Known records from SK (Canada); AZ, CA, CO, NM, NV, OR, TX, UT, WA, WY (United States)

Phidippus johnsoni (Peckham & Peckham, 1883)

Common Name Johnson jumper

Other Colloquial Names Johnson's jumping spider

Total Body Length Female 9.0–14.2 mm, Male 6.2–10.7 mm

female

male

male

male courting female

female and male

Description The prosoma is black. The females have scattered yellow to white hairs. The female clypeus is white, and the male clypeus is black. The chelicerae are metallic green. The opisthosoma is often completely orange to red, or it has some dark markings along the midline. The females often have a white anterior band and white markings. The legs are black, with white and yellow hairs; the hairs are more scarce on the males. The males look similar to *Phidippus princeps*, which is an eastern species, but the ranges do not overlap.

Habitat These spiders can be found in habitats ranging from beaches to forests and in and around human structures. They have been found at elevations up to 12,000 ft.

Known records from AB, BC, NT, SK (Canada); AZ, CA, CO, ID, MT, NM, NV, OR, SD, UT, WA, WY (United States)

Phidippus mystaceus (Hentz, 1846)

Common Name This species does not have a common name.

Other Colloquial Names High eyelashed jumping spider

Total Body Length Female 6.8–13.3 mm, Male 4.9–8.6 mm

Description The female prosoma is gray. A set of three white markings is outlined in black in the eye region; they resemble a hair bow on the top of her head. The clypeus is white to very pale yellow. Two pairs of hair tufts are near the anterior lateral eyes: the first points nearly horizontally, while the other points upward at about a 45° angle. The chelicerae are metallic green. The female opisthosoma is gray. A white to cream anterior band that is outlined in black often continues most of the way down the lateral edges. Two transverse bands run from the lateral band toward the midline but do not meet. The dorsal opisthosoma has paired white to cream spots outlined in black. The legs are black, with white to pale yellow hairs creating a banded look. The male prosoma is very dark gray, almost black, and lighter gray in the eye region. A red band between the posterior lateral eyes is often broken; in some cases, there is just a pair of red dots. The clypeus is white, and the chelicerae are clothed in long pale gray hairs. Two pairs of hair tufts are near the anterior lateral eyes: the first points nearly horizontally, while the other points upward at about a 45° angle. The male opisthosoma is rust colored and mottled. There is a cream anterior band and a faint impression of paired pale spots down the dorsal surface. The legs are black and have orange to yellow hairs that create a banded and striped look. The male pedipalps are covered with pale hairs. The cymbium of the male palp has a white and yellow dot on the dorsal surface.

female

juvenile

subadult male

male

male

541

male ventral

male

Habitat These spiders are most often found in wooded habitats.

Known records from AL, AR, CO, CT, DC, FL, GA, IL, KS, KY, MD, MO, NC, NJ, NY, OH, OK, SC, TN, TX, VA

Phidippus octopunctatus (Peckham & Peckham, 1883)

Common Name This species does not have a common name.

Total Body Length Female 10.2–18.9 mm, Male 9.4–13.2 mm

Description The female prosoma and opisthosoma are mottled with gray tan. Tufts of hairs are near the posterior median eyes. The clypeus is cream to tan. There are long cream-colored hairs over the chelicerae. The legs are brown, with bands of tan hairs. The male prosoma and opisthosoma are blue gray. Tufts of hairs are near the posterior median eyes. The clypeus has a fringe of white hairs. The legs are black, with a few scattered longer white hairs.

Habitat These spiders are often found in grasslands, fields, and prairies.

Known records from AZ, CA, CO, IA, ID, KS, MO, MT, NE, NM, NV, SD, TX, UT, WA

female

female

male

male male

male

Phidippus otiosus (Hentz, 1846)

Common Name This species does not have a common name.

Other Colloquial Names Canopy jumping spider

Total Body Length Female 7.6–17.1 mm, Male 6.3–13.7 mm

Description The female prosoma is black, with pale cream hairs. The area between the posterior eyes is usually black and somewhat iridescent. There may be a white line, or a broken white line, between the posterior lateral eyes (and in some cases, just a dot). The clypeus is white. The chelicerae are metallic green. Two pairs of thin hair tufts are near the

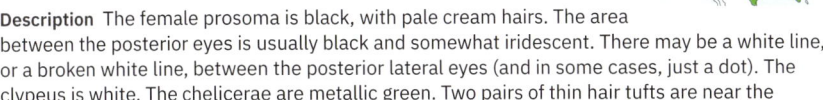

female

female

female

female
ventral

juvenile

male

male

male

egg sac
with young

anterior lateral eyes: the first pair stands erect, while the other pair projects horizontally. The female opisthosoma is black, with an iridescent sheen. A cream to pale tan anterior band breaks into spots on the lateral edges. Tan to cream hairs create a series of stripes around the dark median stripe. There are paired cream dashes in the median stripe. The legs are dark brown, with white to cream hairs. The male prosoma is black and slightly iridescent. There is a white to cream band around the posterior. The clypeus is black. The chelicerae are metallic green. Two pairs of thin hair tufts are near the anterior lateral eyes: the first pair stands erect, while the other pair projects horizontally. The male opisthosoma is black and has an iridescent sheen. There is a cream anterior band. Paired white dashes are on the dorsal opisthosoma. The legs are black, with white hairs. In Florida, specimens of males and females are yellow to orange, rather than white to cream, as seen in other locales. This may be due to introgression with *Phidippus regius*. Hybridization between these two species has been documented.

Habitat This is an arboreal species, and spiders are often found in the canopies of trees.

Known records from AL, AR, AZ, CO, DC, FL, GA, KY, LA, MD, MO, MS, NC, OH, OK, SC, TN, TX, VA

Phidippus phoenix Edwards, 2004

Common Name This species does not have a common name.

Other Colloquial Names Phoenix jumping spider

Total Body Length Female 7.7–12.2 mm, Male 7.4–9.2 mm

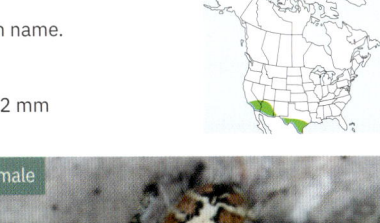

Description The prosoma is yellow to cream and darker from the posterior lateral eyes to the posterior. Some males have a black prosoma. There may be a dark line above the anterior eyes. Tufts of hairs are above the anterior median eyes and laterally out from the anterior lateral eyes. The clypeus is white, with a fringe of long white to yellow hairs over the metallic green chelicerae. The opisthosoma is brick red, with a white anterior band. There is a white median stripe, bordered in black, that may be in the form of spots. Three pairs of transverse white bands do not meet in the middle. Some males have a bright red opisthosoma with no other markings. The legs are brown, with copious white and tan hairs. The males have white fringes on leg I.

female

female

male

male

male

Habitat Juveniles are often found on shrubby vegetation, whereas the adults tend to be found on or near the ground.

Known records from AZ, CA, TX

Phidippus princeps (Peckham & Peckham, 1883)

Common Name This species does not have a common name.

Other Colloquial Names Grayish jumping spider

Total Body Length Female 6.0–11.5 mm, Male 4.4–8.6 mm

Description The female prosoma and opisthosoma are covered in pale gray hairs. The prosoma has a thin white line around the edge. The chelicerae are metallic green. Hair tufts are above the anterior lateral eyes. An indistinct anterior band and scattered longer white hairs are on the opisthosoma. The legs are banded in brown and black and have copious gray hairs. The male prosoma is black. The chelicerae are metallic blue to green. The opisthosoma is covered in red to tan hairs, occasionally with a black midline. The legs are black, with white tufts on the prolateral femur I and the dorsal pedipalp femur and patella. The males look similar to *Phidippus johnsoni*, but note that *P. princeps* is an eastern species, and the ranges do not overlap.

Habitat These spiders are found in forested habitats and old fields.

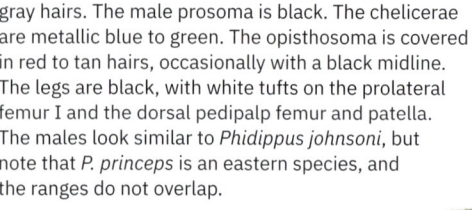

female

female

female

female
ventral

juvenile

male

male

male

Known records from MB, NS, ON, QC,
SK (Canada); AL, AR, CT, GA, IA, IL, IN,
KS, KY, LA, MA, MD, ME, MI, MN, MO, MS,
NC, NE, NH, NJ, NY, OH, OK, A, RI, SC, TN,
TX, UT, VA, VT, WI, WV (United States)

Phidippus putnami (Peckham & Peckham, 1883)

Common Name This species does not have a common name.

Other Colloquial Names Putnam's jumping spider

Total Body Length
Female 8.5–11.2 mm, Male 7.3–9.0 mm

Description The female prosoma is black,
with a slight iridescent sheen. There is
a pale triangle that may be indistinct.
Its base is oriented posteriorly, and the
point is between the posterior lateral
eyes. The clypeus has a thin white to pale
yellow band. The chelicerae are metallic
green to blue. There are three pairs of
hair tufts: the first thin pair is
horizontal, near the anterior

female

female

female

547

female ventral

male

male

male

lateral eyes; the second pair is erect, near the anterior lateral eyes; and the third pair is erect, above the anterior median eyes. The female opisthosoma is black and has an iridescent sheen. There is a white to cream anterior band. Two to three pairs of transverse lateral bands are confined to the lateral edges. There are paired spots and a central fused spot in white to cream. Tan to pale yellow hairs are on the opisthosoma between the dark median stripe and the lateral markings. The legs are dark brown, with cream to yellow hairs creating stripes and bands. The female pedipalps have long cream hairs. The male prosoma is black, with a white triangle. Its base is oriented posteriorly, and the point is between the posterior lateral eyes. White spots are between the posterior lateral and posterior median eyes on each side of the prosoma. Just below the anterior lateral eyes is a blue-gray area. The clypeus is reddish, with pale gray stripes. The base of the chelicerae is covered with white to cream hairs with longitudinal brown stripes, giving the appearance almost of human teeth. The tips of the chelicerae have blue-gray hairs. The male opisthosoma is black and has an iridescent sheen. A white anterior band continues down the lateral edges to the spinnerets. There are paired white spots and a fused white central spot. The legs are black, with cream to yellow hairs creating stripes and bands. The male pedipalp femurs have white hairs, and the cymbium is covered with blue-gray hairs.

Habitat These spiders are often found in forested habitats.

Known records from AL, AR, AZ, DC, GA, IA, ID, IL, IN, KS, KY, LA, MD, MO, MS, NC, NE, NJ, OH, OK, PA, SC, TN, TX, UT, VA, WV

Phidippus regius C. L. Koch, 1846

Common Name Regal jumper

Other Colloquial Names Regal jumping spider

Total Body Length Female 14.0–23.0 mm, Male 6.0–18.0 mm

Description The female prosoma is black, has a white to orange median spot, and is white to orange around all the edges. The females have two

tufts of hairs that sit erect above the anterior lateral eyes. The chelicerae are metallic green to red violet. The female opisthosoma is black but heavily marked in white to orange. There is a central triangular white to pale yellow mark. The male prosoma is black, rarely with a white marking around the posterior end. A white spot may be between the posterior lateral eyes. The chelicerae have a metallic blue-green-violet coloration. The male opisthosoma is black, with a white central mark and one or two pairs of smaller white markings on either side of the midline posteriorly. The legs are black, with white hairs that create some banding and, in the males, white tufts on the prolateral surface of femur I. The palps of the males have white markings on the dorsal femurs. This species, especially the males, looks similar to *Phidippus audax*. You can distinguish the two by the lack of iridescent scales on the anterior opisthosoma, which *P. audax* has, and the lack of matt-black markings on the posterior opisthosoma, which *P. audax* has.

male and female mating

male and female mating

Habitat These spiders are often found in old fields. This species is becoming common in the pet trade and may be seen outside its native range when pets escape or are released.

Known records from AL, FL, GA, MS, NC, SC, VA

Phidippus tigris Edwards, 2004

Common Name This species does not have a common name.

Total Body Length Female 8.5–10.7 mm, Male 6.2–9.0 mm

Description The prosoma is dark brown, almost black. There are white to mottled cream dashes behind the posterior lateral eyes. Three stripes on the anterior start near the posterior lateral eyes: the middle stripe ends at the anterior median eyes, while the other two continue under the anterior lateral eyes and then along the edge of the prosoma in a posterior direction for a short distance. The face pattern looks almost like tiger stripes; this is most contrasting in the males. The metallic green chelicerae are covered with golden hairs in the males and cream hairs in the females. The opisthosoma is dark brown, with a somewhat jagged, mottled, tan anterior band that continues down the lateral edges to just above the spinnerets. A median stripe of mottled tan is broken into dashes and spots.

male

male

male

The legs are dark brown, with tan banding. The male pedipalps have tan femurs and tibiae and black cymbiums.

Habitat These spiders are often found in oak-conifer forests.

Known records from AZ

Phidippus tux Pinter, 1970

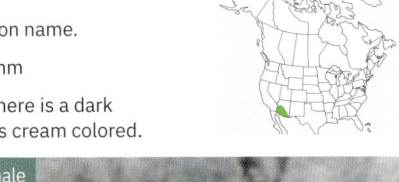

Common Name This species does not have a common name.

Total Body Length Female 11.0 mm, Male 7.8–9.4 mm

Description The prosoma is pale yellow to peach. There is a dark brown band around the anterior eyes. The clypeus is cream colored. The opisthosoma is pale yellow to peach. There is a slightly paler anterior band. There may be a slightly darker heartmark. The median stripe is the same yellow to peach color and may be bordered in thin to broad black stripes that have an iridescent sheen and form a U before reaching the spinnerets. The females often have pale transverse bands that do not touch in the middle and a pair of pale spots anterior to the spinnerets. The legs are pale yellow to peach. The males have a dark band on the tibiae of legs I and IV.

female

Habitat These spiders are found in mesquite and grasses along streams.

Known records from AZ

female

female

male

Phidippus tyrrelli Peckham & Peckham, 1901

Common Name This species does not have a common name.

Other Colloquial Names Tyrell's tufted jumping spider

Total Body Length Female 7.8–12.6 mm, Male 6.9–9.0 mm

Description The prosoma is black, with a white band that runs through the anterior eye row and posteriorly just beyond the posterior lateral eyes. A slightly darker band runs under the anterior eyes; this is most contrasting in the males. The clypeus is white. In the females, tan hairs cover the chelicerae. In the males, alternating tan and white hairs cover the chelicerae. The chelicerae are metallic green. There is a thin white band on the very edge.

The opisthosoma is red and usually somewhat dully colored. A white anterior band stops about halfway down the lateral edge. Iridescent black bands may be on either side of the midline. There are usually three pairs of white spots; the first pair is often fused. There may be transverse white bands that do not meet in the middle.

male courting female

The legs are dark brown, almost black, with yellow to white banding. The male pedipalps have white femurs and banded black and white cymbiums.

Habitat These spiders can be found under rocks or on coniferous plants. They are usually found at elevations between 7,000 ft and 10,000 ft.

Known records from BC (Canada); AZ, CA, CO, ID, MT, NM, OR, UT, WY (United States)

Phidippus whitmani Peckham & Peckham, 1909

Common Name This species does not have a common name.

Other Colloquial Names Whitman's jumping spider

Total Body Length Female 5.2–11.6 mm, Male 4.1–9.6 mm

Description The female prosoma is black, with orange hairs between the posterior eyes. The female opisthosoma is covered with orange hairs. There is a white anterior band, and a pair of transverse white bands are on the lateral edges. A pair of iridescent black

female

juvenile

juvenile

male

male

male

stripes are usually on the dorsal surface, one on either side of the midline. There are paired white markings along the black stripes. The legs have dark femurs but are otherwise brown, with pale banding. The male prosoma is red, with black in the eye region. The opisthosoma is red, with a white anterior band and faint lateral bands. The legs are dark, with light-colored hairs creating stripes and bands. The male pedipalps have yellow hairs on the dorsal surfaces and a white spot on the cymbium.

Habitat These spiders are usually found in leaf litter or low herbaceous vegetation.

Known records from MB, NL, NS, NT, ON, QC (Canada); AL, AR, CT, DC, DE, FL, GA, IA, IL, IN, KS, KY, LA, MA, MD, ME, MI, MN, MO, MS, NC, ND, NE, NH, NJ, NY, OH, OK, PA, RI, SC, TN, TX, VA, WI (United States)

Platycryptus arizonensis (Barnes, 1958)

Common Name This species does not have a common name.

Total Body Length Female 5.9–9.2 mm, Male 5.5–6.5 mm

Description The prosoma is mottled with tan, cream, and golden brown. Two pairs of hair tufts are near the anterior lateral eyes: the first pair points horizontally, while the other pair points upward at about a 45° angle. The female clypeus is white; the male clypeus is tan to light brown. The opisthosoma is a palette of similar colors, with a pale median stripe that undulates and is often edged in cream to white. The

female

female

legs are dark, with light hairs creating the look of bands, and usually with a darker overall appearance in the males. This is the smallest species of *Platycryptus* in the United States and is usually paler than the other species.

female

Habitat These spiders are often found on tree bark or wooden structures.

Known records from AZ, CA, CO, NM, NV, UT

Platycryptus californicus (Peckham & Peckham, 1888)

Common Name This species does not have a common name.

Other Colloquial Names California flattened jumping spider

Total Body Length Female 9.4–10.6 mm, Male 6.2–7.1 mm

Description The prosoma is mottled with gray, brown, and tan, with dark sides. Two pairs of hair tufts are near the anterior lateral eyes: the first pair points horizontally, while the other pair points upward at about a 45° angle. The opisthosoma is mottled with tan, gray, and brown, and a series of chevron shapes runs down the dorsal surface. The lateral edges are darker, creating a strong contrast with the chevron-shaped markings. The legs are tan, brown,

female

female

female

female

and gray, with banding. This species looks very similar to *Platycryptus undatus*, but its colors contrast more strongly, and note that the range does not overlap with *P. undatus*.

Habitat These spiders are often found on the bark of trees and on and around human structures.

Known records from BC (Canada); AR, AZ, CA, CO, ID, MT, NM, NV, OR, TX, UT, WA, WY (United States)

Platycryptus undatus (DeGeer, 1778)

Common Name This species does not have a common name.

Other Colloquial Names Tan jumping spider, Lorax spider

Total Body Length Female 6.0–13.0 mm, Male 6.7–9.5 mm

Description The prosoma is mottled with gray, brown, and tan. The clypeus is mahogany on the males and brilliant white on the females. The male chelicerae are covered in long white hairs. Two pairs of hair tufts are near the anterior lateral eyes: the first pair points horizontally, while the other pair points upward at about a 45° angle. The opisthosoma is mottled with tan, gray, and brown, and a series of chevron shapes runs down the dorsal surface. The legs are tan, brown, and gray, with banding. This species looks very similar to *Platycryptus californicus*. Note that the range of *P. undatus* does not overlap with the other species in the genus *Platycryptus*.

Habitat These spiders are often found on the bark of trees and on and around human structures (including wooden fences).

female

female

female on quarter
for scale

male

male

male ventral

male with prey

Known records from MB, NS, ON, QC (Canada); AL, AR, CT, DC, FL, GA, IA, IL, IN, KS, KY, LA, MA, MD, MI, MN, MO, MS, NC, NE, NJ, NY, OK, PA, SC, TN, TX, VA, WI, WV (United States)

Plexippus paykulli (Audouin, 1826)

Common Name Pantropical jumper

Other Colloquial Names Pantropical jumping spider

Total Body Length Female 10.0–12.0 mm, Male 9.5 mm

Description The female prosoma is light brown, with a tan median stripe that ends just before the eye region. The female opisthosoma is light brown, with a tan median stripe. The anterior portion has a dark longitudinal stripe. Two pairs of pale dots are on the edge of the median stripe: the first is just over halfway down, while the second is just

female

female

female

juvenile

juvenile

juvenile

male

male

male

above the spinnerets. The legs are tan to brown. The male prosoma is dark brown to almost black. There is a cream median stripe, and cream bands run around the edges. The opisthosoma is dark brown to almost black, with a cream median stripe that darkens slightly toward the posterior end. Two pairs of white dots are on the edges of the median stripe at the posterior end. The lateral edges are cream. The legs are tan, with dusky longitudinal dashes.

Habitat These spiders are often found in and around human structures, especially greenhouses.

Known records from CO, FL, GA, LA, NM, SC, TX
This is an introduced species.

Salticus austinensis Gertsch, 1936

Common Name This species does not have a common name.

Total Body Length Female 4.9 mm, Male 4.9 mm

Description The prosoma is black, with a white band around the edge and two white bands bordered with pale yellow across the prosoma. One is near the anterior eyes, while the other is behind the eye region.

female

female female

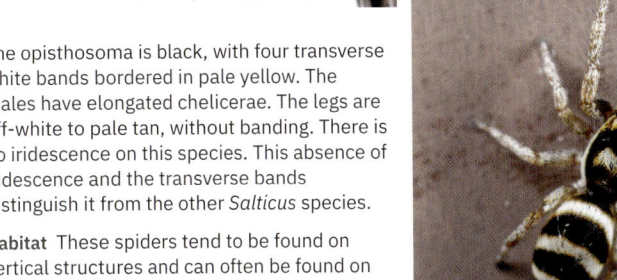

male

The opisthosoma is black, with four transverse white bands bordered in pale yellow. The males have elongated chelicerae. The legs are off-white to pale tan, without banding. There is no iridescence on this species. This absence of iridescence and the transverse bands distinguish it from the other *Salticus* species.

Habitat These spiders tend to be found on vertical structures and can often be found on tree trunks. They can also be found on and around human structures.

Known records from KS, OK, TX

Salticus palpalis (Banks, 1904)

Common Name This species does not have a common name.

Total Body Length Female 4.5 mm, Male 4.5 mm

Description The prosoma is black, with iridescent scales. There are white bands on the lateral edges, and a white band is just above the anterior eyes. The males have elongated chelicerae. The opisthosoma is dark rust brown to brick red, with four white transverse bands. The legs have iridescent femurs and are otherwise dark, with white hairs. The iridescence on the prosoma and the transverse red and white bands on the opisthosoma distinguish this species from the other *Salticus* species.

Habitat These spiders tend to be found in open habitats and seem to like vertical surfaces.

female

female

Known records from AZ, CA

Salticus scenicus (Clerck, 1757)

Common Name Zebra jumper

Other Colloquial Names Zebra jumping spider

Total Body Length Female 4.3–6.4 mm, Male 4.0–5.5 mm

Description The prosoma is black, with a thin white border. A pair of triangular white markings are just posterior of the eye region, and a white marking is just above the anterior eyes. There are iridescent scales in the eye region. The males have elongated chelicerae. The opisthosoma is black, with four white transverse bands that are slightly oblique; the middle two do not quite meet in the middle. The legs are black and banded in white. The oblique bands and the limited iridescence distinguish *S. scenicus* from the other *Salticus* species.

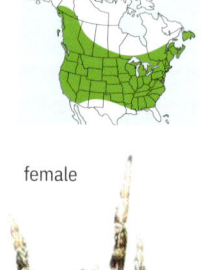

female

female

male

male
ventral

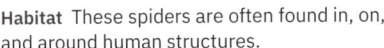

male and female mating

male and female mating

male and female mating

Habitat These spiders are often found in, on, and around human structures.

Known records from AB, BC, MB, NB, NL, NS, ON, PE, QC, SK (Canada); AR, CA, CO, CT, GA, IA, IL, IN, KS, MA, ME, MI, MN, NH, NJ, NY, OH, OR, PA, TX, UT, WI, WY (United States)

Sassacus cyaneus (Hentz, 1846)

Common Name This species does not have a common name.

Other Colloquial Names Iridescent leaf-beetle jumping spider

Total Body Length Female 3.3–4.6 mm, Male 2.4–4.0 mm

Description The entire spider is black, with small iridescent scales. The prosoma is almost square and slightly wider at the posterior end. A bristle of hairs is on the anterior of the opisthosoma. The legs are dark, with yellow to pale yellow bands on the distal segments.

Habitat These spiders are often found in shrubs and other low vegetation.

Known records from NB (Canada); AL, AR, CT, FL, GA, IA, IL, MA, MO, MS, NC, NE, NJ, NY, OH, PA, SC, TX, VA, WI, WV (United States)

juvenile

juvenile

male ventral

male

Sassacus papenhoei Peckham & Peckham, 1895

Common Name This species does not have a common name.

Other Colloquial Names Common leaf-beetle jumping spider

Total Body Length Female 4.4–5.5 mm, Male 2.8–4.8 mm

Description The prosoma is black and covered in small iridescent scales. There are white bands on the lateral edges, which are wider in the females. The opisthosoma is covered in iridescent scales and has a white anterior band that continues down the lateral edges to the spinnerets. The legs are dark with longitudinal light-colored stripes.

Habitat These spiders are often found in prairies, on low vegetation, in sagebrush, and in agricultural fields.

female

female

Known records from BC (Canada); AL, AR, AZ, CA, CO, DC, IA, ID, IL, IN, KS, LA, MD, MI, MN, MO, MS, MT, NC, NE, NM, NV, OH, OK, OR, SC, TN, TX, UT, VA, WA, WI, WY (United States)

Sassacus vitis (Cockerell, 1894)

Common Name This species does not have a common name.

Other Colloquial Names Buttonhook leaf-beetle jumping spider

Total Body Length Female 4.0–4.9 mm, Male 3.6–4.8 mm

Description The prosoma is orange brown and covered with golden scales. A white patch is often just behind the posterior lateral eyes. The females have a white band around the edges that goes through the anterior eye row. The edge is white. The opisthosoma is covered in golden scales. A white to cream anterior band continues down the sides almost to the

juvenile

male

male

male

male
ventral

spinnerets. One to three pairs of dark spots are often highlighted with white on the dorsal edges posteriorly. The legs are yellow, with white hairs creating bands or stripes. The males tend to be a little bit darker and have robust chelicerae.

Habitat These spiders are often found on herbaceous vegetation, including in agricultural fields.

Known records from AB, BC (Canada); AZ, CA, CO, FL, ID, KS, MT, NM, NV, OK, OR, TX, UT, WA, WY (United States)

Synageles bishopi Cutler, 1988

Common Name This species does not have a common name.

Other Colloquial Names Bishop's ant-mimic jumping spider

Total Body Length Female 3.2 mm, Male 2.6 mm

Description The prosoma is dark golden brown. There is a thin white line of hairs between the posterior lateral eyes. The eye region occupies less than half of the prosoma. The prosoma slopes gently behind the posterior lateral eyes. The opisthosoma is dark golden brown, slightly lighter anteriorly. A pair of white dots are on either side less than halfway down, and there is a constriction (more pronounced in the males) in the opisthosoma at this point. A small scutum is on the dorsal anterior. The legs are pale to golden brown, with a somewhat darker line on the prolateral surfaces of the tibiae and patellae. The first pair of legs of the males is very robust. These spiders often raise their second pair of legs in an antenna-like pose. Species in the genus *Synageles* look very similar to those in *Peckhamia*, and both use the second pair of legs to

female

female

male

male

mimic antennae. The slope behind the posterior lateral eyes is steeper in *Peckhamia* species, and the constriction on the opisthosoma is usually less pronounced in *Synageles*.

Habitat These spiders are often found on the bark of trees, on fences, or in grassy habitats.

Known records from AR, CT, FL, KS, KY, MA, MD, MO, MS, NC, NJ, NM, NY, OH, PA, SC, TN, TX, VA

Synemosyna formica
Hentz, 1846

Common Name This species does not have a common name.

Other Colloquial Names
Slender ant-mimic jumping spider

Total Body Length
Female 4.7–5.7 mm, Male 4.2–5.0 mm

Description The prosoma is golden to dark brown. There are constrictions just behind the eyes and just before the pedicle. The opisthosoma is dark golden brown, and a well-defined constriction is about halfway down. A small scutum is on the dorsal opisthosoma. The legs are yellow to light brown. Legs I have longitudinal black stripes on the prolateral and retrolateral surfaces and are often held above the prosoma, resembling ant antennae. These spiders usually do not jump and often move in a manner similar to that of ants.

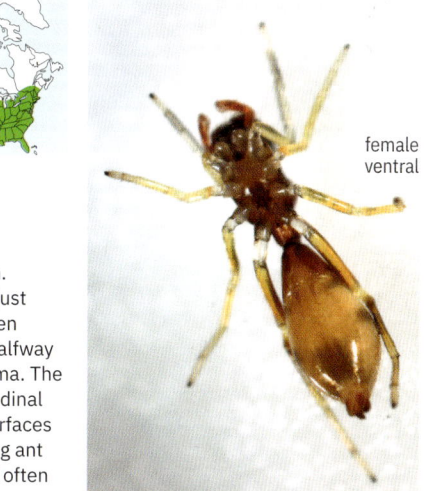

female
ventral

female

female

male

male

young spiderling

egg sac with young

Habitat These spiders are often found on low vegetation in humid environments and are often mistaken for ants.

Known records from ON (Canada); AR, AZ, CT, FL, GA, IA, IL, IN, KS, MA, MN, NC, NH, NJ, PA, SC, TX, WI (United States)

Talavera minuta (Banks, 1895)

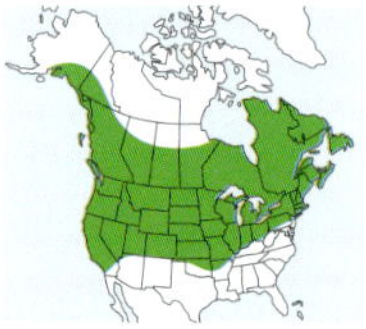

Common Name This species does not have a common name.

Other Colloquial Names Minute jumping spider

Total Body Length Female 2.0 mm, Male 2.5–2.7 mm

Description The prosoma and opisthosoma are mottled with dark brown, golden brown, and pale gray. Indistinct chevrons are at the posterior end of the opisthosoma. The legs are heavily banded in brown and yellow. The pedipalps are bright yellow.

Habitat These spiders are often found in leaf litter.

female

female

female

female
ventral

male

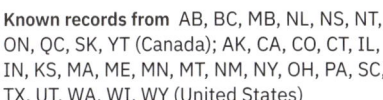

male on penny for scale

Known records from AB, BC, MB, NL, NS, NT, ON, QC, SK, YT (Canada); AK, CA, CO, CT, IL, IN, KS, MA, ME, MN, MT, NM, NY, OH, PA, SC, TX, UT, WA, WI, WY (United States)

Tutelina elegans (Hentz, 1846)

Common Name This species does not have a common name.

Other Colloquial Names Thin-spined jumping spider

Total Body Length Female 5.5–7.0 mm, Male 4.0–4.5 mm

Description The prosoma is covered in iridescent scales. The males have very conspicuous erect black hairs from the anterior median eyes to the posterior lateral eyes, creating the look of stiff erect eyebrows. The opisthosoma is covered in iridescent scales and has a white to cream anterior band. The legs are yellow to pale orange, with black longitudinal lines on the dorsal surface. The males have a bristle of black hairs on the tibiae of leg I.

female

female

female

female
ventral

male

male

male

Habitat These spiders are often found in prairies, fields, and herbaceous vegetation.

Known records from NS, ON (Canada); AR, CO, CT, IA, IL, IN, KS, MI, MN, MO, MS, NJ, OH, OK, PA, SC, TX, WY (United States)

Tutelina harti (Emerton, 1891)

female

Common Name This species does not have a common name.

Other Colloquial Names Hart's jumping spider

Total Body Length Female 4.0–7.0 mm, Male 4.0–5.5 mm

Description The prosoma is dark gray, sometimes with scattered light gray markings and a thin white border on the edge. The opisthosoma is dark gray, sometimes with scattered light gray markings and a white anterior band. Overall a slight metallic sheen can be detected. The legs are banded in black and tan, and both the males and females have a bristle of hairs on the robust tibiae I.

Habitat These spiders are often found on the bark of trees.

female

female

female

male ventral

male

male

male

Known records from MB, ON, QC (Canada); CO, CT, FL, IL, IN, KS, MI, MN, NE, OH, PA, SC (United States)

Zygoballus rufipes Peckham & Peckham, 1885

Common Name Hammerjawed jumper

Other Colloquial Names Hammer-jawed jumping spider

Total Body Length Female 4.3–6.0 mm, Male 3.0–4.0 mm

Description The prosoma is dark brown to almost black and has an iridescent sheen; it is usually darker in the males. The females are sometimes covered with pale hairs. The opisthosoma is reddish brown, with a thin anterior band and two oblique transverse bands that are restricted to the lateral edges. The females often have two pairs of white markings on either side of the midline. The opisthosoma is less iridescent. The legs are pale yellow, with darker, sometimes red, femurs, metatarsi, and tarsi on leg I; dark spots on femur II; and dark markings on leg IV. The females may have banding on some or all legs. The males' chelicerae are somewhat elongated and possess a spur that gives the spider the common name "hammerjawed."

female

female

male

male

male near
penny for scale

male

male

Habitat These spiders are usually found in open habitats on herbaceous vegetation.

Known records from ON (Canada); AR, AZ, CT, FL, IL, IN, KS, MA, ME, MI, MN, MO, MS, NM, NY, OH, PA, SC, TX (United States)

male
ventral

Zygoballus sexpunctatus (Hentz, 1845)

Common Name This species does not have a common name.

Other Colloquial Names Six-spotted hammer-jawed jumping spider

Total Body Length Female 3.0 mm, Male 3.0–4.5 mm

Description The prosoma is dark brown to black. There are scattered iridescent scales and a white spot on the posterior surface where the prosoma slopes down toward the pedicel. White scales are along the lateral edges under the posterior median and anterior lateral eyes. The very edge of the prosoma has a fringe of white scales. The male has robust chelicerae. The opisthosoma is bronze to black and slightly iridescent. There is a white anterior band. Two oblique transverse bands do not meet in the middle; these bands tend to be less conspicuous on the females. The first band starts to curve toward the posterior end as it approaches the midline. In *Zygoballus rufipes,* this first band is truncated and does not curve in this way. All femurs are suffused with black, and femur I is more robust than the others. The other leg segments are yellow to tan. The males' chelicerae are somewhat elongated and possess a spur.

female

juvenile

juvenile

male

male

male

male

male

Habitat These spiders are usually found on herbaceous vegetation.

Known records from IL, FL, LA, MS, NC, OH, OK, SC, TX

SCYTODIDAE ▸ SPITTING SPIDERS

Genera in This Family *Scytodes* (10)

Total Body Length 3.5–10.0 mm

Description These unique spiders have a highly domed prosoma. They are haplogyne and ecribellate. They have six eyes that are in three diads. They have a pair of book lungs and a single tracheal spiracle located near the spinnerets. They are usually tan to brown, often with darker markings. They capture prey by spitting silk and venom from their fangs, sticking their prey to the surface it is on. Many scytodids are synanthropic and are found in and around human structures.

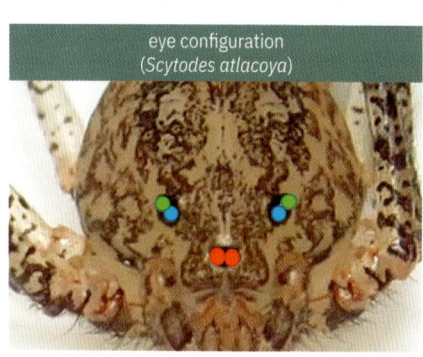

eye configuration
(*Scytodes atlacoya*)

Similar Families Diguetidae, Mimetidae, Pholcidae, Sicariidae

How to distinguish Scytodidae from similar families:

Diguetids, sicariids, and pholcids all can have six eyes, and the eye configuration of diguetids and sicariids is similar to that of scytodids, but neither of these species has the highly domed prosoma that is seen in scytodids. Mimetids have eight eyes and characteristic supination on legs I and II, which is not seen in scytodids. None of the similar families spits silk and venom to secure its prey to the surface it is on.

Scytodes atlacoya
(Rheims, Brescovit & Durán-Barrón, 2007)

Common Name This species does not have a common name.

Other Colloquial Names Spotted spitting spider

Total Body Length Female 6.3–10.1 mm, Male 4.8–8.7 mm

Description The prosoma is tan to brown, with dark brown markings that are somewhat like squiggly lines. The prosoma is highly domed, especially posteriorly. There are six eyes arranged

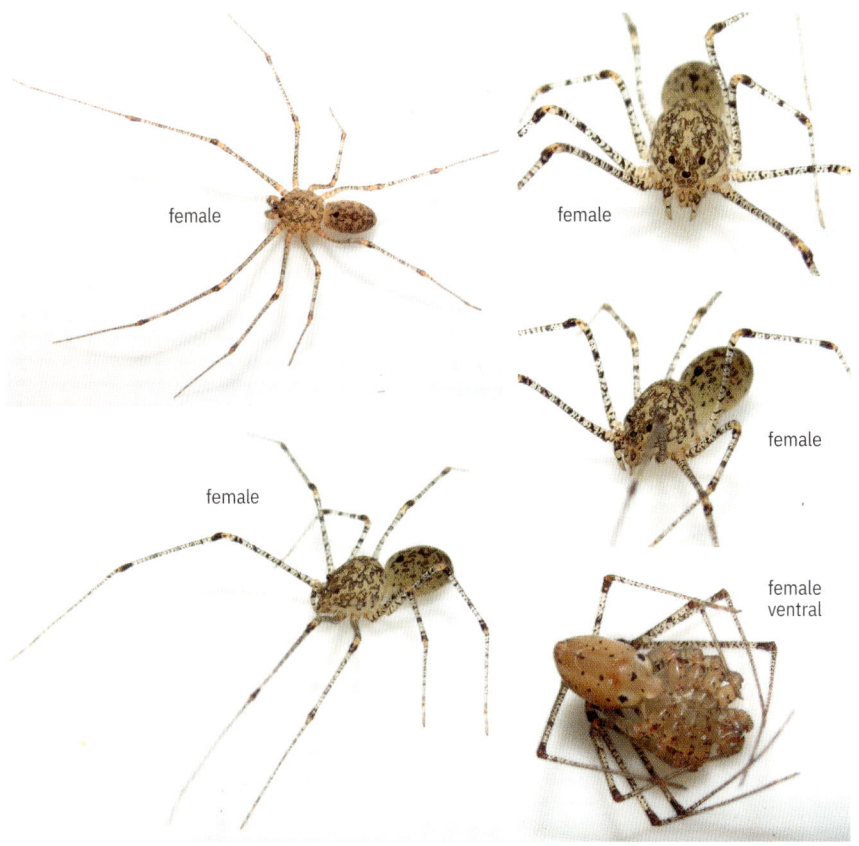

female

female

female

female

female
ventral

in three pairs. There is an outlined teardrop to diamond shape on the chelicerae. The opisthosoma is tan to brown, with dark brown spots that are longer than they are wide and the darkest of which is the heartmark. The legs are very pale, with lots of dark spots and dark bands at the joints.

Habitat These spiders are often found in and around human structures.

Known records from FL, GA, MS, TX

Scytodes fusca Walckenaer, 1837

Common Name This species does not have a common name.

Other Colloquial Names Brown spitting spider

Total Body Length Female 4.4–6.0 mm, Male 4.0–5.5 mm

Description The prosoma is dark reddish brown, with a slightly darker median stripe. Radial lines may be visible. (They are usually more visible on the males.) There are six eyes arranged in three pairs. The prosoma is highly domed, especially toward the posterior. The opisthosoma is dark gray and covered with fine hairs, giving it a pale sheen. The legs have reddish-brown femurs, and the distal segments are yellow. The females carry their egg sac, and the young are highly striped when they emerge.

female

female

female

female

juvenile

juvenile

Habitat These spiders seem to be associated with human structures, man-made tropical environments, or areas where tropical plants have been imported.

Known records from QC (Canada); IN, FL, LA, NM, OH, TX (United States)
This is an introduced species.

Scytodes thoracica (Latreille, 1802)

Common Name This species does not have a common name.

Other Colloquial Names Common spitting spider

Total Body Length Female 4.0–5.5 mm, Male 3.5–4.0 mm

Description The prosoma is tan to pale orange. The prosoma is highly domed, especially posteriorly. A pair of brown stripes are on either side of the midline, creating an arrow pointing toward the opisthosoma. There is a brown submarginal stripe that is somewhat wavy. The opisthosoma is tan to light brown and usually has four transverse bands that are broken into spots or dashes. The legs are yellow, with brown banding.

female

female

female with egg sac

male

Habitat These spiders are often found in and around human structures.

Known records from ON (Canada); CO, CT, FL, GA, MA, ME, MS, OH, TX (United States) This is an introduced species.

SPARASSIDAE ▶ GIANT CRAB SPIDERS

Other Colloquial Names Huntsman spiders

Genera in This Family *Curicaberis* (4), *Decaphora* (1), *Heteropoda* (1), *Macrinus* (1), and *Olios* (2)

Total Body Length 7.0–50.0 mm

Description The large entelegyne spiders are ecribellate. They have eight eyes in two rows. They are somewhat dorsoventrally flattened and have laterigrade legs; these features allow them to fit in small cracks and crevices. They have a pair of book lungs and a single tracheal spiracle located near the spinnerets. They tend to be tans, browns, or gray in color.

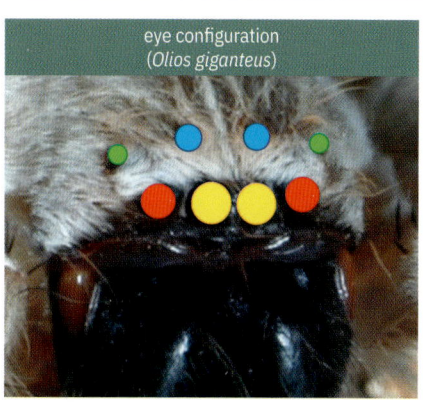
eye configuration
(*Olios giganteus*)

Similar Families Anyphaenidae, Cheiracanthiidae, Clubionidae, Homalonychidae, Liocranidae, Miturgidae, Philodromidae, Selenopidae, Zoropsidae

How to distinguish Sparassidae from other similar families:

Sparassids have laterigrade legs, which distinguishes them from anyphaenids, cheiracanthiids, clubionids, homalonychids, liocranids, miturgids, and most of the zoropsids. Sparassids can be distinguished from philodromids by their overall larger size (7.0–50.0 mm total body length), as philodromids tend to be fairly small (3.0–8.2 mm total body length). The eye configuration can distinguish sparassids from selenopids: In selenopids, the posterior eye row is so strongly procurved that it appears they have six eyes in the anterior row and only two eyes

dorsoventrally flat, laterigrade legs (*Heteropoda venatoria* with egg sac)

in the posterior row. In contrast, sparassids have their eyes in two rows of four eyes. Spiders in the genus *Lauricius* (Zoropsidae) are dorsoventrally flattened and have laterigrade legs, and they tend to be smaller (8.0–18.0 mm total body length) than most sparassids (7.0–50.0 mm total body length). *Lauricius* spiders tend to be found in leaf litter and caves, whereas sparassids are usually found up on vegetation and other structures. *Lauricius* also have stockier shorter legs than most sparassids.

Curicaberis abnormis (Keyserling, 1884)

Common Name This species does not have a common name.

Total Body Length Female 8.2–9.1 mm, Male 7.4–9.0 mm

Description The prosoma is pale orange and mostly covered with off-white hairs. There are scatted dark spots just behind the eye region. The opisthosoma is mottled with tan and off-white, and a slightly darker lobed median stripe may be broken and may look like small chevron shapes. The legs are mottled with tan and off-white and speckled with abundant darker spots.

Habitat These spiders have been found in rocky limestone thornscrub desert habitats. Occasionally, they can be found hunting on human structures, especially at night.

Known records from AZ, NM

female

male

Decaphora cubana
(Banks, 1909)

Common Name This species does not have a common name.

Total Body Length Female 6.6–10.2 mm, Male 7.5–10.0 mm

female

female with prey

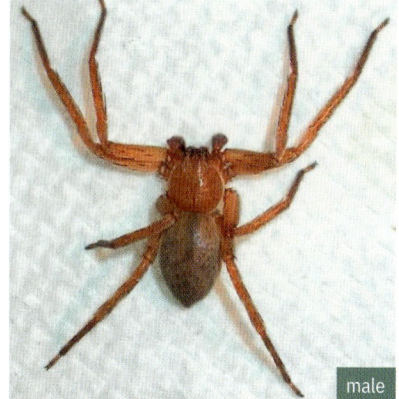

male

Description The prosoma is orange and slightly darker toward the edges. The opisthosoma is cream to pale orange, with a faint chevron pattern toward the posterior end. The females have a much more mottled appearance. The legs are tan to pale orange. The male pedipalp cymbium is covered with gray hairs.

Habitat Not much is published about this species' preferred habitats, but the spiders can be found in and around human structures.

Known records from FL
This is likely an introduced species. It is native to Cuba.

Heteropoda venatoria
(Linnaeus, 1767)

Common Name
Huntsman spider

Other Colloquial Names
Giant crab spider

Total Body Length Female 20.0–30.0 mm, Male 15.0–25.0 mm

female with egg sac

Description The female prosoma is brown, with a white clypeus. There is a dark line around the edge. Faint radial lines may be visible. The male prosoma is black, with a tan border and a white clypeus. A tan median line starts at the fovea and forks widely to the lateral eyes. The eye

female with egg sac

female with egg sac

female ventral with egg sac

male

region is paler. There is a dark band around the posterior. The opisthosoma for both the male and the female is brown, with a dark heartmark and a few indistinct black spots. The legs are brown, with black spots at the base of the spines. The female carries her flat egg cases with her pedipalps.

Habitat These spiders are often found in urban areas, and humans seem to transport them to new locations easily.

Known records from AL, FL, GA, LA, MS, NM, OH, PA, SC, TX
This is an introduced species.

Olios giganteus Keyserling, 1884

Common Name Golden huntsman spider

Other Colloquial Names Giant crab spider

Total Body Length Female 14.6–48.0 mm, Male 11.3–29.4 mm

Description The prosoma is orange to orange brown and slightly darker on the clypeus. The opisthosoma is orange to brownish gray. There is a slightly darker heartmark. At the posterior end, it is bordered in black, and the border joins to form a thin median line.

Habitat These spiders are usually found in dry areas and in and around human structures.

Known records from AZ, CA, NM, TX, UT

female

female

female

male

TRECHALEIDAE > LONGLEGGED WATER SPIDERS

Genera in This Family *Trechalea* (1)

Total Body Length 13.0–20.0 mm

Description These spiders are entelegyne and ecribellate. They have eight eyes in two rows, both of which are recurved. They have somewhat laterigrade legs, with flexible tarsi. They have a pair of book lungs and a single tracheal spiracle near the spinnerets. They have three tarsal claws. They create a flattened egg sac that is attached to spinnerets. The young ride on the egg sac for a short period after emerging before dispersing.

Similar Families Lycosidae, Pisauridae

How to distinguish Trechaleidae from other similar families:

Pisaurids and lycosids can be distinguished from trechaleids by their prograde legs, as trechaleids have somewhat laterigrade legs. Additionally, trechaleids do not carry their egg sac in their chelicerae (the egg sacs are attached to the spinnerets), and they do not build a nursery web (the young ride on the egg sac for a period before dispersing), as pisaurids do. Trechaleids can be distinguished

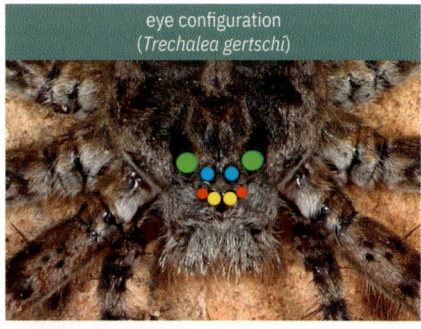
eye configuration
(*Trechalea gertschi*)

flexible tarsi
young ride on egg sac
Trechalea gertschi

from lycosids by their eye configuration. In lycosids, the four small anterior eyes are in a nearly straight row across the front of the prosoma. The posterior median eyes are usually quite large. The posterior lateral eyes are placed back on the prosoma, giving the appearance of three eye rows. In trechaleids, the eyes are in two rows, both of which are recurved. Their eyes may be subequal in size, but their posterior median eyes are not enlarged, as they are in lycosids. Trechaleids are also represented in North America north of Mexico by only one species, and it is found only in Arizona.

Trechalea gertschi Carico & Minch, 1981

Common Name This species does not have a common name.

Total Body Length Female 16.4–20.0 mm, Male 13.0–15.0 mm

Description The prosoma is dark gray, with a pale gray border. The posterior eye row is recurved. The opisthosoma is mottled with browns and grays. The legs are somewhat laterigrade and banded in gray and brown. They have long flexible tarsi that are often held in a curved position. The females attach their egg sac to their spinnerets, and the young ride on the egg sac when they emerge. *T. gertschi* looks very similar to *Dolomedes* species from the Pisauridae family. To distinguish the two, note the somewhat laterigrade legs, the flexible tarsi, and the fact that the females attach the egg sacs to their spinnerets. (In *Dolomedes* species, the egg sacs are held in the chelicerae.)

Habitat These spiders are found in the Gila River drainage area. They are often found on the rocks above the water.

Known records from AZ

female with egg sac

female with young on egg sac

male

juvenile

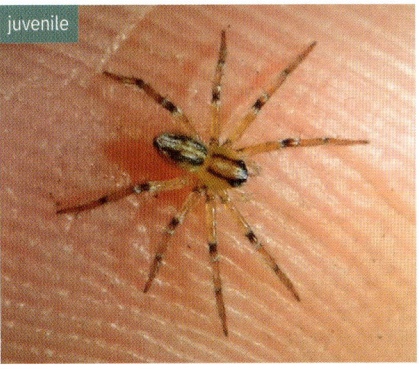

juvenile

ZOROPSIDAE ▷ FALSE WOLF SPIDERS

Other Colloquial Names Wandering spiders

Genera in This Family *Anachemmis* (5), *Lauricius* (2), *Liocranoides* (5), *Socalchemmis* (14), *Titiotus* (16), *Zorocrates* (5), and *Zoropsis* (1)

Total Body Length 5.0–20.0 mm

Descriptions These large spiders are entelegyne. This family contains both cribellate and ecribellate species. They have many ventral tibial spines. The legs are prograde in some species and laterigrade in others. They have a pair of book lungs and a single tracheal spiracle near the spinnerets. They have eight eyes in two rows. They are mostly found in the Southwest.

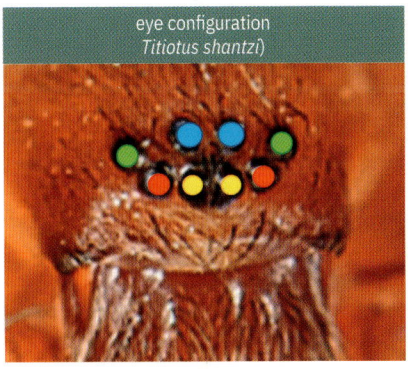

eye configuration
Titiotus shantzi)

Similar Families Amaurobiidae, Ctenidae, Desidae, Filistatidae, Lycosidae, Selenopids, Sparassidae, Titanoecidae

How to distinguish Zoropsidae from other similar families:

Many zoropsids are ecribellate, which easily distinguishes them from the amaurobiids, desids, filistatids, and titanoecids. Two genera of zoropsids have a cribellum: *Zoropsis* and *Zorocrates*. Spiders in *Zoropsis* can be distinguished by the fact that the posterior eyes are recurved, while they are procurved or straight in amaurobiids, procurved in desids, in a central cluster on a mound in filistatids, and nearly straight in titanoecids. Spiders in *Zorocrates* have fine fuzzy hairs on the prosoma that are not found in amaurobiids or desids. Zoropsids can be distinguished from lycosids, as they lack the enlarged posterior median eyes that are diagnostic of that family, and the eyes are in two rows. (In lycosids, the posterior lateral eyes are far back on the prosoma, so it appears that the eyes are in three rows.) *Lauricius* species have laterigrade legs, which could cause them to be confused with selenopids or sparassids. Selenopids are extremely dorsoventrally flat, and the posterior eyes are so strongly recurved that it appears that six eyes are in the anterior row and only two eyes are in the posterior row. In zoropsids, the eyes are in two rows of four. Sparassids tend to have much longer legs than the stout legs seen in *Lauricius*. Spiders in the genus *Acanthoctenus* (Ctenidae) are superficially similar to those in *Zoropsis* and *Zorocrates*, but in *Zoropsis* and *Zorocrates*, the anterior eye row is straight (it is recurved in ctenids), and the anterior lateral eyes are not significantly reduced in size (which is a trait seen in ctenids).

Lauricius hooki Gertsch, 1941

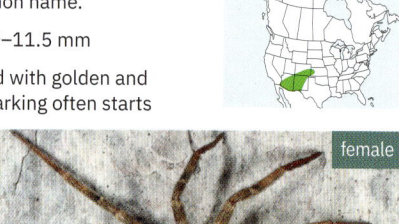

Common Name This species does not have a common name.

Total Body Length Female 10.6–18.0 mm, Male 8.0–11.5 mm

Description The prosoma is reddish brown, covered with golden and black hairs, and somewhat flattened. A visible V marking often starts at the fovea and opens widely to the lateral eyes. There is usually a slightly darker broken submarginal band, and the very edges may be somewhat darker. The opisthosoma is tan to light brown. The heartmark is brown. Just posterior to the heartmark, a series of chevron markings may lead to the spinnerets. The sides are usually somewhat mottled with brown and tan. The legs are orange to brown, are either unbanded or have very faint dusky bands, and are laterigrade. The female attaches her cream to white-colored egg sacs to the surface and guards them by sitting over them.

female

female

female

Habitat These spiders are often found in leaf litter, in caves, and in and around human structures. They have been found at elevations up to 8,900 ft.

Known records from AZ, CO, KS, NM, TX

Titiotus shantzi Platnick & Ubick, 2008

Common Name This species does not have a common name.

Total Body Length Female 11.0–14.0 mm, Male 8.0–10.0 mm

Description The prosoma is golden brown, with a slightly darker area, almost triangular, from the fovea to the eye region. The opisthosoma is silvery gray. The legs are golden brown and unbanded. In this species, the middle prong on the male pedipalp tibial apophysis is short, compared to other similar-looking *Titiotus* species, and the females are very difficult to identify to species. *Titiotus* species can be difficult to distinguish from some other zoropsids without viewing a specimen under a microscope. They have three claws and claw tufts, and the male pedipalp has a tibial apophysis consisting of two or three prongs, which is diagnostic of the genus. This genus often seems to be confused with *Loxosceles reclusa* (brown recluse), but *T. shantzi* is much larger than *L. reclusa*, it lacks the violin-shaped marking, it has eight eyes, and it has spines on its legs.

Habitat These spiders are often found in forested habitats and caves, and in and around human structures.

Known records from CA

Zoropsis spinimana
(Dufour, 1820)

Common Name This species does not have a common name.

Other Colloquial Names
Mediterranean spiny false wolf spider

Total Body Length Female 10.0–19.0 mm, Male 10.0–13.0 mm

Description The prosoma is tan gray to golden. The median stripe is flared and bordered by a slightly darker flared band with radial lines. The very edge has a thin dark line. The opisthosoma is tan gray to golden and somewhat variegated. The heartmark is usually dark brown. Sometimes just brown spots are on the outer edge, and there are short transverse lines on the dorsal surface posterior to the heartmark. The legs are tan gray to golden and spotted. These spiders are often mistaken for lycosids (wolf spiders), but note eye configuration.

Habitat In North America, these spiders have often been found in and around human structures.

Known records from CA This is an introduced species.

THE SPIDER HUNTER GUILD

The spiders of the Spider Hunter Guild usually do not build a web of their own but prey upon other spiders, sometimes invading the web of their prey. All spiders in this guild are Araneomorphae. This guild includes the following families: Caponiidae, Cithaeronidae, and Mimetidae.

KEY TO FAMILY

1a. First two pairs of legs longer than others with characteristic spination, 2.5–7.0 mm total body length, sometimes found in web of host spider -- **Mimetidae** (page 588)

1b. Legs without characteristic spination --- **2**

Mimetus notius

2a. Two eyes or eight eyes in three rows, lacks book lungs (two pairs of tracheal spiracles), robust legs, 3.0–6.0 mm total body length ---------- **Caponiidae** (page 585)

2b. Eight eyes in two rows, pair of book lungs and a single tracheal spiracle near spinnerets, long thin legs, 3.0–8.0 mm total body length ------- **Cithaeronidae** (page 587)

Calponia harrisonfordi

Orthonops icenoglei

Cithaeron praedonius

CAPONIIDAE ▶ BRIGHT LUNGLESS SPIDERS

Genera in This Family *Calponia* (1), *Orthonops* (7), and *Tarsonops* (4)

Total Body Length 3.0–6.0 mm

Description These small haplogyne spiders are ecribellate. They lack book lungs and have two pairs of tracheal spiracles near the epigastric furrow. Many have only two eyes (the anterior median eyes); in those with eight eyes, they are clustered together. These spiders specialize in preying upon other spiders but may not be strictly araneophagous.

Similar Families Dysderidae, Gnaphosidae, Oonopidae, Plectreuridae, Trogloraptoridae

How to distinguish Caponiidae from other similar families:

Caponiids can be distinguished from dysderids, gnaphosids, oonopids, plectreurids, and trogloraptorids by their lack of book lungs, which all similar families have. Additionally, they can be distinguished by their eye configuration. In most caponiids, only the anterior median eyes are present, or, if they have eight eyes, they are clustered together. Dysderids have six eyes in a cluster (the anterior median eyes are missing); gnaphosids and plectreurids have eight eyes in two rows; oonopids have six eyes in a cluster; and trogloraptorids have six eyes, with the lateral eyes adjacent to one another and the posterior median eyes between them.

eye configuration
(*Calponia harrisonfordi*)

eye configuration
(*Tarsonops systematicus*)

Calponia harrisonfordi Platnick, 1993

Common Name This species does not have a common name.

Total Body Length Female 5.2 mm, Male 5.0 mm

Description The prosoma is reddish orange, with a thin pale band around the border. It is somewhat flattened and narrows anterior of the eyes. These spiders have eight eyes in a cluster on the front of the prosoma on a dark spot. The opisthosoma is pale orange to reddish orange and has long pale hairs. The legs are pale yellow to orange and unbanded. The first two pairs of legs have robust femurs.

Habitat These spiders are often found in leaf litter or under rocks.

Known records from CA

female

Orthonops icenoglei Platnick, 1995

Common Name This species does not have a common name.

Total Body Length Female 5.0 mm, Male 3.3–3.4 mm

Description The prosoma is orange and narrows anterior of the eyes. This species' two eyes are placed on a black spot. The opisthosoma is tan to pale brown, with long brown hairs. The legs are yellow brown. The tarsi are subsegmented (that is, with a groove in the cuticle, making it appear there are two segments). The metatarsi are not subsegmented, which distinguishes them from *Tarsonops* species. The legs are also stouter than those of *Tarsonops*. The first two pairs of legs have robust femurs.

Habitat These spiders are often found under rocks. They have been observed feeding on filistatid spiders.

Known records from AZ, CA

Tarsonops systematicus
Chamberlin, 1924

Common Name
This species does not have a common name.

Total Body Length
Female 3.0 mm, Male 3.0 mm

Description The prosoma is pale orange and narrows anterior of the eyes. This species' two eyes are placed on a black spot. The opisthosoma is pale orange, with long brown hairs. The legs are pale orange. The first pair of legs is longer and slenderer than those of *Orthonops* species. They have subsegmented metatarsi.

Habitat These spiders are often found in desert habitats.

Known records from AZ, CA, TX

CITHAERONIDAE ▸ RUNNING SPIDER HUNTERS

Other Colloquial Names Swift ground spiders

Genera in This Family *Cithaeron* (1)

Total Body Length 3.0–8.0 mm

Description These spiders are entelegyne and ecribellate. They have eight eyes in two rows. They have a pair of book lungs and a single tracheal spiracle near the spinnerets. They have unusually long thin legs; the tarsi are pseudosegmented and have two claws. These are quick little spiders. They are nocturnal and can be found during the day in silk retreats under rocks and debris. Only one species, which is introduced, is found in North America.

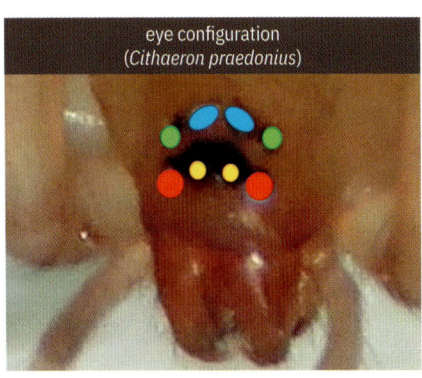

eye configuration
(*Cithaeron praedonius*)

Similar Families Anyphaenidae, Cheiracanthiidae, Clubionidae, Gnaphosidae

How to distinguish Cithaeronidae from other similar families:

The presence of the pseudosegmented tarsi distinguishes cithaeronids from all similar families, as none of them has this trait. Cithaeronids can be distinguished from anyphaenids by the placement of the tracheal spiracle: it is placed anteriorly in anyphaenids (between the spinnerets and the epigastric furrow), whereas it is placed near the spinnerets in cithaeronids. Cithaeronids have conical spinnerets that are not widely spaced; this distinguishes them from gnaphosids, which have widely spaced, cylindrical spinnerets.

Cithaeron praedonius O. Pickard-Cambridge, 1872

Common Name This species does not have a common name.

Total Body Length Female 5.0–8.0 mm, Male 3.0–4.0 mm

Description The prosoma is pale yellow to pale orange. The posterior median eyes are oval and oblique. The eye region is slightly darker. The opisthosoma is pale yellow to pale orange, and the heartmark is slightly darker. The legs are pale yellow to dusky yellow and lack spines. The tarsi are pseudosegmented. The spinnerets are conical and not widely spaced. There is evidence that these spiders specialize in preying upon other spiders.

female

female

female ventral (in retreat)

female creating an egg sac

female with prey

female creating an egg sac

Habitat These spiders tend to be found in dry habitats and are often seen in and around human structures.

Known records from FL
This is an introduced species.

MIMETIDAE ▶ PIRATE SPIDERS

Other Colloquial Names Cannibal spiders

Genera in This Family *Ero* (4) and *Mimetus* (10)

Total Body Length 2.5–7.0 mm

Description These spiders are entelegyne and ecribellate. They have eight eyes in two rows. They have a pair of book lungs and a single tracheal spiracle located near the spinnerets. The first two pairs of legs tend to be long and have a diagnostic spination pattern. They have three tarsal claws. These spiders specialize in preying upon other spiders. They tend to target web-building spiders, taking over the host spider's web and opportunistically eating insects caught in the web. They have also been

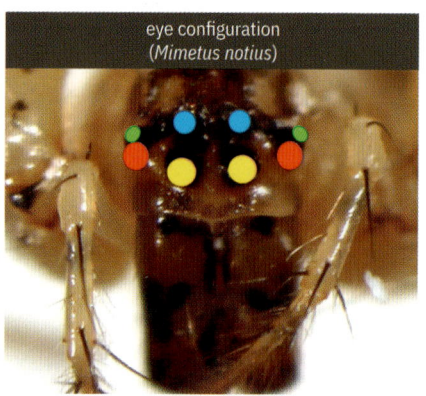

eye configuration
(*Mimetus notius*)

observed eating the egg sacs of their host spiders.

Similar Families Araneidae, Theridiidae, Scytodidae

How to distinguish Mimetidae from other similar families:

The diagnostic leg spination of mimetids distinguishes them from similar families.

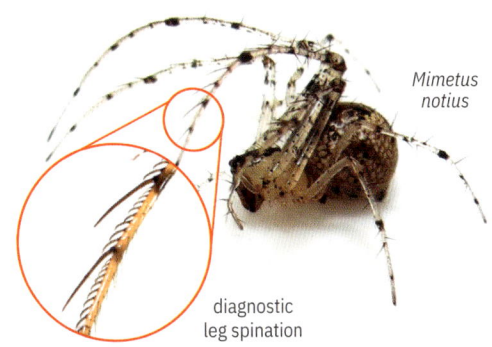

Mimetus notius

diagnostic leg spination

Ero canionis
Chamberlin & Ivie, 1935

Common Name This species does not have a common name.

Total Body Length
Female 3.0 mm, Male 2.2–2.5 mm

eyes (female)

Description The prosoma is tan to light brown. A thin irregular median stripe starts at the fovea and proceeds to the eye region. Dark bands from the eye region start to circle back toward the fovea but stop short. There is a thin dark band around the border. The clypeus is high. The opisthosoma is mottled with tan, brown, dark brown, red, cream, and yellow. Two small, often pale, tubercles are on the dorsal anterior. Sometimes the markings give the appearance of a second pair of tubercles, but there is only one pair in this species. The legs are banded in brown, tan, and dark brown. The first pair of legs is much longer than the others and has the characteristic leg supination of a mimetid. The outward appearance of this species is very similar to that of *Ero leonina* (which had previously been misidentified as *Ero furcata*, a

female

female

female

egg sac

male

male ventral

male

European species) and can be distinguished only by the genitalia. The egg sacs of *Ero* species are suspended on a stalk of silk and covered in a tangle of thicker darker silk.

Habitat These spiders tend to be found on the understory vegetation of forested habitats.

Known records from AB, BC, MB, NB, NL, NS, ON, QC, SK (Canada); IL, IN, ME, OH, TX, UT, WA, WI (United States)

Ero pensacolae
Ivie & Barrows, 1935

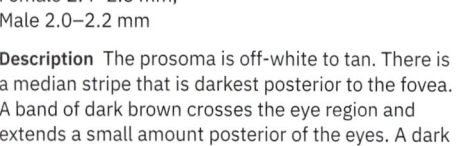

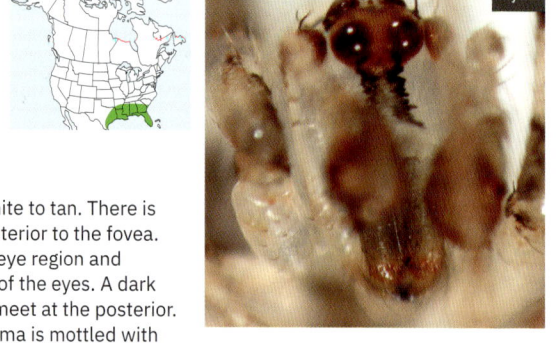

eyes

Common Name This species does not have a common name.

Total Body Length
Female 2.4–2.5 mm,
Male 2.0–2.2 mm

Description The prosoma is off-white to tan. There is a median stripe that is darkest posterior to the fovea. A band of dark brown crosses the eye region and extends a small amount posterior of the eyes. A dark band around the border does not meet at the posterior. The clypeus is high. The opisthosoma is mottled with white, tan, brown, and pale yellow. There are two pairs of tubercles, usually pale, in a transverse row across the anterior. The legs are off-white and banded in dark brown. The first pair of legs is

male

male

male

juvenile

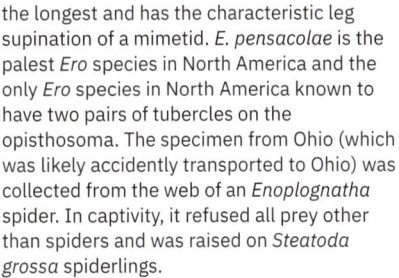

juvenile

the longest and has the characteristic leg supination of a mimetid. *E. pensacolae* is the palest *Ero* species in North America and the only *Ero* species in North America known to have two pairs of tubercles on the opisthosoma. The specimen from Ohio (which was likely accidently transported to Ohio) was collected from the web of an *Enoplognatha* spider. In captivity, it refused all prey other than spiders and was raised on *Steatoda grossa* spiderlings.

Habitat These spiders are often found in woody habitats.

Known records from FL, OH, TX

Mimetus notius Chamberlin, 1923

eyes

Common Name
This species does not have a common name.

Other Colloquial Names
Reticulated pirate spider

Total Body Length Female 4.2–5.0 mm, Male 3.5 mm

Description The prosoma is off-white to pale yellow. An irregular dark marking that resembles an ink blot reaches from the fovea to the eye region. There may be other scattered small dark markings.

female

female

female
ventral

female

A short dark dash is on the very edge of the prosoma near coxae I, and a dark line is under the eyes. The clypeus is low. The chelicerae are dark, often with a hint of red at the distal end. The opisthosoma is round, with a hump or shoulders about a quarter of the way back, but this may not be visible in well-fed or gravid individuals. The dorsal surface is covered with a variegated red, brown, yellow, and orange marking. The heartmark is usually black and highlighted in white, although it is not particularly well defined. The lateral edges are cream to pale yellow. The legs are off-white to cream, with scattered dusky bands and dark spots. An orange to brown band is often at the distal end of the femurs and tibiae of legs I and II. This species looks very similar to *Mimetus puritanus*. To confirm an identification, one would need to look at the genitalia. The egg sacs are a white sphere covered in curly silk.

egg sac

Habitat These spiders tend to be found in dry areas, where they will invade the webs of other spiders. They are sometimes seen guarding egg sacs of other spiders, presumably waiting for the young to emerge so they can feed on them.

Known records from ON, QC (Canada); CO, FL, IL, LA, MA, MI, NH, NC, NY, OH, PA, SC, TX, WI (United States)

Mimetus puritanus Chamberlin, 1923

Common Name
This species does not have a common name.

Other Colloquial Names Common pirate spider

Total Body Length
Female 5.0–5.6 mm, Male 4.0–4.5 mm

Description The prosoma is off-white to
pale orange. An irregular dark marking that
resembles an ink blot reaches from the fovea
to the eye region. Some people see a double
V marking. Sometimes a pair of dots, one on
each side, is about halfway down. The clypeus
is low. The chelicerae tend to be lighter on the
basal half and darker below but are sometimes
completely dark. There is a spot on the face of
the chelicerae about a quarter of the way down;
the spot can be difficult to see if the chelicerae
are darker. The opisthosoma is round, with a
hump or shoulders about a quarter of the way
back; this may not be visible in well-fed or gravid individuals.
The dorsal surface is covered by a variegated red, brown,
black, and orange marking. There is white marking at the
posterior of the heartmark. A broken transverse band
of white is often at the peak of the opisthosoma hump
or shoulders. The legs are off-white to pale orange.
An orange to brown band is often at the distal end
of the femurs and tibiae of legs I and II. There
are other scattered dusky bands and spots. This
species looks very similar to *Mimetus notius*.
To confirm an identification, one would need
to look at the genitalia. The egg sacs are a
golden sphere covered with curly silk.

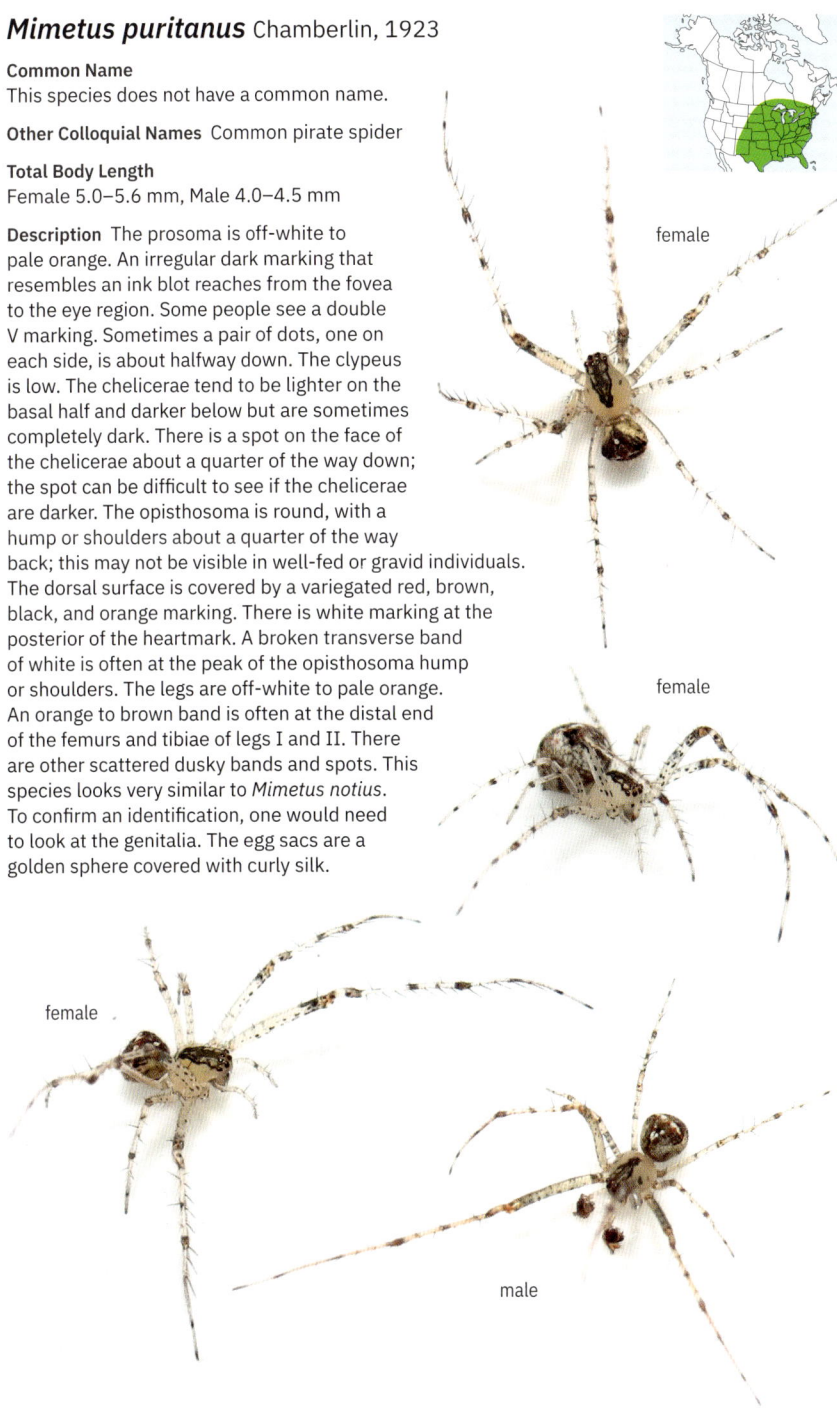

female

female

female

male

male

male

juvenile

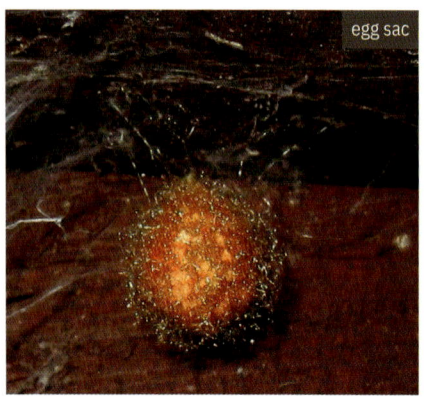

egg sac

female with *Parasteatoda* sp. as prey

Habitat These spiders tend to be found in dry areas, where they will invade the webs of other spiders. They are sometimes seen guarding egg sacs of other spiders, presumably waiting for the young to emerge so they can feed on them.

Known records from ON, QC (Canada); AL, CT, FL, GA, IL, IN, KY, MA, ME, MI, ND, NH, NM, NY, OK, PA, SC, TN, TX, VA (United States)

Anal tubercle	A tubercle that is small (in most spiders) and sits close to the anus
Antenniform	Modified for sensory purposes
Anterior	Toward the front of the animal
Apodeme	A depression in the exoskeleton where muscles attach (also called *sigilla*)
Apophysis	A process or extension, usually on the male pedipalp tibia

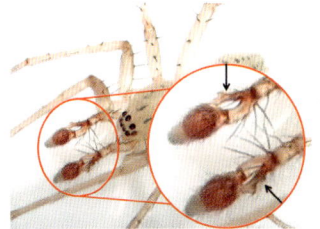

Wulfila
saltabundus

Arachnida	A class of arthropods that includes mites, spiders, scorpions, and several other orders
Araneae	The order to which spiders belong
Araneophagous	Feeding on spiders
Autospasy	The act of separating an extremity with a mechanism that prevents blood loss
Ballooning	A technique whereby a spider produces long strands of silk that allow the spider to fly using air and electrical currents
Bifurcate	Divided into two sections, forked
Book lungs	Respiratory organ found in most spiders
Calamistrum	One or two rows of bristles on metatarsi IV that are used for combing cribellate silk

Amaurobius
ferox

Carina	A ridge located below the anterior eye row

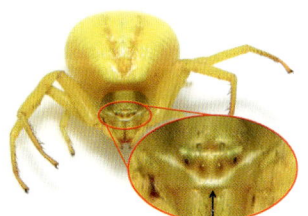

Misumenoides
formosipes

Chelicera	The first pair of appendages on the anterior of the prosoma; composed of a basal segment and the fang
Clypeus	The area between the bottom of the eyes and the top of the chelicerae
Coxa	The base segment of the legs
Cribellate	Possessing a cribellum
Cribellum	A modified silk-producing plate derived from the first pair of spinnerets; the cribellum produces cribellate silk

Amaurobius ferox

Cuccullus	The hoodlike structure found on Ricinulei

Ricinulei

Cymbium	The dorsal covering of the terminal segment of the male pedipalp
Diads	Pairs, in twos
Diaxial	At an angle to the central line of the body
Distal	Away from the body core
Dorsal	On the upper side
Ecdysis	The process of shedding the exoskeleton (also used to refer to the shedding of skin in some animals)
Ecribellate	Not possessing a cribellum
Endites	The first segment of the pedipalps (similar to leg coxae)
Entelegyne	Spiders that have more complex genitalia (female with three genital openings—difficult to see—and a sclerotized epigynum, and males with complex palp structures)
Epigastric furrow	A transverse fold that contains the opening to the reproductive organs on the ventral side of the opisthosoma

Epigynum	The external female reproductive opening
Femur	The third segment of the leg and pedipalp
Fovea	A depression on the dorsal prosoma usually where muscles attach
Habitus	The general outward appearance
Haplogyne	Spiders that have simple genitalia (females with a single genital opening and lacking a sclerotized epigynum—although sclerotized structures may be present—and males with a simple pedipalps bulb)
Introduced species	A species that is not native to an area and was likely introduced by humans
Invasive species	A species that is harming the environment, usually due to an unnatural increase in the population in an area due to human activity; most cases involve species that are introduced into an area to which they are not native
Kleptoparasitism	A life strategy of feeding upon the prey captured by another animal
Labium	A segment at the lower border of the mouth
Labrum	A segment at the upper border of the mouth
Marginal bands	Markings that run along the outer edge of the prosoma
Mastidion	A tooth or projection, usually on the chelicerae

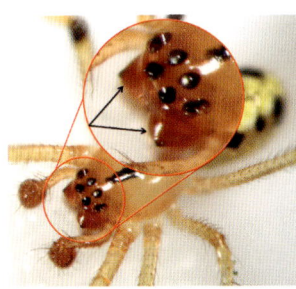

Theridion frondeum

Matriphagy	The consumption of the mother by her young
Median strip or band	A marking that runs longitudinally along the midline of the prosoma and/or opisthosoma
Medically significant	In regard to spiders, a species whose venom has been shown to cause adverse reactions in humans that may require medical attention
Metatarsus	The sixth segment of the legs (not found on the pedipalps)
Molting	The act of shedding the exoskeleton
Opisthosoma	The abdominal body region or posterior tagmata of a spider
Patella	The fourth segment of the legs
Pedicel	The narrow stalk that connects the prosoma and opisthosoma
Pedipalp	Small leglike appendages on the anterior of the prosoma; male spiders have modified pedipalps, as they are used for sperm transfer
Phoresy	The act of attaching to another animal for dispersal

Poison	A biological toxin that is ingested, inhaled, or absorbed and that is delivered passively
Pooter	An aspirator, a tool used to collect small animals
Porrect	Extended forward, like the chelicerae of *Bagheera prosper*

Bagheera prosper

Posterior	Toward the rear of the animal
Precoxal triangles	Triangular structures on the sternum near the coxae
Procurved	Position of the eyes where the lateral eyes are situated more anteriorly than the median eyes of that eye row
Prolateral	On the forward-facing side
Prosoma	The head region or anterior tagmata of a spider
Proximal	Toward the body core
Pseudosegmented	Constricted so it appears to be a joint forming another segment

Trogloraptor marchingtoni

Radial lines	Markings that run from the median area to the marginal area, similar to the spokes of a wheel
Raptorial	Modified for grasping, like the pedipalps of Amblypigi
Recurved	Position of the eyes where the lateral eyes are situated more posteriorly than the median eyes of that eye row
Retrolateral	On the backward-facing side

Scape   A projection that covers the copulatory area of the female epigynum

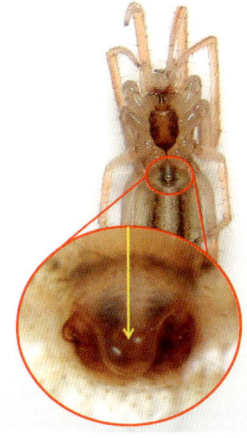

Larinia directa

Sclerite   A sclerotized area (See *tergites* for an example)

Sclerotized   An area that is hardened into sclerotin

Scopulae   Hair tufts that are on the distal ends of the legs and provide adhesion

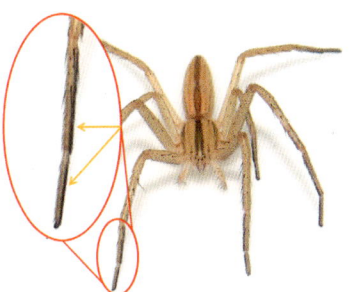

Tibellus oblongus

Scutum   A sclerotized plate, usually on the opisthosoma

Nodocion eclecticus

Serulla   A row of small teeth on the endites

*Elaver
excepta*

Sigilla   An apodeme

Spinnerets   Modified appendages with spigots for dispensing silk

Stridulation   Sounds generated by strumming or rubbing things together

Stridulatory file   A series of ridges, usually on the edge of the chelicerae or on the book lung
                           covers, that are used to create sounds by being rubbed

*Mermessus
fradeorum*

Submarginal bands   Markings that fall between the median and marginal areas of the prosoma

Synanthropic   Associated with humans

Tagmata   A distinct body region; in Arachnida, the prosoma and opisthosoma

Tapetum   An inner layer of the eyes that causes them to reflect light

Tarsal claws   Small claws at the end of the tarsi

Tarsus   The seventh segment of the legs and the sixth segment of the pedipalps

Telson   The terminal segment in arthropods; in Scorpiones, it is modified to have a venom
                           bulb and stinger

Scorpiones

Tergites	An area of cuticle that is sclerotized usually on the dorsal opisthosoma

*Antrodiaetus
unicolor*

Thanatosis	The act of feigning death
Thoracic furrow	See *fovea*
Tibia	The fifth segment of the legs
Toxungen	A biological toxin that is intentionally delivered without a wound
Tracheal spiracle	An opening to the breathing tubes
Trichobothria	Fine sensory hairs

*Leucauge
venusta*

Trochanter	The second segment of the legs or pedipalps
Troglobites	Species that are known to occupy caves or other underground structures
Troglophiles	Cave dwellers
Urticating hair	A patch of hairs on the dorsal opisthosoma of Theraphosidae that can be kicked off when the spider is disturbed; the barbed shape of the hairs causes irritation in other animals

*Aphonopelma
eutylenum*

Venom	A biological toxin that is delivered through an intentionally caused wound
Ventral	On the underside
Xeric	Of an environment that contains little moisture or an arid habitat

REFERENCES

Arnold, Kathryn E., Scot L. Ramsay, Christine Donaldson, and Aileen Adam. 2007. "Parental Prey Selection Affects Risk-Taking Behaviour and Spatial Learning in Avian Offspring." *Proceedings of the Royal Society B: Biological Sciences* 274, no. 1625: 2563–69.

Ballesteros, Jesús A., and Prashant P. Sharma. 2019. "A Critical Appraisal of the Placement of Xiphosura (Chelicerata) with Account of Known Sources of Phylogenetic Error." *Systematic Biology* 68, no. 6: 896–917.

Bradley, Richard A. 2013. *Common Spiders of North America.* University of California Press.

Brady, A. R. 1987. "Nearctic Species of the New Wolf Spider Genus *Gladicosa* (Araneae: Lycosidae)." *Psyche* 93: 285–319.

Cardoso, Pedro, Stano Pekár, Rudy Jocqué, and Jonathan A. Coddington. 2011. "Global Patterns of Guild Composition and Functional Diversity of Spiders." *PloS ONE* 6, no. 6: e21710.

Carlin, Albert S., Hunter G. Hoffman, and Suzanne Weghorst. 1997. "Virtual Reality and Tactile Augmentation in the Treatment of Spider Phobia: A Case Report." *Behaviour Research and Therapy* 35, no. 2: 153–58.

Forster, R. R. 1971. "Notes on an Airborne Spider Found in Antarctica." *Pacific Insects Monograph* 25: 119–20.

Glenday, Craig. 2009. *Guinness World Records 2010.* Guinness World Records.

Han, S. I., H. C. Astley, D. D. Maksuta, and T. A. Blackledge. 2019. "External Power Amplification Drives Prey Capture in a Spider Web." *Proceedings of the National Academy of Sciences* 116, no. 24: 12060–65.

Hentz, N. M. 1821. "A Notice Concerning the Spiders Whose Web Is Used in Medicine." *Journal of the Academy of Natural Sciences of Philadelphia* 2: 53–55.

Hodkinson, Ian D., Stephen J. Coulson, Joanna Harrison, and Nigel R. Webb. 2001. "What a Wonderful Web They Weave: Spiders, Nutrient Capture and Early Ecosystem Development in the High Arctic—Some Counter-Intuitive Ideas on Community Assembly." *Oikos* 95, no. 2: 349–52.

Isbister, G. K., and M. R. Gray. 2004. "Black House Spiders Are Unlikely Culprits in Necrotic Arachnidism: A Prospective Study." *Internal Medicine Journal* 34, no. 5: 287–89.

Johnson, J. Chadwick, Patricia Trubl, Valerie Blackmore, and Lindsay Miles. 2011. "Male Black Widows Court Well-Fed Females More than Starved Females: Silken Cues Indicate Sexual Cannibalism Risk." *Animal Behaviour* 82, no. 2: 383–90.

Marx, Michael Thomas, Patrick Guhmann, and Peter Decker. 2012. "Adaptations and Predispositions of Different Middle European Arthropod Taxa (Collembola, Araneae, Chilopoda, Diplopoda) to Flooding and Drought Conditions." *Animals* 2, no. 4: 564–90.

Murphy, John A., and Michael J. Roberts. 2015. *Spider Families of the World and Their Spinnerets*. British Arachnological Society.

Nelsen, David R., Zia Nisani, Allen M. Cooper, Gerad A. Fox, Eric C. K. Gren, Aaron G. Corbit, and William K. Hayes. 2014. "Poisons, Toxungens, and Venoms: Redefining and Classifying Toxic Biological Secretions and the Organisms That Employ Them." *Biological Reviews* 89, no. 2: 450–65.

Nentwig, Wolfgang, ed. 2011. *Ecophysiology of Spiders.* Springer.

Nyffeler, Martin, and Klaus Birkhofer. 2017. "An Estimated 400–800 Million Tons of Prey Are Annually Killed by the Global Spider Community." *The Science of Nature* 104: 30.

Řezáč, Milan, Stanislav Pekár, and Yael Lubin. 2008. "How Oniscophagous Spiders Overcome Woodlouse Armour." *Journal of Zoology* 275, no. 1: 64–71.

Roberts, Michael J. 1995. *Spiders of Britain and Northern Europe.* HarperCollins.

Schwartz, Steven K., William E. Wagner Jr., and Eileen A. Hebets. 2014. "Obligate Male Death and Sexual Cannibalism in Dark Fishing Spiders." *Animal Behaviour* 93: 151–56.

Stoecker, William V., Richard S. Vetter, and Jonathan A. Dyer. 2017. "NOT RECLUSE—A Mnemonic Device to Avoid False Diagnoses of Brown Recluse Spider Bites." *JAMA Dermatology* 153, no. 5: 377–78.

Thornton, Ian. 1997. *Krakatau: The Destruction and Reassembly of an Island Ecosystem.* Harvard University Press.

Vetter, Richard S., Leonard S. Vincent, Douglas W. R. Danielsen, Kathryn I. Reinker, Daniel E. Clarke, Amelia A. Itnyre, John N. Kabashima, and Michael K. Rust. 2012. "The Prevalence of Brown Widow and Black Widow Spiders (Araneae: Theridiidae) in Urban Southern California." *Journal of Medical Entomology* 49, no. 4: 947–51.

Wise, David H. 1993. *Spiders in Ecological Webs.* Cambridge University Press.

INDEX

This index includes the *scientific name* (in *italics*) and the common name (as approved by the American Arachnological Society) of each species in this book. Family scientific names are capitalized, and the guilds are capitalized and in **bold**. Page numbers are followed by figure numbers (in parenthesis and in *italics*) when referring to a photograph. The page numbers in **bold** highlight the main written account.

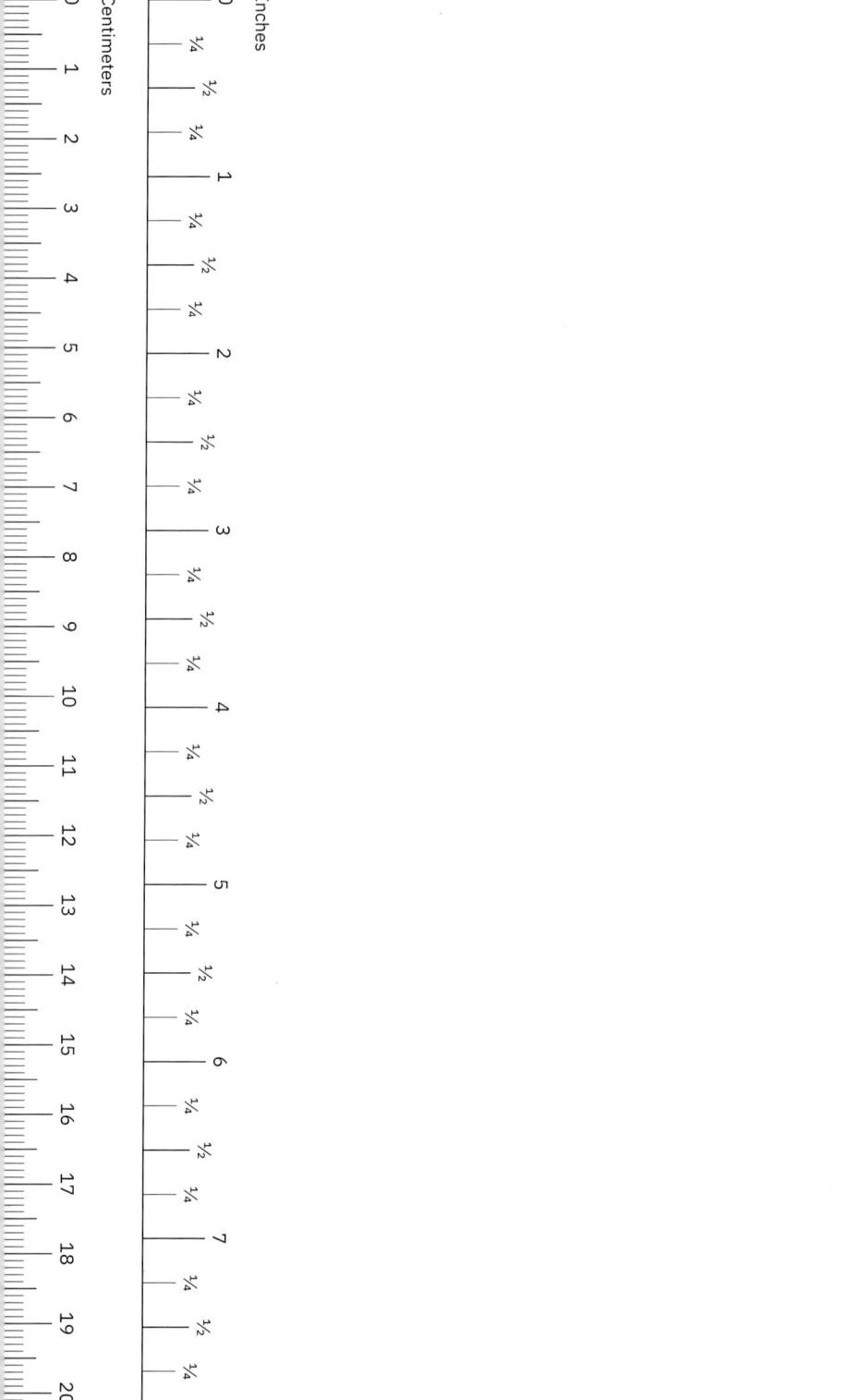